Lecture Notes in Computer Science 3479

Commenced Publication in 1973
Founding and Former Series Editors:
Gerhard Goos, Juris Hartmanis, and Jan van Leeuwen

Thomas Strang   Claudia Linnhoff-Popien (Eds.)

Location- and Context-Awareness

First International Workshop, LoCA 2005
Oberpfaffenhofen, Germany, May 12-13, 2005
Proceedings

 Springer

Volume Editors

Thomas Strang
German Aerospace Center (DLR), Institute of Communications and Navigation
82234 Wessling/Oberpfaffenhofen, Germany
E-mail: thomas.strang@dlr.de

Claudia Linnhoff-Popien
University of Munich (LMU), Mobile and Distributed Systems Group
Oettingenstr. 67, 80538 Munich, Germany
E-mail: linnhoff@ifi.lmu.de

Library of Congress Control Number: 2005925756

CR Subject Classification (1998): H.3, H.4, C.2, H.5, K.8

ISSN   0302-9743
ISBN-10   3-540-25896-5 Springer Berlin Heidelberg New York
ISBN-13   978-3-540-25896-4 Springer Berlin Heidelberg New York

Springer is a part of Springer Science+Business Media

springeronline.com

© Springer-Verlag Berlin Heidelberg 2005
Printed in Germany

Typesetting: Camera-ready by author, data conversion by Scientific Publishing Services, Chennai, India
Printed on acid-free paper   SPIN: 11426646   06/3142   5 4 3 2 1 0

Preface

Context-awareness is one of the drivers of the ubiquitous computing paradigm. Well-designed context modeling and context retrieval approaches are key prerequisites in any context-aware system. Location is one of the primary aspects of all major context models — together with time, identity and activity. From the technical side, sensing, fusing and distributing location and other context information is as important as providing context-awareness to applications and services in pervasive systems.

The material summarized in this volume was selected for the 1st International Workshop on Location- and Context-Awareness (LoCA 2005) held in cooperation with the 3rd International Conference on Pervasive Computing 2005. The workshop was organized by the Institute of Communications and Navigation of the German Aerospace Center (DLR) in Oberpfaffenhofen, and the Mobile and Distributed Systems Group of the University of Munich.

During the workshop, novel positioning algorithms and location sensing techniques were discussed, comprising not only enhancements of singular systems, like positioning in GSM or WLAN, but also hybrid technologies, such as the integration of global satellite systems with inertial positioning. Furthermore, improvements in sensor technology, as well as the integration and fusion of sensors, were addressed both on a theoretical and on an implementation level.

Personal and confidential data, such as location data of users, have profound implications for personal information privacy. Thus privacy protection, privacy-oriented location-aware systems, and how privacy affects the feasibility and usefulness of systems were also addressed in the workshop.

A total of 84 papers from 26 countries were submitted to LoCA 2005, from which 26 full and 7 short papers were selected for publication in the proceedings. The overall quality of the submissions was impressive, demonstrating the importance of this field. The Program Committee did an excellent job — all papers were reviewed by at least 3 referees, which left each member with up to 18 papers to be reviewed within a very tight schedule.

May 2005

Thomas Strang
Claudia Linnhoff-Popien

Organization

Program Committee

Alessandro Acquisti	Carnegie Mellon University (USA)
Victor Bahl	Microsoft Research (USA)
Christian Becker	University of Stuttgart (Germany)
Anind K. Dey	Carnegie Mellon University (USA)
Thomas Engel	University of Luxemburg (Luxemburg)
Dieter Fensel	DERI (Austria/Ireland)
Jens Grossklags	University of California, Berkeley (USA)
Mike Hazas	Lancaster University (UK)
Jeffrey Hightower	Intel Research Seattle (USA)
Jadwiga Indulska	University of Queensland (Australia)
John Krumm	Microsoft Research (USA)
Axel Küpper	University of Munich (Germany)
Gerard Lachapelle	University of Calgary (Canada)
Marc Langheinrich	ETH Zurich (Switzerland)
Claudia Linnhoff-Popien	University of Munich (Germany)
Jussi Myllymaki	IBM Almaden Research Center (USA)
Harlan Onsrud	University of Maine (USA)
Aaron Quigley	University College Dublin (Ireland)
Kay Römer	ETH Zurich (Switzerland)
Albrecht Schmidt	University of Munich (Germany)
Stefan Schulz	Carleton University (Canada)
Frank Stajano	University of Cambridge (UK)
Thomas Strang	German Aerospace Center (Germany)

Additional Reviewers

Henoc Agbota	Lancaster University (UK)
Michael Angermann	German Aerospace Center (Germany)
Stavros Antifakos	ETH Zurich (Switzerland)
Trent Apted	National ICT Australia (Australia)
Martin Bauer	University of Stuttgart (Germany)
Alastair Beresford	University of Cambridge (UK)
Jan Beutel	ETH Zurich (Switzerland)
Jürgen Bohn	ETH Zurich (Switzerland)
Thomas Buchholz	University of Munich (Germany)
Jim Campbell	University of Maine (USA)
Nicolas Christin	University of California, Berkeley (USA)

Table of Contents

Positioning Sensor Systems III

From Location to Context

Bayesian Networks

Context Inference

Privacy

Location- and Context-Aware Applications

Hybrid Positioning and User Studies

Location Awareness:
Potential Benefits and Risks

Vidal Ashkenazi

Nottingham Scientific Ltd, United Kingdom
vidal.ashkenazi@nsl.eu.com

Abstract. The development of the European Satellite Navigation System, Galileo, the modernisation of GPS, and the recent advances in High Sensitivity GNSS technology have opened up new horizons, leading to new location and context based applications. Nevertheless, these satellite based technologies may not always deliver the necessary navigation and positioning information in a number of difficult environments, as well as when there is accidental or intentional interference with the satellite signals.

Underground car parks and railway tunnels are two examples of difficult environments, where the reception of satellite signals is affected. Similarly, the malicious jamming of satellite signals near landing sites at airports, or the intentional or unintentional uploading of incorrect orbit predictions will render the satellite derived navigation and positioning information unusable. The risk of such interference may be low, but difficult environments are always present in a number of safety-critical transport applications, as well as in a variety of commercial location-based-services, involving the continuous tracking of goods or individuals.

This is when there is a requirement to combine satellite derived navigation and positioning data with other positioning technologies, such as inertial navigation, cellular telephone networks, such as GSM/GPRS, and Wireless Local Area Networks (WLAN). Clearly, the specific combination of a hybrid system will depend on the required accuracy, integrity and extent of geographical coverage of the corresponding application.

The wide variety of tracking applications, involving persons, vehicles, devices or merchandise, for safety, convenience, security, marketing and other purposes, presents multiple challenges, not only with respect to technology development and service provision, but also in terms of what is legally and ethically acceptable. Many of the proposed commercial applications would create few problems regarding general public acceptance. These include the tracking of motor vehicles for congestion monitoring, taxation and insurance purposes, and the tracking of vulnerable individuals, such as the very young or individuals suffering from a debilitating infirmity such as Alzheimer.

Some of these technologies could also be exploited not only by governments for national and internal security purposes, but also by criminals. Clearly, there is a fine boundary between what is ethically acceptable and what is not. Therefore, there is a need for raising public awareness of these issues and starting a debate involving the public at large as well the relevant government, legal and political institutions.

T. Strang and C. Linnhoff-Popien (Eds.): LoCA 2005, LNCS 3479, p. 1, 2005.
© Springer-Verlag Berlin Heidelberg 2005

Context Modelling and Management in Ambient-Aware Pervasive Environments

Maria Strimpakou[1], Ioanna Roussaki[1], Carsten Pils[2], Michael Angermann[3], Patrick Robertson[3], and Miltiades Anagnostou[1]

[1] School of Electrical and Computer Engineering, National Technical University of Athens, 9 Heroon Polytechneiou Street, 157 73, Athens, Greece
{mstrim, nanario, miltos}@telecom.ntua.gr
[2] Department of Computer Science 4, Ahornstr. 55, 52056 Aachen, Germany
pils@i4.informatik.rwth-aachen.de
[3] Institute of Communications and Navigation, German Aerospace Center, D-82234 Wessling/Oberpfaffenhofen, Germany
{michael.angermann, Patrick.Robertson}@dlr.de

Abstract. Services in pervasive computing systems must evolve so that they become minimally intrusive and exhibit inherent proactiveness and dynamic adaptability to the current conditions, user preferences and environment. Context awareness has the potential to greatly reduce the human attention and interaction bottlenecks, to give the user the impression that services fade into the background, and to support intelligent personalization and adaptability features. To establish this functionality, an infrastructure is required to collect, manage, maintain, synchronize, infer and disseminate context information towards applications and users. This paper presents a context model and ambient context management system that have been integrated into a pervasive service platform. This research is being carried out in the DAIDALOS IST Integrated Project for pervasive environments. The final goal is to integrate the platform developed with a heterogeneous all-IP network, in order to provide intelligent pervasive services to mobile and non-mobile users based on a robust context-aware environment.

1 Introduction

Pervasive computing is about the creation of environments saturated with computing and communication capabilities, yet having those devices integrated into the environment such that they disappear [1]. In such a pervasive computing world [2], service provisioning systems will be able to proactively address user needs, negotiate for services, act on the user's behalf, and deliver services anywhere and anytime across a multitude of networks. As traditional systems evolve into pervasive, an important aspect that needs to be pursued is context-awareness, in order for pervasive services to seamlessly integrate and cooperate in support of user requirements, desires and objectives. Context awareness in services is actually about closely and properly linking services, so that their user is relieved from submitting information that already exists in other parts of the global system. In this manner services are expected to act

T. Strang and C. Linnhoff-Popien (Eds.): LoCA 2005, LNCS 3479, pp. 2–15, 2005.

in a collaborative mode, which finally increases the user friendliness. Yet, context awareness cannot be achieved without an adequate methodology and a suitable infrastructure. In this framework, the European IST project DAIDALOS[1] works on the adoption of an end-to-end approach for service provisioning, from users to service providers and to the global Internet. DAIDALOS aims to design and develop the necessary infrastructure and components that support the composition and deployment of pervasive services. The middleware software system currently being developed provides, amongst others, the efficient collection and distribution of context information. Eventually, DAIDALOS will offer a uniform way and the underlying means that will enable users to discover and compose services, tailored to their requirements and context, while also preserving their privacy.

Quite a few articles in the research literature on context awareness have outlined the benefits of using context and have proposed various, albeit similar to each other, context definitions. Most popular is the definition of Dey [3]: "Context is any information that can be used to characterise the situation of an entity." Evidently, context comprises a vast amount of different data sources and data types. Thus, collection, consistency, and distribution of context data is challenging for the development of context aware systems. This paper presents how the DAIDALOS middleware addresses these challenges in pervasive computing environments. The rest of the paper is structured as follows. Section 2 gives a short overview of context-aware approaches known from research literature and outlines their advantages and shortcomings. Subsequently, in Section 3 the DAIDALOS context data model is described and Section 4 proposes a context management infrastructure. Sections 5 and 6 present how DAIDALOS captures context information and ensures the consistency of the data. Finally, Section 7 concludes the paper and gives an outlook towards future work.

2   Context-Aware Systems Overview

In this section, a short overview of the most important context-aware systems is provided, while the wide variety of fields where context can be exploited is identified.

Early work in context awareness includes the Active Badge System developed at Olivetti Research Lab to redirect phone calls based on people's locations [4]. Subsequently, the ParcTab system developed at the Xerox Palo Alto Research Center in the mid 90's could be considered as an evolution of the active badge that relied on PDAs to support a variety of context-aware office applications [5]. A few years later, Cyberdesk [6] built an architecture to automatically integrate web-based services based on virtual context, or context derived from the electronic world. The virtual context was the personal information the user was interacting with on-screen including email addresses, mailing addresses, dates, names, URLs, etc. While Cyberdesk could handle

[1] This work has been partially supported by the Integrated Project DAIDALOS ("Designing Advanced network Interfaces for the Delivery and Administration of Location independent, Optimised personal Services"), which is financed by the European Commission under the Sixth Framework Programme. However, this paper expresses the authors' personal views, which are not necessarily those of the DAIDALOS consortium.

limited types of context, it provided many of the mechanisms that are necessary to build generic context-aware architectures. The Cyberguide application [7] enhanced the prevailing services of a guidebook by adding location awareness and a simple form of orientation information. The context aware tour guide, implemented in two versions for indoor and outdoor usage, could give more precise information, depending on the user's location. In general, tour guidance is a popular context-aware application, which has been explored by several research and development groups.

The Ektara architecture [8] is a distributed computing architecture for building context-aware ubiquitous and wearable computing applications (UWC). Ektara reviewed a wide range of context-aware wearable and ubiquitous computing systems, identified their critical features and finally, proposed a common functional architecture for the development of real-world applications in this domain. At the same time, Mediacup [9] and TEA [10] projects tried to explore the possibility of hiding context sensors in everyday objects. The Mediacup project studied capture and communication of context in human environments, based on a coffee cup with infrared communication and multiple sensors. Using the Mediacup various new applications were developed exploiting the context information collected. The TEA project investigated "Technologies for Enabling Awareness" and their applications in mobile telephony, building a mobile phone artefact.

Other interesting examples of context management are provided by Owl context service, Kimura System and Solar. Owl [11] is a context-aware system, which aimed to gather, maintain and supply context information to clients, while protecting people's privacy through the use of a role-based access control mechanism. It also tackled various advanced issues, such as historical context, access rights, quality, extensibility and scalability [12]. Kimura System [13], on the other hand, tried to integrate both physical and virtual context information to enrich activities of knowledge workers. Finally, Solar [14] is a middleware system designed in Dartmouth College that consists of various information sources, i.e., sensors, gathering physical or virtual context information, together with filters, transformers and aggregators modifying context to offer the application usable context information.

A quite promising approach to context-awareness was introduced by the Context Toolkit [3] that isolated the application from context sensing. The proposed architecture was based on abstract components named context widgets, interpreters and aggregators that interact, in order to gather context data and disseminate it to the applications. On the other hand, the Aura Project [15] at Carnegie Mellon University (CMU) investigated how applications could proactively adapt to the environment in which they operated. A set of basic "contextual services" was developed within Aura, in order to provide adaptive applications with environmental information. While the Context Toolkit focused on developing an object oriented framework and allowed use of multiple wire protocols, Aura focused on developing a standard interface for accessing services and forced all services and clients to use the same wire protocol. This sacrificed flexibility, but increased interoperability.

HotTown [16] is another project that developed an open and scalable service architecture for context-aware personal communication. Users and other entities were represented by mobile agents that carried a context knowledge representation reflecting the capabilities of the entity and associated objects and relations between them. In HotTown, entities could exchange context knowledge, merge it with existing knowledge, and interpret context knowledge in the end devices. The Cooltown project by HP labs attempted to solve the problems of representing, combining and exploiting context information, and by introducing a uniform Web presence model for people, places and things [17]. Rather than focusing on creating the best solution for a particular application, Cooltown concentrated on building general-purpose mechanisms common to providing Web presence for people, places, and things. The Cooltown architecture was deployed for real use within a lab.

Other current research activities include the CoBrA, SOCAM, CASS and CORTEX projects. CoBrA (Context Broker Architecture) [18] is an agent based architecture supporting context-aware computing in intelligent spaces populated with intelligent systems that provide pervasive computing services to users. CoBrA has adopted an OWL-based ontology approach and it offers a context inference engine that uses rule-based ontology reasoning. The SOCAM (Service-oriented Context-Aware Middleware) project [19] is another architecture for the rapid prototyping and provision of context-aware mobile services. It is based on a central server that retrieves context data from distributed context providers, processes it and offers it to its clients. Third party mobile services are located on top of the architecture and use the different levels of available context information to adapt their behavior accordingly. SOCAM also uses ontologies to model and manage the context data and has implemented a context reasoning engine. Another scalable server-based middleware for context-aware mobile applications on hand-held and other small mobile computers is designed within the CASS (Context-awareness sub-structure) project [20]. CASS enables developers to overcome the memory and processor constraints of small mobile computer platforms while supporting a large number of low-level sensor and other context inputs. It supports high-level context data abstraction and separation of context-based inferences and behaviours from application code, thus opening the way for context-aware applications configurable by users. The CORTEX project has built context-aware middleware based on the Sentient Object Model [21]. It is suitable for the development of context-aware applications in ad-hoc mobile environments and allows developers to fuse data from disparate sensors, represent application context, and reason efficiently about context, without the need to write complex code. It provides an event-based communication mechanism designed for ad-hoc wireless environments, which supports loose coupling between sensors, actuators and application components.

Finally, one of the most recent projects that focused on context-awareness is the IST CONTEXT project [22]. Its main objective is the specification and design of models and solutions for an efficient provisioning of context-based services making use of active networks on top of fixed and mobile infrastructure. CONTEXT has proved that active networks are also powerful in context-aware systems for tackling the issues of context distribution and heterogeneity. Furthermore, via the CONTEXT platform it has been demonstrated that gathering and disseminating context using active networks is efficient with regards to traffic and delay parameters.

3   A Context Model for Pervasive Systems

An efficient context model is a key factor in designing context-aware services. Relevant research efforts have indicated that generic uniform context models are more useful in pervasive computing environments, in which the range and heterogeneity of services is unique. In [23] a survey of the most important context modelling approaches for pervasive computing is provided. These approaches have been evaluated against the requirements of pervasive services and the relative results are presented in the following table [23].

Table 1. Evaluation of the existing context modelling schemes against the pervasive computing requirements

Context Modelling Approach \ Pervasive Computing Req/ments	distributed composition	partial validation	richness & quality of information	incompleteness and ambiguity	level of formality	applicability to existing environments
Key-Value Models	-	-	- -	- -	- -	+
Mark-up Scheme Models	+	+ +	-	-	+	+ +
Graphical Models	- -	-	+	-	+	+
Object Oriented Models	+ +	+	+	+	+	+
Logic Based Models	+ +	-	-	-	+ +	- -
Ontology Based Models	+ +	+ +	+	+	+ +	+

In DAIDALOS, we developed a context model adequate for the pervasive service platform designed, which demonstrates features of both the graphical and the object oriented models, while it can be extended to incorporate a context ontology. The DAIDALOS Context Model (DCM) addresses, to some degree, all the pervasive computing requirements in Table 1.

A simplified class diagram of the DCM is depicted in Figure 1. This approach is based on the object-oriented programming principles. The DCM is built upon the notion of an **Entity**, which corresponds to an object of the physical or conceptual world. Each Entity instance is associated with a specific **EntityType**, e.g., person, service, place, terminal, preferences, etc, modelled by the homonymous class. Entities may demonstrate various properties, e.g., "height", "colour", "address", "location", etc, which are represented by **Attributes**. Each Attribute is related to exactly one Entity, and belongs to a specific **AttributeType**. An Entity may be linked to other Entities via **DirectedAssociations** (DirAs), such as "owns", "uses", "located in", "student of", etc, or **UndirectedAssocations** (UndirAs), such as "friends", "team-mates", etc. The DirAs are relationships among entities with different source and target roles. Each DirA originates at a single entity, called the parent entity, and points to one or more entities, called child entities. The UndirAs do not specify an owner entity, but form generic associations among peer entities. All Entities, Attributes and Associations are marked with a timestamp indicating their most recent update time. The Attributes and Associations also have an activation status boolean parameter, which indicates whether or not these instances are currently activated. In the set of Attributes of the same type that belong to a specific Entity, only one may be activated at a given time. All Entities, Attributes and Associations implement an XML interface, which enables them to be (de)serialised for the purpose of remote communication.

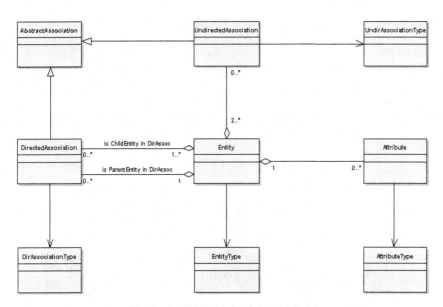

Fig. 1. The DAIDALOS Context Model

The advantages of the DCM are outlined in the following. First, it is sufficiently generic and extendable in order to capture arbitrary types of context information. Second, via the timestamp property, it addresses the temporal aspects of context, a quite critical feature in pervasive computing systems as context information may change constantly and unpredictably. Third, it supports activation/deactivation functionality to indicate the current context value or user selection. Fourth, it addresses the fact that context information is usually highly interrelated, introducing various types of associations between entities. Fifth, it could be seen as a distributed model, where each entity contains its own information and the type of relationship with other entities. Sixth, it is flexible and scalable, as the addition/deletion of an entity does not require the modification of the content or status of the existing entities.

Future plans involve the extension of the DCM in order to support multiple representations of the same context in different forms and different levels of abstraction, and should be able to capture the relationships that exist between the alternative representations. New classes will be added, derived from the Attribute base class, which will be used to add logic to the context data. These classes will serve as "context data translators", e.g., a class TemperatureAttribute will provide methods for accessing temperature information encoded either in Celsius or Fahrenheit degrees. Furthermore, in future versions the DCM will be extended to express the fact that context may be imperfect, i.e., outdated, inaccurate, or even unknown. To this effect, context quality parameters will be included and stochastic modelling principles will be used.

4 A Context Awareness Enabling Architecture

The Context Management Subsystem (CMS) establishes the context-awareness related functionality in the DAIDALOS pervasive services platform thoroughly described in [24]. It provides context consumers with a uniform way to access context information, thus hiding the complexity of context management. Furthermore, it offers streamlined mechanisms for large-scale distribution and synchronization of personal repositories across multiple administrative domains. The set of the software components that constitute the CMS and offer the above functionality are depicted in Figure 2 and are briefly described in the following paragraphs.

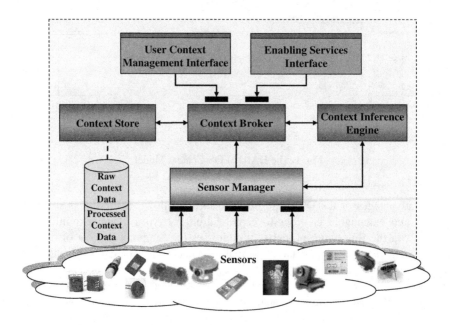

Fig. 2. The Context Management Subsystem architecture

The *Context Broker* (CoB) answers context requests and serves as the access point to the context data. A context consumer requests context from the broker that handles the subsequent negotiation. This negotiation is based on the requestor's context requirements (e.g., type, accuracy, cost, time of update, priorities), the context owner's authorization settings (e.g., access rights for privacy) and the available context and its quality parameters. Based on this information, it decides on the most appropriate context source among multiple sources and providers. It may retrieve the context information from the local Context Store, from the inference engine, from peer CoBs or from another appropriate context provider. CoBs are also responsible for maintaining the consistency of replicated context entities, an issue further examined in a subsequent section. Finally, upon addition, update or removal of context data, the CoB triggers the pertinent subscribed context events using the designed notification mechanisms.

The *Inference Engine* (IE) infers additional and not directly observable context information. It enhances the intelligence of the system as it produces useful information and extracts patterns that have high added value for both consumers and providers. A chain of components is set up to allow the enrichment of context, by combining, converting, or refining context. The functionality of the IE is very important for pervasive services provision, as it greatly contributes to the minimisation of user interaction with the pervasive computing system. A typical example of inference is that of estimating the activity of a person given some lower level sensor data.

The *Context Store* retrieves context from the Sensor Manager (raw data) and the Inference Engine (context data inferred), processes this data, stores it to the appropriate repositories and updates the context databases. Context retrieval implies pulling data or receiving pushed data and converting the data in the context system's uniform format. The Context Store provides context to the CoB on demand.

The *Sensor Manager* keeps track of the sensors which are associated with or attached to the local host, configures them according to the device's requirements, and updates new sample data in the context system. Since the functionality of the Sensor Manager is quite important for the support of context-awareness in pervasive environments, this entity will be studied in more detail in the following section.

Two main interfaces are offered to actors or other entities outside the CMS by the CoB: the User Context Management Interface and the Enabling Services Interface. The *User Context Management Interface* allows a context consumer (user, service, content and context provider) to configure any data previously defined. The *Enabling Services Interface* is offered to the rest of the components in the pervasive system architecture in order to support and control access to the context data maintained by the platform. The designed CMS interacts and cooperates with peer systems in foreign domains via service level agreements and federation mechanisms negotiating the provision/acquisition of additional context information.

Both the presented DCM and a simplified version of the described CMS architecture have already been evaluated in a prototype implementation of the DAIDALOS pervasive service platform. This prototype was built on an OSGi Service Platform [25]. The OSGi™ specifications define a standardized, component oriented, computing environment for networked services. For remote communication SOAP [26] was used, while the discovery of services and components was based on the SLP protocol [27].

5 Sensor Management Supporting Context-Aware Services

Primarily, dynamic information is collected by sensors that monitor the state of resources or the environment at regular intervals. In general, sensors are either attached to a device (e.g., a mobile terminal's CPU load sensor), or are at least associated with a device (e.g., a location or temperature sensor). Since devices may have different configuration requirements with respect to sensors, each CMS has a Sensor Manager (SM). The SM keeps track of the sensors which are associated with or attached to the local host, configures them according to the device's requirements, and updates new sample data in the CMS. Consequently, when a new sensor is activated it must first register with a SM. On receiving a registration request, the SM adds the sensor's identification to the set of active sensors. Subsequently, it retrieves a suitable sensor configuration from the CMS and forwards it to the sensor. A sensor configuration comprises the required sample rate, the accepted sample types (given that the sensor samples different types of data), or some thresholds. Generally, the SM requires that sensors send sample data at regular intervals, allowing the SM to keep track of the sensor state. When the SM does not receive sample data within a certain time frame, it considers the sensor to be inactive and updates its state in the CMS database accordingly. In that sense, the SM is also a data source and can be considered to be a virtual sensor. On receiving sample data updates the SM inspects the sample source and its type and stores it in the CMS database. To this end, each sensor has a corresponding entity data object in the CMS database. Moreover, the SM fires an event and dispatches it to the Inference Engine to inform it about the update. On receiving the event the Inference Engine must decide whether other entity objects must be updated as a consequence.

In general, a sensor (e.g., temperature or GPS sensor) is attached and thus associated with a single device. Obviously, there is also a strong semantic and functional association between the device and the sensor. That is, the sample data is used predominantly by the local CMS and rarely retrieved by remote ones. Yet there is another kind of sensor which is characterised by its association with multiple devices. A location sensor located in the supporting environment which samples location information of mobile devices based on signal strength measurement or on the cell of origin is associated with multiple devices. Such a sensor is also registered with multiple SMs and has thus multiple configurations. To this end, this sensor implements handlers which keep track of the state information of individual SMs. When a device location should be determined with the help of a sensor which serves multiple devices, the SM must look-up a suitable location sensor. Sensor look-up is supported by the DAIDALOS service discovery mechanism. Once a suitable sensor reference has been determined, the SM registers with it. Based on the two sensor concepts, the SM gains the flexibility required for capturing the different data types from various information sources.

6   Consistency of Context Data

The research activity in the field of context awareness is very intense, while various context-aware application paradigms have been introduced to the wide market. The supply and demand for context-aware services are now considerable and are estimated to increase significantly in the years to come. Due to the existence of geographically wide business domains and the extended inter-domain activities, context management systems are expected to collect, store, process, and disseminate context information from data generated at geographically dispersed sites. In such large-scale distributed systems the cost of remote communication is sufficiently high to discourage the system relying on a context entity hosted at just a single store. Thus, the maintenance of multiple context entity replicas across several administrative domains may improve the system performance with regards to the context retrieval delay. Nevertheless, the existence of entity replicas requires the adoption of appropriate synchronisation mechanisms to ensure data consistency. To enforce concurrency control, update transactions must be processed on all entity replicas upon the update of any single replica.

The problem of strong [28] or weak [29] replica consistency maintenance has been studied in various domains since the introduction of digital information. The algorithms and protocols designed to solve this problem may be classified in two main categories: the pessimistic and optimistic ones [30]. Pessimistic strategies prevent inconsistencies between replicas by restraining the availability of the replicated system, making worst-case assumptions about the state of the remaining replicas. If some replica operation may lead to any inconsistency, that operation is prevented. Thus, pessimistic strategies provide strong consistency guarantees, while the replicated system's availability is reduced, as any failure of some replica broker or node freezes the distributed (un)locking algorithms [31]. On the other hand, optimistic strategies do not limit availability. All requests are served as long as the replica broker is accessible, as the requestor does not have to wait for all replicas to be locked before performing the update operation. However, this results in weaker consistency guarantees, and in order to restore replica consistency, replica brokers must detect and resolve conflicts [32].

The DAIDALOS CMS prototype deals with update control of context information replicas across multiple remote repositories. The mechanism used to synchronise context information in the distributed CMS databases (DBs) is based upon the employment of "locks", and bears close resemblance to the "single writer multiple readers" token protocol [33], a quite popular pessimistic strategy. In the designed scheme, a single master copy of each context entity exists, while every CMS has to store a list of references to the entity replicas of all the master entity copies it maintains. Each time a request for a non-locally available context entity is performed, the local CMS contacts the affiliated site holding the master entity, which then stores the requestor CMS's address and delivers a replicated copy. Following the GSM principles [34], the home location (HL) concept has been used to signify the CMS of the master entity. The HL address (i.e., the CMS URL) has been encapsulated in the context information identifier. Thus, the HL of the master copy is obtained by interrogating the context identifier. Each context entity has a single associated token or lock, which allows only the replica holding it to update the entity's master copy, while CoB's of several CMSs can simultaneously access a particular entity master copy (get/read operation) without any restriction. If many CoBs are interested in updating a specific context entity, they have to wait until the entity lock is released. This locking mechanism is essential in distributed context management systems, as it ensures the concurrent synchronisation of context updates originating from various and heterogeneous sources such as sensors, network, users and applications. The lock management current implementation is quite simple, and the update mechanism addresses all replicas irrespective of the context properties and context consumers/sources profiles. We are currently working on exploiting consistency control and replica dissemination algorithms originating in the distributed DB domain, but considering the specific nature of context information.

Maintaining entity replicas in distributed context aware systems involves tradeoffs between storage, processing, communication, and outdated information costs, and it should consider the anticipated rate of incoming context update & retrieval requests. Designing competent schemes for consistent replication of context entities distributed throughout the network becomes imperative, considering the dynamic nature and other inherent features of context. For instance, the location information of people travelling on high speed trains is so frequently updated, that the cost of constantly updating the relative entity replicas is prohibitively high. To reduce the respective overall context update & retrieval cost, efficient algorithms should be employed in distributed context management systems, before they are introduced in the wide market.

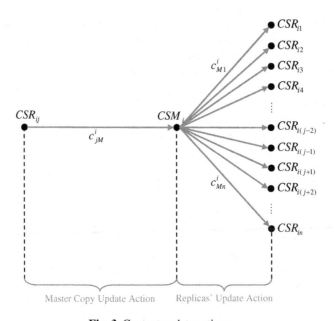

Fig. 3. Context update actions

In our framework, two optimisation problems can be distinguished, that aim to minimise the distributed context management costs (Figure 3). The first one deals with the decision making in the entity-replica-holding CMS, where the context update request is originated. If an update request of context entity i is triggered in the jth CMS domain (CSR_{ij}) that holds a replica of i, the problem is reduced to deciding whether or not it is more efficient to propagate the updated value of i to the master copy (CSM_i), given the estimated retrieval and update requests per time unit, the communication cost, the outdated information cost, and the current and former values of context information (*Master Copy Selective Update Problem*). Given the same parameters, the second problem is reduced to the minimisation of cost during the dissemination of updated context information by the master-copy-holding CMS (CSM_i), to the CMSs maintaining the entity replicas (CSR_{ik}) (*Replica's Selective Update Problem*). In future versions of the CMS prototype, we plan to study, design, evaluate and enhance algorithms that solve the combination of the aforementioned problems (*End-to-End Selective Update Problem*). To this respect, epidemic propagation [35], lazy update techniques [36], gossip algorithms [37], and antientropy schemes [38] will be considered.

7 Conclusions and Future Work

In this paper a context management architecture has been presented, adequate for large scale ambient systems that have a strong relation to heterogeneous and distributed networks. Since the system is embedded in a larger framework comprising service discovery and composition, configurable networks, distributed event handling and sensor networks, we have aligned the architecture to provide accessible interfaces to both the context consuming services and the context providers. An object oriented context model has been developed, which addresses the main requirements for context representation, distribution and validation, as well as requirements for the provision of pervasive services and applications. The architecture supports context refinement, uniform access to context data by context consumers, finely grained privacy restrictions on context access, as well as federation of context repositories.

Future plans mainly focus on two research areas. The first will address the context model itself and the inclusion of suitable context ontologies. On the one hand we will need to provide the flexibility to include any kind of domain specific ontologies on behalf of the system operator and/or end-users. On the other hand the scenarios we intend to demonstrate at a project wide level need the support of suitable ontologies which have to be selected and possibly modified. We also plan to extend the model to support multiple representations of essentially the same context information and to treat different levels of abstraction. This is an essential element to support features such as inference. Finally, the model needs to be enhanced so as to represent uncertain, partial or outdated context information, also an important precondition for inference. Our second field will be that of enhancing the interrelation and functions of the context management systems, such as improved and cost aware context synchronisation, issues on the performance of large scale context deployment serving potentially millions of active entities and countless sensors. This task aims to provide a solid base on which future operators can build robust and scalable context systems. As indicated above, the topic of context inference will be addressed allowing existing and future inference algorithms to be employed uniformly.

References

[1] Gupta, P., Moitra, D.: Evolving a pervasive IT infrastructure: a technology integration approach. Personal and Ubiquitous Computing Journal. Vol. 8. No. 1. (2004) 31–41

[2] Satyanarayanan, M.: Pervasive Computing: Vision and Challenges. IEEE Personal Communications Magazine. Vol. 8. No. 4. (2001) 10–17

[3] Dey, A., Salber, D., Abowd, G.: A Conceptual Framework and a Toolkit for Supporting the Rapid Prototyping of Context-Aware Applications. Human-Computer Interaction Journal, Vol. 16. No. 2-4. (2001) 97–166

[4] Want, R., Hopper, A., Falcao, V., Gibbons, J.: The active badge location system. ACM Transactions on Information Systems. Vol. 10. No 2. (1992) 91–102

[5] Schilit, B., Adams, N., Want, R.: Context-Aware Computing Applications. Proc. of the Workshop on Mobile Computing Systems and Applications, IEEE Computer Society, Santa Cruz (1994) 85–90

[6] Dey, A., Abowd, G., Wood, A.: CyberDesk: A Framework for Providing Self–Integrating Context–Aware Services. Knowledge Based Systems, Vol. 11. No. 1. (1998) 3–13

[7] Abowd, G.D., Atkeson, C.G., Hong, J., Long, S., Kooper, R., Pinkerton, M.: Cyberguide: A mobile context-aware tour guide. ACM Wireless Networks. Vol. 3. No 5. (1997) 421–433

[8] DeVaul, R., Pentland, A.: The Ektara Architecture: The Right Framework for Context-Aware Wearable and Ubiquitous Computing Applications. The Media Laboratory, MIT (2000)

[9] Beigl, M., Gellersen, H.W., Schmidt, A.: Mediacups: Experience with design and use of computer-augmented everyday objects. Computer Networks, Vol. 35. No. 4. Elsevier (2001) 401–409

[10] Technology for Enabling Awareness (TEA), http://www.teco.edu/tea/.

[11] Lei, H., Sow, D.M., Davis II, J.S., Banavar, G., Ebling. M.R.: The design and applications of a context service. ACM SIGMOBILE Mobile Computing and Communications Review, Vol. 6. No. 4. (2002) 45–55

[12] Henricksen, K., Indulska, J., Rakotonirainy, A.: Modeling context information in pervasive computing systems. 1st International Conference on Pervasive Computing, Springer, Lecture Notes in Computer Science. Springer-Verlag, Zurich, Switzerland (2002) 167–180

[13] Voida, S., Mynatt, E., MacIntyre, B., Corso, G.: Integrating virtual and physical context to support knowledge workers. IEEE Pervasive Computing, Vol. 1. No. 3. (2002) 73–79

[14] Chen, G., Kotz, D.: Context-sensitive resource discovery. First IEEE International Conference on Pervasive Computing and Communications. (2003) 243–252

[15] Garlan, D., Siewiorek, D., Smailagic, A., Steenkiste, P.: Project Aura: Towards Distraction-Free Pervasive Computing. IEEE Pervasive Computing, Vol. 1. No. 2. (2002) 22-31

[16] Kanter, T.: Attaching Context-Aware Services to Moving Locations. IEEE Internet Computing, Vol. 7. No. 2. (2003) 43–51

[17] Kindberg, T, Barton, J., Morgan, J., Becker, G., Caswell, D., Debaty, P., Gopal, G., Frid, M., Krishnan, V., Morris, H., Schettino, J., Serra, Spasojevic, M.: People, Places, Things: Web Presence for the Real World. ACM MONET (Mobile Networks & Applications Journal) Vol. 7. No. 5. (2002) 365–376

[18] Chen, H., Finin, T., Joshi, A.: An ontology for context-aware pervasive computing environments. Special Issue on Ontologies for Distributed Systems, Knowledge Engi-neering Review. Vol. 18. No. 3. (2004) 197–207

[19] Gu, T., Pung, H.K., Zhang, D.Q.: A Service-Oriented Middleware for Building Context-Aware Services. Journal of Network and Computer Applications (JNCA). Vol. 28. No. 1. (2005) 1–18

[20] Fahy, P. and Clarke, S. CASS: Middleware for Mobile Context-Aware Applications. ACM MobiSys Workshop on Context Awareness, Boston, USA (2004).

[21] Biegel, G., Cahill, V.: A Framework for Developing Mobile, Context-aware Applications. 2nd IEEE Conference on Pervasive Computing and Communications, Orlando, FL, USA (2004)

[22] Xynogalas, S., Chantzara, M., Sygkouna, I., Vrontis, S., Roussaki, I., Anagnostou, M.: Context Management for the Provision of Adaptive Services to Roaming Users, IEEE Wireless Communications. Vol. 11. No. 2. (2004) 40–47

[23] Strang, T., Linnhoff-Popien C.: A Context Modeling Survey. Workshop on Advanced Context Modelling, Reasoning and Management as part of UbiComp 2004 - The Sixth International Conference on Ubiquitous Computing, Nottingham/England (2004)

[24] B. Farshchian, B., Zoric, J., Mehrmann, L., Cawsey, A., Williams, H., Robertson, P., Hauser, C.: Developing Pervasive Services for Future Telecommunication Networks. IADIS International Conference WWW/Internet, Madrid, Spain (2004) 977–982

[25] OSGi Alliance, http://www.osgi.org/.

[26] Simple Object Access Protocol (SOAP), http://www.w3.org/TR/soap/.

[27] Guttman, E.: Service Location Protocol: Automatic Discovery of IP Network Services. IEEE Internet Computing. Vol. 3. No. 4. (1999) 71–80

[28] Triantafillou, P., Neilson, C.: Achieving Strong Consistency in a Distributed File System. IEEE Transactions on Software Engineering, Vol. 23. No. 1. (1997) 35–55

[29] Lenz, R.: Adaptive distributed data management with weak consistent replicated data. ACM symposium on Applied Computing, Philadelphia, USA (1996) 178–185

[30] Ciciani B., Dias, D.M., Yu, P.S.: Analysis of Replication in Distributed Database Systems. IEEE Transactions on Knowledge and Data Engineering, Vol. 2. No. 2. (1990) 247–261

[31] Wiesmann, M., Pedone, F., Schiper, A., Kemme, B., Alonso, G.: Understanding replication in databases and distributed systems. 20th International Conference on Distributed Computing Systems, Taipei, Taiwan (2000) 264–274

[32] Parker, D.S., Popek, G.J., Rudisin, G., Stoughton, A., Walker, B., Walton, E., Chow, J., Edwards, D., Kiser, S., Kline, C.: Detection of mutual inconsistency in distributed systems. IEEE Transactions on Software Engineering. Vol. 9. No. 3. (1983) 240–247

[33] Li, K., Hudak, P.: Memory coherence in shared virtual memory systems. ACM Transactions on Computer Systems. Vol. 7. No. 4. (1989) 321–359

[34] Mouly, M., Pautet, M.B.: The GSM System for Mobile Communications. Telecom Pub (1992)

[35] Holliday, J., Steinke, R., Agrawal, D., Abbadi A.-El.: Epidemic Algorithms for Replicated Databases. IEEE Transactions on Knowledge and Data Engineering, Vol. 15. No. 5. (2003) 1218–1238

[36] Ladin, R., Liskov, B., Shrira, L., Ghemawat, S.: Providing High Availability Using Lazy Replication. ACM Transactions on Computer Systems, Vol. 10. No. 4. (1992) 360–391

[37] Schótt, T., Schintke, F., Reinefeld, A.: Efficient Synchronization of Replicated Data in Distributed Systems. International Conference on Computational Science ICCS 2003, Springer LNCS 2657, (2003) 274–283

[38] Petersen, K., Spreitzer, M.J., Terry, D.B., Theimer, M.M., Demers, A.J.: Flexible Update Propagation for Weakly Consistent Replication. 16th ACM Symposium on Operating Systems Principles, Saint-Malo, France (1997) 288–301

A Context Architecture for Service-Centric Systems

Johanneke Siljee, Sven Vintges, and Jos Nijhuis

University of Groningen, Department of Computer Science,
PO Box 800, 9700 AV Groningen, the Netherlands
{b.i.j.siljee, j.a.g.nijhuis}@cs.rug.nl
s.vintges@home.nl

Abstract. Service-centric systems are highly dynamic and often complex systems, with different services running on a distributed network. For the design of context-aware service-centric systems, paradigms have to be developed that deal with the distributed and dynamic nature of these systems, and with the unreliability and unavailability problems of providing information on their context. This paper presents a context architecture for the development of context-aware service-centric systems that provides the context information and deals with these challenges.

1 Introduction

Building software from services is emerging as the new software paradigm [1], [2], [3]. These service-centric systems consist of multiple services, possibly from different service providers. Service-centric systems are highly dynamic systems, as service discovery and binding happen at runtime. To exploit the full potential of software services, we need to be able to rapidly develop flexible, dependable service-centric systems that can adapt dynamically to their *context* [4]. The context of a software system is any environment information that influences the functional and non-functional behavior of the system.

A service provider deploys services, describes their functionality and Quality of Service (QoS), and publishes the description to enable its discovery. When developing service-centric systems, a choice has to be made to determine which services to integrate in the system. Often different service providers offer services that all provide the required functionality, but have different QoS characteristics. A choice for a certain service will therefore not only depend on the functional requirements for that service, but also on the QoS requirements. Furthermore, service-centric systems live in a dynamic environment, with different services of the same system distributed over a network. Changes in the system's context (like a change in bandwidth) will often occur. To keep fulfilling the system's requirements, a changing context causes the need for finding new services – optimal for the new context – resulting in a reconfiguration of the system. Thus, service-centric systems need to be context-aware.

The development of context-aware software systems is not a new research area [5], [6]. However, creating context-aware *service-centric systems* provides some new challenges:

T. Strang and C. Linnhoff-Popien (Eds.): LoCA 2005, LNCS 3479, pp. 16–25, 2005.

Distribution: the services of which the service-centric system is composed are highly distributed. Each service may have a different context, and the context of the entire system is highly distributed. It is not possible to locate the sensors that monitor the system context at one central location; the sensors may have to be placed all around the world (or even further away, when considering space systems). This is the case for all context-aware distributed systems, of which service-centric systems are a specific type.

Dynamism: Service-centric systems are highly dynamic, frequently unbinding services and binding with new services, provided by different providers and with different contexts. Services built after deployment of the system are just as easily incorporated in the system, making it impossible to have sensors in place to measure the context of every possible service the system may bind.

To cope with the distribution and dynamism problems, in a service-oriented world a system's context information will often be entirely or partly provided by other services.

Reliability: How can we know if the context information provided by the services is always reliable? Simply trusting a service to always provide correct context information may not be safe. Reliability of context information is a much bigger issue for service-centric systems than for systems that have all their parts combined and tested by one and the same party.

Availability: Since a service-centric system is dependent on other services to provide context information, the availability of the context information cannot be controlled by the system, as would be the case if the context information provision is part of the system itself. Relying on a service to always send the context information in time will cause major problems when the service fails.

A developer of a context-aware service-centric system will have to deal with these extra challenges.

A software architecture provides a global perspective on the software system in question. Architecture-based design of software systems provides major benefits [7], as it shifts the focus away from implementation details towards a more abstract level. Most research on context-aware systems, however, focuses on ad-hoc solutions at the implementation level, as designers lack abstractions to reason about context-aware systems [6], [8]. This hinders the reuse of design paradigms and ideas. The part of the software architecture that obtains context information and presents it to the application is the *context architecture*. It should relieve the application developer of the problem of retrieving context information [9].

In this paper we present a context architecture for service-centric systems. This architecture explicitly deals with distribution, dynamism, reliability and availability of context information, which are typical for service-centric systems.

The remainder of the paper is organized as follows: in the next section we present related work. The elements of the context architecture are discussed in Section 3, and further research and a summary of the paper is given in the last section.

2   Related Work

Examples of application-specific context-aware systems are the Active Badge project [10], which provides a phone redirection service based on the location of a person in an office, the Cyberguide [11], which is a context-aware tour guide to visitors, and the Context Awareness Subsystem (CAS) [12], where the mobile phone is used to gather data about the context.

The Context Toolkit [6] consists of domain-independent, reusable context widgets, interpreters, aggregators, and context-aware services. In [13] an architecture is discussed that uses probes, gauges and gauge consumers. The probes are deployed in the 'target' system, they measure and report data to the gauges. Gauges interpret the data and translate the data into a higher level syntax. The gauge consumer uses the data from the gauges to update models, alert the system, etc. One of the biggest disadvantages about this system is the lack of control on the measurements of the probes; no information is kept about the time of measurement, the error probability, etc.

The Service-oriented Context-Aware Middleware (SOCAM) [14] architecture focuses on providing context to different services. A service can send a request for specific context information to the service locating service, and the service locating service matches the provided context information with the needed context information. Then a reference to the context provider is returned to the requesting service.

None of the above approaches provide support for the unreliability and unavailability of context information of service-centric systems.

3   Context Architecture

In this section we present the context architecture for service-centric systems. The architecture provides separation of concerns while developing context-aware service-centric systems. Fig. 1 shows the complete context architecture for service-centric systems. The architecture contains collectors for measuring the context, a context source for the storage of the context information, a context engine for the analysis of the context information, and an actor for enacting changes in the application system.

3.1   Collectors

Collectors are responsible for retrieving data about the context of the system. They can be located anywhere, and in the case of service-centric systems will be highly distributed. Collectors can be *intrinsic, extrinsic,* or *foreign*.

Intrinsic collectors determine the local context of a software system, by measuring information available inside the system itself (e.g. memory consumption). Extrinsic collectors measure the context outside the system. Foreign collectors do not belong to the system itself, but to a different context-aware system or to external services providing context information. The context information gathered by foreign collectors is available through external context information sources (CISs), which are described further on.

Two types of collectors exist: simple and intelligent collectors. Simple collectors do not perform any processing on the data, and immediately send the data to the context source. Using simple collectors results in a higher bandwidth and memory usage, because the data is not filtered. Furthermore, all the interpretation needs to be performed by the context engine, placing a larger burden upon that part of the system.

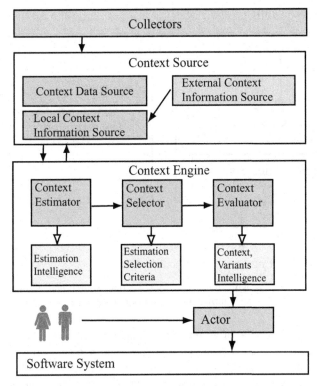

Fig. 1. The context architecture for service-centric systems consists of collectors, a context source, a context engine, and an actor

Intelligent collectors do analyse the context data they measure. Due to this filtering intelligent collectors produce less data; this is useful in systems with limited bandwidth or memory. Furthermore, the context engine has less processing to perform. A disadvantage of intelligent collectors is that (valuable) information may be lost.

When the collector sends the context data to the context source, the collector provides additional information. This additional information consists of the identity of the collector and the time the measurement took place, and may be extended with domain-specific information.

3.2 Context Source

The context source stores the context data from the collectors. The context engine analyzes the data from the context source, and in certain situations adds data to the context source. The context source with all its components is shown in Fig. 2.

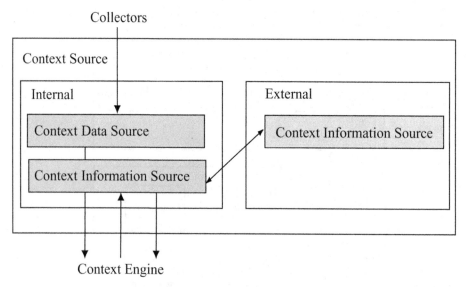

Fig. 2. The context source, with internally a context data source and a context information source, and a subscription to an external context information source

The context source exists of an internal and an external part. The internal part on its turn consists of a context data source (CDS) and a context information source (CIS). The external part consists of one or more external CISs, which belong to the internal parts of context architectures of different systems. The context information stored in these external CISs is retrievable through a subscription.

The CDS is a database that contains the measurements of the collectors. This information is accessible by the context engine. The context engine analyses the data from the CDS, and the results of the analysis are stored in the CIS. The CDS is only locally available and cannot be imported from other systems, nor can it be exported to other systems. This is because the data in the CDS of other systems is analyzed by their context engines, and placed in their CIS. If every system would analyse the same context data, a lot of redundant data transport and processing would be the result, causing unnecessary system load. Therefore, the data in the CDS is only locally available, and the information placed in the CIS is publicly available.

The CIS is used to make context information publicly available and to store context information imported from other systems. This way the context data from foreign collectors can be used. Access to certain information on an external CIS is implemented through a subscription policy. Whenever the CIS is updated, it broadcasts the new information to all its subscribers. This way the system can use the context information measured by other systems or special context information services. This relieves the burden of having to place and move many collectors to deal with the dispersal and dynamism of service-centric systems.

A problem with storing context data is to determine when to remove data. To implement garbage collection different policies can be used:

1. FIFO: the oldest data is removed first.
2. Longest unused: the data that has not been used for the longest period is removed first.
3. Least referenced: the data that has been referenced least (maybe in a certain period of time) is removed first.

Each data element stored in the CDS or in the CIS is called a context element. Context elements have the following properties:

- *Collector ID:* The ID of the collector responsible for fetching the data this element contains. This information is used to discover which collector is responsible for measuring the data and what its type is (i.e. simple or intelligent).
- *Related Context Elements:* Description of the context elements containing the same type of context information. If the data represented in one context element contains errors or is too old, another context element with possibly better data can be found. Related collectors can be defined in multiple ways, two of which are:
  - *Reference*; the collected data elements have a reference to other data elements providing the same data. This can be, for example, a linked list or a web where each data elements points to all other date elements providing the same data.
  - *Record with multiple collectors;* records that contain a collection of data elements representing the same information.
- *Update:* This flag indicates that the item has been updated. This way the context engine can determine if it will be useful to read the context element. After a read from the context engine the flag set to false.
- *Error probability:* This field indicates the probability that the measured data is faulty. When a selector in the context engine finds an error or determines that the measurement was wrong, it increases the probability that the element contains errors. If a selector deems a specific element to be correct, it decreases the error probability. Depending on the intensity of errors or the error probability can be increased or decreased in bigger steps.

3.3 Context Engine

The context engine is responsible for analyzing the context data stored in the CDS. After analysis the resulting information on the context is stored in the CIS and used to adapt the application system to its current context.

Context Estimator

In the CDSs the measurements of the context collectors are saved. The context estimators use these measurements to estimate the context situation. This means that an estimator translates the measurement data from the collectors into more useful information. For example, an estimator translates (e.g. by using fuzzy logic) the light intensity measured in lux to values like "day", "sunny", "cloudy", "night", "clear night" etc. Other techniques for translating the data are decision trees, neural networks, and ontologies. The knowledge that a context estimator uses for translation is represented by the Estimation Intelligence (see Fig. 3). The estimator can also pass through the data from the context source without translation.

The estimator collects data from the CDS by requesting for a specific type of context information. Multiple ways of requesting the data are:

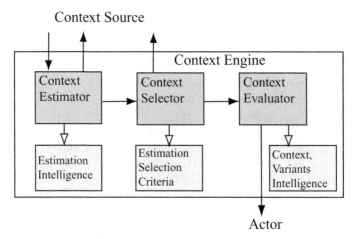

Fig. 3. The data from the context source is processed by the context estimator, the context selector, and the context evaluator, before it is sent to the actor

- *Request by collector:* The estimator request data from a specific collector. The CDS returns the context elements provided by the requested collector.
- *Request by context type:* The estimator requests a specific type of context on a certain moment (in the past or the present). It receives all the context elements for the specific context type.

The estimator can request multiple types of context elements, and aggregating these elements results in a more abstract representation of the context. By aggregating for example the time of day, the processor load of the system, and the number of processes running, the conclusion can be reached that it is night and that the system has enough processing power left for a full system scan or backup.

The abstracted context information is saved in the CIS, as back-up and to provide this information to other context-aware systems.

Context Selector
The purpose of the context selector is to make sure that only the right context information is used by the context evaluator. To make sure that the context evaluator receives the right information, the selector determines which context estimator performs

the best estimation for a specific context type. When the evaluator requests the best estimator for a specific type of context; the selector looks up the possible estimators and passes through the best context estimation to the evaluator.

The alternatives for determining the best estimator are:

- *Error probability:* estimators can output the error probability of the context elements they use; the best choice is the estimator with the lowest error probability.
- *Combine context estimations:* Another method is to combine context estimations. One way to combine context estimations is to compare the different estimations and choose the one that occurs most (i.e. count how many estimators produce the same value). Another way is to determine the current context from different types of estimators.

The strategy for the selection of the best estimator is stored in the Estimation Selection Criteria. The above alternatives can be used to create a list of preferred estimators, reducing the number of estimators used.

When the selector notices errors coming from a certain context element, it will provide feedback to the context source, which adjusts the error probability of the context element (and its representative context estimators) accordingly.

Context Evaluator

After the selector has provided the context evaluator with the best context estimation, the evaluator will use the information to determine the best configuration of the application system. When the context has changed, the evaluator decides to either keep the system running as it is or to reconfigure the system. If the latter is the case, then it is also the evaluator's task to decide what to change. This decision is based on the current context and on the current configuration.

The first necessity for making such a choice is the knowledge of which choices are available. Therefore the evaluator needs to know all the possible variations of the system, and have a mapping with how suitable they are for the given context situation. A first mapping will have to be made during development, but it can evolve dynamically during operation. Due to the fact that this lies beyond the scope of this research we will not go into more detail on the methods for doing so.

After the evaluator chooses the best new configuration for the new context, it makes sure that the changes are enacted by invoking the actor.

3.4 Actor

The actor is responsible for enacting changes in the running application system. The actor performs as the link between the implementation of the context architecture, the application system, and possibly the user. As most users will first want to control the changes in the system, it is possible to let the context evaluator propose changes, which the user can then overrule if desired. After a while the involvement of the user can be decreased until finally the context-aware system is completely self-adaptive.

It is also possible to combine multiple implementations of the context architecture at the actor, see Fig. 4. The actor uses a weight factor to decide which of the propositions for change will be invoked and which will be ignored.

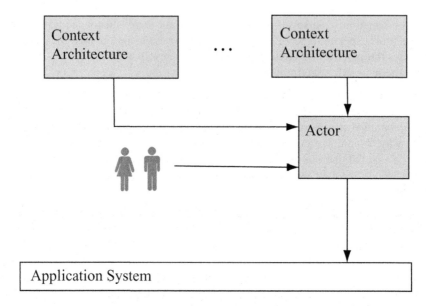

Fig. 4. The actor uses the input of one or more implementations of the context architecture and possibly the user to enact changes in the application system

4   Conclusion and Further Research

The context architecture proposed in this paper is tailored to deal with the characteristics of service-centric systems. Context information coming from context providing services or context architectures of other systems can easily be integrated, solving the distribution and dynamism problems of service-centric systems. Unreliable context information can be detected and ignored by using the error probability and related context element properties of the context elements, and by combining context estimations. Possible unavailability of context information will not cause system crashes, as the context information is stored as back-up and related context elements providing the same type of context information are listed.

As the architecture is very general, it can be developed separately from the software system, enabling separation of concerns. The few points where direct intervention with the software system is needed are the collectors monitoring the internal context of the application system, and the actor, which is responsible for reconfiguration.

The next challenge is to further define the context evaluator, i.e. the mapping between the context information and the possible configurations of the application system. Currently, this mapping is always application-specific, defined by the systems developer.

The actor allows the connection of multiple context architectures to the software system. Another interesting research topic is the policy for conflicting advices from the different evaluators.

Acknowledgements

This research has been sponsored by the SeCSE project, under contract no. IST-511680. The primary goal of the SeCSE project is to create methods, tools and techniques for system integrators and service providers and to support the cost-effective development and use of dependable services and service-centric applications.

References

1. Duggan J., "Simplify Your Business With an SOA Approach", AV-18-6077, Gartner (2002)
2. Microsoft, Service-Oriented Architecture: Implementation Challenges, http://msdn.microsoft.com/library/en-us/dnmaj/html/aj2soaimpc.asp (2004)
3. Schmelzer R., "Service-Oriented Process Foundation Report", ZTR-WS108, ZapThink (2003)
4. SeCSE, Service Centric Systems Engineering, IST project, contract no. IST-511680
5. Horn P., Autonomic Computing: IBM's Perspective on the State of Information Technology, IBM Corporation (2001)
6. Dey A. K., Abowd G. D., Salber D., "A conceptual framework and a toolkit for supporting the rapid prototyping of context-aware applications", *Human-Computer Interaction*, vol. 16 (2-4) (2001) 97-166
7. Shaw M., Garlan D., Software Architecture: Perspectives on an Emerging Discipline, Prentice Hall, Upper Saddle River, New Jersey (1996)
8. Brown P. J., Bovey J. D., Chen X., "Context-Aware Applications: from the Laboratory to the Marketplace", *IEEE Personal Communications*, vol. 4 (5) (1997) 58-64
9. Ensing J., "Software architecture for the support of context aware applications", NL-UR 2002-841, Philips (2002)
10. Want R., Hopper A., Falcao V., Gibbons J., "The Active Badge Location System", *ACM Transactions on Information Systems*, vol. 10 (1) (1992) 91-102
11. Long S., Kooper R., Abowd G. D., Atkeson C. G., "Rapid Prototyping of Mobile Context-Aware Applications: The Cyberguide Case Study", *Proceedings of the 2nd ACM International Conference on Mobile Computing and Networking (MobiCom'96)*, ACM, Rye, New York (1996)
12. Lonsdale P., Baber C., Sharples M., Arvanitis T., "A Context Awareness Architecture for Facilitating Mobile Learning", *Proceedings of MLEARN 2003: Learning with Mobile Devices*, London, UK (2003)
13. Garlan D., Schmerl B., Chang J., "Using Gauges for Architecture-Based Monitoring and Adaptation", *Working Conference on Complex and Dynamic Systems Architecture*, Brisbane, Australia (2001)
14. Gu T., Pung H. K., Zhang D. Q., "A Service-Oriented Middleware for Building Context-Aware Services", *Proceedings of IEEE Vehicular Technology Conference (VTC 2004)*, Milan, Italy (2004)

Towards Realizing Global Scalability in Context-Aware Systems

Thomas Buchholz and Claudia Linnhoff-Popien

Ludwig-Maximilian-University Munich,
Mobile and Distributed Systems Group, Institute for Informatics,
Oettingenstr. 67, 80538 Munich, Germany
{thomas.buchholz, claudia.linnhoff-popien}@ifi.lmu.de,
www.mobile.ifi.lmu.de

Abstract. Context-aware systems are systems that use context information to adapt their behavior or to customize the content they provide. Prior work in the area of context-aware systems focused on applications where context sources, Context-Aware Services (CASs), and users are in each other's spatial proximity. In such systems no scalability problem exists. However, other relevant CASs are subject to strong scalability problems due to the number of users, the geographical distribution of the context sources, the users, and the CAS as well as the number of organizations that participate in the provision of the context-aware system.

In this paper, we classify CASs according to their scalability needs and review context provision and service provision infrastructures with regard to their scalability. For building large-scale context-aware systems it is not sufficient to Zuse large-scale service provision and context provision infrastructures. Additionally an integration layer is needed that makes the heterogeneous access interfaces of the context provision infrastructures transparent for the CASs. We outline our proposal for such an integration layer and describe how it can be combined with an infrastructure that allows handhelds themselves to gather context information in their immediate vicinity via wireless technologies.

1 Introduction

Context-aware systems are systems that use context information to adapt their behavior or to customize the content they provide. In these systems three main types of entities can be distinguished: Context Sources (CS), Context-Aware Services (CASs) and users.

Early research in ubiquitous computing focused on applications where all entities involved in a user session are located in each other's spatial proximity like figure 1(a) illustrates. In such a setting there is no scalability problem. [1] coined the term "localized scalability" for this principle. Localized scalability is reached "by severely reducing interactions between distant entities" [1] or by locally restricting the distribution of information [2].

However, there are applications where the context sources, the CASs and the users are not colocated. Thus, interactions between distant entities are needed. In these situations the question arises how global scalability in context-aware systems can be reached. The key to this lies in replicating and distributing CASs as well as context information.

T. Strang and C. Linnhoff-Popien (Eds.): LoCA 2005, LNCS 3479, pp. 26–39, 2005.

The paper is structured as follows: In section 2 we argue under which conditions a need for global scalability exists and give examples for applications. Section 3 reviews approaches for context provision as well as for service provision. In section 4 we discuss how the existing approaches can be combined and show which components are still missing in order to build large-scale context-aware systems. We describe our proposals for these components in section 4.1 and 4.2. In section 5 we outline directions for future research.

2   From Local Towards Global Scalability

In this section we will define classes of CASs with different scalability needs and will provide examples for these classes. To be able to discuss the scalability of the different application classes we first need to define what scalability is. According to [3] scale consists of a numerical, a geographical, and an administrative dimension. The numerical dimension refers to the number of users, deployed objects like e.g. context sources and services in a system. The geographical dimension means the distance between the farthest nodes, while the administrative dimension consists of the number of organizations that execute control over parts of the system. A large administrative dimension in general leads to heterogeneity. The various organizations are likely to use different hardware, software, protocols, and information models. In such an environment service interoperability becomes a problem. The interoperability problem is classically divided into a signature level, a protocol level, and a semantic level [4]. The syntax of the interface of a service is described on the signature level. It includes in general the name of any operation and the type and sequence of any parameter of the interface. On the protocol level the relative order in which methods of a service are called is specified. The problem of divergent understanding and interpretation is dealt with on the semantic level. [3] defines a system to be scalable if users, objects and services can be added, if it can be scattered over a larger area, and if the chain of value creation can be divided among more organizations without the system "suffering a noticeable loss of performance or increase in administrative complexity."

Locally scalable CASs (figure 1(a)): Most CASs that were built so far colocated the CAS, the context sources, and the users (see e.g. [5] for a survey). Typical examples are shopping and conference assistants. In these systems scalability is not an issue. All ways between the entities are short. The number of context sources and users is low. Moreover, the physical context sources were known when the CAS was developed. Thus, the CASs as well as the context sources were designed to interoperate.

CASs with heightened scalability needs (figure 1(b)-(d)): Recently, some CASs have been built where only two of the three types of entities are colocated and the third is far away. For example, many health surveillance applications [6] assume that sensors are attached to the body of persons with health problems. The gathered data is transmitted via the mobile Internet to a central database where it is analyzed by a physician or by a data-mining process that alerts a doctor if something is wrong (see figure 1(b)). Other applications colocate the user and the CAS (figure 1(c)), but receive context information from distant locations. An example are on-board navigation systems in cars that receive

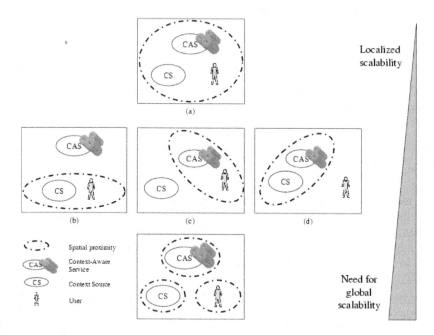

Fig. 1. Depending on the application context-aware services, context sources and users may be colocated or far apart

information about traffic jams via radio technologies. Surveillance of remote areas with the help of sensor networks is an application where in general the context sources and the CAS are colocated. When something interesting happens the user that is far away is informed (figure 1(d)). These types of CASs cause moderate scalability problems.

CASs with potentially global scalability needs (figure 1(e)): The largest scalability problems arise when the context sources, the CASs, and the users are far apart from each other. Examples are context-based push information services, contextualized yellow pages services, and server-based navigation applications. A context-based push information services for example is a friend-finder (cp. [7]). A friend-finder constantly tracks all members of a community within a large area. Whenever two members with similar interests enter each others vicinity, they are informed if their current activity does not forbid this (if e.g. one must not be disturbed because he is in a meeting). Another example is an application that reminds people to buy e.g. milk when they are near a supermarket. Contextualized yellow pages services (cp. [7]) help people to find nearby sights, restaurant, gas stations etc.. Scalability problems arise especially if dynamic properties of target objects (target context) and of objects between the user's current position and the target's position (transition context) are included into the recommendations given to the user. For example, a tour guide might incorporate the current waiting times at major sights into the proposed tour or a restaurant finder might suggest only places that can be reached within 15 minutes and still have seats available. To determine which restaurants are reachable within a certain time, the CAS finds out which bus stops are near the user and when the next bus will arrive including all known current delays. If the user is within a car, the current traffic situation is taken into account. A server-based navigation system keeps all maps on the server and calculates routes based on the user's position, the point he wants to reach and dynamic information like traffic jams, weather condition, and road works.

The scalability problem of these applications is mainly caused by three properties of the overall system: 1. The CAS can be invoked from within a very large area. 2. Many persons might want to use the CAS. Thus, it must be ready to cope with a high workload. Moreover, the context of a plethora of entities must be accessible for the CAS. 3. The context sources will have different access interfaces in different regions and will be operated by different organizations. For example, a navigation system that uses traffic information for its recommendations will have to address that this information will be in a different language and format in Germany and France. Standards can partly solve this problem. However, many competing standards exist. For example, a traffic information system in the USA will express distances in miles and feet, while in Europe kilometers are used. For all these problems solutions must be found to allow for scalable context-aware services.

After having outlined for which classes of CASs global scalability is needed, we will review approaches for context provision and CAS provision with regard to their suitability for building globally scalable context-aware systems.

3   Approaches for Context Provision and CAS Provision

In order to build globally scalable context-aware systems the context provision process as well as the provision of the CAS must be scalable. We will address both subsystems in turn.

3.1   Scalable Context Provision

Scalability research in ubiquitous computing so far has mainly focused on the question how the context provision process can be made scalable. Figure 2 classifies the context provision approaches.

Context Provision approaches can be divided into infrastructureless and infrastructure-based approaches. If no infrastructure is employed, the CAS retrieves information from the context sources directly via wireless links or a multi-hop routing scheme is used. Multi-hoping is used in Wireless Sensor Networks (WSN) [8] where context sources collaborate and in Data Diffusion Systems [9, 10] where context sinks interact to distribute information. These technologies are only suitable if the distance between the context source and the context sink is not too large (limited geographical scalability).

Early infrastructure-based systems directly linked CASs with sensors that were attached to a Local-Area Network (tight coupling). To loosen the coupling two main approaches can be distinguished. Service-centric approaches provide a homogeneous interface for context retrieval from sensors. The Context Toolkit [11] (CTK) was the first representative of this school of thought. It encapsulates sensors in so called widgets that provide a unified interface to the CAS. The heterogeneity is diminished, but application-specific code still needs to interact directly with context sources, i.e. sen-

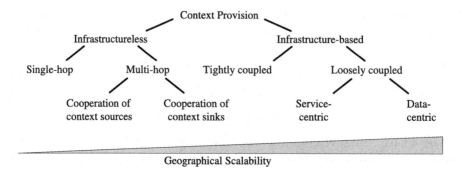

Fig. 2. Classification of context provision approaches

sors plus widgets. Other representatives are Solar [12] and iQueue [13, 14]. Data-centric systems introduced a data repository (e.g. a tuple-space) between the context sources and the CASs (e.g. [2]). The context sources provide context information to the data repository. CASs can query the repository via a query language. In these systems the CASs only need to comply to the service interface of the data repository and the used information model of the context sources.

Many proposals were made how geographically scalable context provision infrastructure can be built using the loose coupling paradigm, e.g. ConFab [7, 15], Solar [12], GLobal Smart Space (GLOSS) [16], Strathclyde Context Infrastructure (SCI) [17, 18, 19], IrisNet [20, 21], and iQueue [13, 14]. All these approaches connect the sensors to a local server that either acts as a data repository (ConFab, GLOSS, IrisNet), as an execution environment (Solar, iQueue), or as both (SCI). Some systems assign sensors to local servers merely based on geographical proximity (Solar, GLOSS, IrisNet, iQueue). Others make the data repositories a representative of entities in an environment (like people, places, and things) (ConFab, SCI). Data repositories are in general based on tuple-space technology (ConFab, GLOSS, SCI), but also XML-databases are employed (IrisNet). The large-scale context provision infrastructures must be able to answer queries that require data from many local servers. Thus, the local servers need to be confederated in some way. Many systems use Peer-to-Peer routing technology for this purpose (Solar, GLOSS, SCI). Others use XML hyperlinks between the data repositories (ConFab) or employ the Domain Name System (IrisNet). The information models that the approaches specify to express queries are very heterogeneous, though most of them are XML-based. Some describe context sources and chains of operators for the generation of context information. Others define functions that calculate the requested context information based on context data that is described in an attribute-based manner.

3.2 Scalable CAS Provision

Scalable context provision infrastructures are a very important building block for scalable context-aware systems. However, even if perfectly scalable context provision were given, there would still be some challenges left to allow for completely scalable context-aware systems:

1. **Numerical dimension:** Large-scale CASs may have millions of users that collectively cause a workload that is much too high for a single server. Thus, replicating and distributing the CAS must be considered.
2. **Geographical dimension:** Large-scale CASs may have users that are spread over a very large area. In this case also relevant context sources will be strongly distributed. Since many interactions especially between the CAS and the context sources, but also between the user and the CAS will be needed, there is a potential that the long ways will lead to large response times of CASs. A solution is to distribute replicas of the CAS to servers that are close to the user and to the context sources.
3. **administrative dimension:** The large-scale context provision infrastructures possess heterogeneous access interfaces. Especially interoperability on the semantic level is a problem. Either a CAS must be able to deal with all of them or an infrastructure must be in place that makes the heterogeneity transparent for the CAS.

Little work has been done on the question how scalable CASs can be built. The only works we are aware of are [6], [22], and [23]. These works suggest to expose context sources and context repositories to the system as grid services. A grid service is a Web Service that conforms to a set of conventions [24]. By using grid technology interoperability on the signature level is reached and the scalability properties of grids can be reused. Besides grids other technologies for scalable service provision exist. They are classified in figure 3.

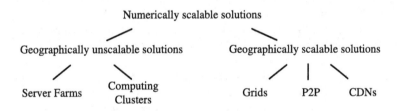

Fig. 3. Classification of CAS provision approaches

If too many requests need to be answered by a web-based application, one of the most often used approaches is to employ a server farm. The application is replicated and installed on many servers that are interconnected by a common high-speed Local-Area Network (LAN). The workload is evenly distributed among the servers by an intelligent switch. Computing clusters are similar to server farms. Computing problems that are too large for a single server are divided into many parts that are solved in parallel by many CPUs. Since the execution of one session of a CAS in general does not overcharge a single server, server farms are more suitable for CASs than computing clusters. Server farms address the numerical dimension, but the geographical dimension of the scalability remains unsolved.

Infrastructures that solve the numerical and geographical scalability problem are grids [25], peer-to-peer (P2P) networks [26], and Content Delivery Networks (CDNs) [27]. All these technologies are able to coordinate globally distributed servers. All three infrastructures have their specific strengths. Grids excel at monitoring resources and balancing load. P2P networks perform content-based routing in a very scalable manner and CDNs are especially good at assigning client requests to nearby servers. The three technologies are converging. Grids will use P2P routing. Recent proposals for P2P networks allow for routing to nearby servers and CDNs use the Apache Tomcat container like the Globus Toolkit [28] (the most widely used toolkit for grid technology) as an execution environment for applications.

To build globally scalable context-aware systems the best choice is to use one of these three numerically and geographically scalable infrastructures to dynamically distribute replicas of CASs and to combine it especially with large-scale context provision infrastructures. However, coupling the CAS provision infrastructures with the various context provision systems poses some challenges like we will discuss in the next section.

4   Components for Scalable Context-Aware Systems

In order to build globally scalable context-aware systems it is necessary to dynamically replicate and distribute CASs on the servers of one of the three numerically and geographically scalable service provision infrastructures that were outlined in the last section. This infrastructure forms the upper layer in figure 4. In this way the numerical and geographical scalability of the CAS is assured.

The replicas of the CASs that are placed on the geographically distributed servers retrieve context information from regional Context Information Services (CIS) (see lower layer in figure 4). A CIS is an abstraction. It is any service that provides context information. In most cases it will be a large-scale context provision infrastructure, but it can also be a single sensor or a user's handheld that provides context information. By combining a large-scale service provision infrastructure with large-scale context provision infrastructures numerical and geographical scalability is reached.

However, the administrative dimension is still a problem: The CISs in the various regions may possess different access interfaces, even if they provide the same type of context information. Figure 4 illustrates this with diversely shaped symbols. Since the developer of a CAS does not know at design time on which servers his CAS will be deployed and which CISs will be used, he cannot anticipate what the access interface of needed CISs will be like. This is a major problem. The replicas of a CAS need an execution environment that is the same on every server. This means that also the access to context information must be identical on every server. Thus, the CAS distribution infrastructure needs to provide an integration layer that binds to the heterogeneous access interfaces of CISs, maps the retrieved information to a standard information model, and provides it to the CASs in a unified way. Such an integration layer is our CoCo infrastructure. It will be outlined in section 4.1.

The CoCo infrastructure resides on each server and binds to local or regional CISs. In general this CISs are large-scale context provision infrastructure. However, for some

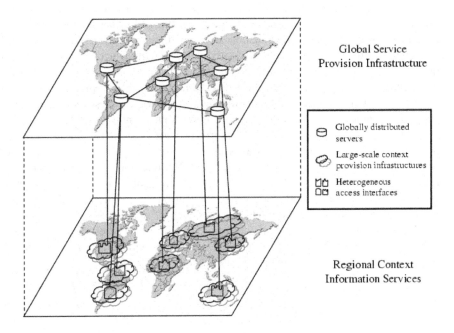

Global Service
Provision Infrastructure

Globally distributed
servers

Large-scale context
provision infrastructures

Heterogeneous
access interfaces

Regional Context
Information Services

Fig. 4. Combining a scalable service provision infrastructure with large-scale context provision infrastructures provides numerical and geographical scalability. However, the administrative dimension of the scalability problem is yet to be solved

applications it might also be useful that the user's handheld itself looks for context information in its immediate vicinity via wireless technologies and passes the context information to the CoCo infrastructure when it invokes the CAS via e.g. UMTS or GPRS. For example, somebody might be looking for restaurants that can be reached within 15 minutes with public transportation means and that are not overly crowded. The information which bus stations are near the user is probably most efficiently retrieved via wireless multi-hoping methods. To find out which candidate restaurants still have free seats available, however, is an information that is generated too far away from the user to be accessible wirelessly. For the wireless gathering of context information in the user's immediate vicinity we developed an infrastructure that will be described in section 4.2.

4.1   Context Composition Infrastructure

The Context Composition (CoCo) infrastructure [29] is a set of collaborating components that generate requested context information. It solves the following problems: It binds to regional CISs and can translate the provided context information into the standard information model. Moreover, it is able to execute workflows that are needed to generate higher-level context information from low-level context information. These workflows and needed pieces of context information are specified in a CoCo document using the CoCo-language which is based on XML. It can be represented as a graph.

Figure 5 shows an example for the generation of context information that a restaurant finder might need. CoCo documents are mainly composed of factory nodes that specify which pieces of context information are needed and operator nodes that describe how context information needs to be adapted, selected, aggregated, or processed in any other way. A directed edge between two nodes stands for a context information flow. In the example the user is looking for nearby restaurants or beer gardens (if the weather is fine) that match his preferences he specified in a profile. He invokes the CAS and passes his phone number. The CAS dynamically adds this phone number to the CoCo document that describes the general context information needs of the restaurant finder. This document is given to the CoCo infrastructure. Based on the phone number the user's profile and his current location can be found out. Not until the user's position has been retrieved, the temperature and the likelihood that it will rain can be requested because these pieces of context information are dependent on the location. Since the CAS is not interested in the temperature and the likelihood of rain directly, an operator node is invoked that uses the former information to determine whether the weather can be considered as good. The retrieved user preferences, the user's location and the decision if the weather is good are given back to the CAS.

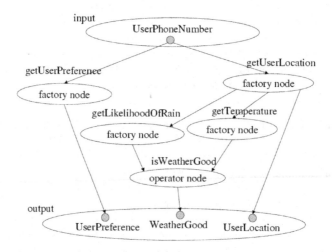

Fig. 5. Example of a CoCo-Graph

The infrastructure that executes the CoCo-Graph consists mainly of a CoCo-Graph Controller, a Context Retriever, the CoCo Processor and two caches like depicted in figure 6.

The CoCo-Graph Controller receives the CoCo document from the CAS and executes the specified steps. Thus, the CoCo-Graph Controller is the component that is in charge of flow control. Whenever it parses a factory node, it sends the corresponding information to the Context Retriever that responds with the respective context information. Operator nodes result in a call to the CoCo Processor. When the output node is reached, the results of the execution of the CoCo document are returned to the CAS.

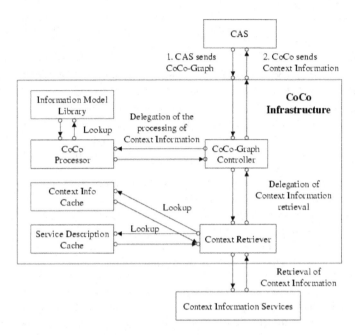

Fig. 6. CoCo Infrastructure

The Context Retriever discovers CISs, selects the most appropriate one and returns the retrieved context information to the CoCo-Graph Controller. Both, the descriptions of discovered CISs and the context information, are cached for later reuse.

The CoCo Processor executes explicit operations that are needed to derive a specific piece of context information from one or many other pieces of context information. For example, the CoCo Processor would process the operator node "isWeatherGood" to decide if the weather is good or bad, based on the retrieved temperature and the likelihood of rain. The CoCo-Graph Controller also delegates implicit instructions to the processor. These are conversion operations between different scales, i.e. formats, of the same context information item. In our scenario, the CoCo Infrastructure locates the user and passes his position to the factory node needed to retrieve the temperature at the user's position. The position might be expressed in WGS84-coordinates, while there might be only CISs available that accept coordinates exclusively in UTM-format. At this point, the CoCo-Graph Controller invokes the CoCo Processor to try to convert the information. For its task the CoCo Processor needs semantic information about the information model and requires conversion operations between the various scales of the same types of context information. This knowledge and functionality is stored in the Information Model Library.

The CoCo infrastructure was conceived as an integration layer between large-scale context provision infrastructures and execution environments for CASs. In the next section we will describe how the CoCo infrastructure can be combined with wireless context provision methods.

4.2   Context Diffusion Infrastructure

We assume that when a user subscribes to a CAS, he downloads a thin client to his hand-held. This thin client uses a Context Diffusion Infrastructure that is already installed on the handheld to gather context information. The main principle of Context Diffusion is that static server nodes announce information via local wireless radio technologies like Bluetooth or IEEE 802.11. Mobile nodes listen to this information and cooperate to disseminate the information by word of mouth.

The core of our Context Diffusion Infrastructure is our data diffusion protocol [30]. It is executed whenever two nodes come into radio range. These can either be two mobile nodes or an immobile and a mobile node. The protocol works as follows:

1. First, profiles are transmitted between the peers. The profiles specify in which types of context information the profile owner is interested.
2. The peer's profile is used to search through the handheld's cache of formerly obtained pieces of context information.
3. All cached pieces of context information that might be of interest for the exchange partner according to his profile are transmitted and are then stored on the receiver's side.

Since each handheld frequently executes a cache invalidation scheme that deletes all pieces of context information that have become outdated or invalid because the user has moved out of the region where the specific piece of context information could be useful, only valid information that refers to the current region is exchanged.

The Context Diffusion Infrastructure offers an API for thin clients to add entries to the profile. Depending on the application's collection strategy, it can either add its profile during its installation phase or when it is invoked. In the former case, the hand-held scans continuously its environment for available context information, in the latter case, context information collection is initiated on demand. The infrastructure allows for push notification and for a request mode. Push notification means that the thin client is informed everytime an interesting piece of context information is received. Alternatively, the thin client can search the cache for relevant entries using the request mode. Our favorite configuration is that the thin client initiates the context retrieval when it is invoked. After a fixed amount of time it searches through the cache for relevant context information, invokes the backend of the CAS via GPRS or UMTS and transfers the context information to it.

Whenever the backend of a CAS is invoked by a client, it analyzes the request to see whether context information was passed. If this is the case, the provided context information is given to the CoCo infrastructure. The CoCo infrastructure receives the information, transforms it into the standard information model if needed, and writes it into the context information cache. The CAS operates in the normal way and requests context information from the CoCo infrastructure with a CoCo document. Since the Context Retriever within the CoCo infrastructure always starts looking in the cache whether needed context information is already stored, the context information provided by the client is very naturally introduced into the standard process of CoCo. The functionality that is in place to check whether cached context information is usable for a certain problem is reused to validate context information that was passed by the client.

Additionally, the information that was gathered by one client is potentially reusable for others. By writing the context information that was gathered by the client into the CoCo cache instead of directly using it, it becomes transparent for the CAS in which way context information was retrieved.

5   Conclusion and Future Work

In this article we have shown that some Context-Aware Services (CASs) need to be globally scalable. We have outlined approaches for context provision and CAS provision and evaluated them with regard to their scalability. Good candidates as a suitable CAS provision infrastructure are grids, P2P networks, and CDNs. These systems need to be coupled with large-scale context provision infrastructures. To provide a homogeneous access interface to context information for replicas of CASs the CAS provision infrastructures need to provide an integration layer that makes the heterogeneity of the access interfaces of the context information services transparent for the CASs. The CoCo infrastructure is such an integration layer. It can be combined with an infrastructure that allows to gather context information in the immediate vicinity of a user. These approaches allow to build globally scalable context-aware services.

Currently, we are working on the development of efficient, but easily applicable heuristics for the distribution of CASs in CDNs. We are developing a simulation that allows to evaluate the resource consumption and improvements of user-perceived latencies that are incurred by the various heuristics. Furthermore, we are refining our infrastructures. Especially, the development of a generic model for context information remains one of the hard research questions.

References

1. Satyanarayanan, M.: Pervasive Computing: Vision and Challenges. IEEE Personal Communications (2001)
2. Schmidt, A.: Ubiquitous Computing - Computing in Context. PhD thesis, Computing Department, Lancaster University, England, U.K. (2002)
3. Neuman, B.C.: Scale in distributed systems. In: Readings in Distributed Computing Systems. IEEE Computer Society, Los Alamitos, CA (1994) 463–489
4. Strang, T., Linnhoff-Popien, C.: Service Interoperability on Context Level in Ubiquitous Computing Environments. In: Proceedings of International Conference on Advances in Infrastructure for Electronic Business, Education, Science, Medicine, and Mobile Technologies on the Internet (SSGRR2003w), L'Aquila/Italy (2003)
5. Chen, G., Kotz, D.: A survey of context-aware mobile computing research. Technical Report TR2000-381, Dept. of Computer Science, Dartmouth College (2000)
6. Barratt et al., C.: Extending the Grid to Support Remote Medical Monitoring. In: 2nd UK eScience All hands meeting, Nottingham, U.K. (2003)
7. Hong, J.I., Landay, J.: An architecture for privacy-sensitive ubiquitous computing. In: Proceedings of The Second International Conference on Mobile Systems, Applications, and Services (Mobisys 2004), Boston, MA, USA (2004)
8. Al-Karaki, J.N., Kamal, A.E.: Routing techniques in wireless sensor networks: A survey. IEEE Wireless Communications (2004)

9. Papadopouli, M., Schulzrinne, H.: Effects of power conservation, wireless coverage and co-operation on data dissemination among mobile devices. In: ACM International Symposium on Mobile Ad Hoc Networking and Computing (Mobihoc), Long Beach, California (2001)

10. Becker, C., Bauer, M., Hähner, J.: Usenet-on-the-fly: supporting locality of information in spontaneous networking environments. In: CSCW 2002 Workshop on Ad hoc Communications and Collaboration in Ubiquitous Computing Environments, New Orleans, USA (2002)

11. Dey, A.: Architectural Support for Building Context-Aware Applications. PhD thesis, College of Computing, Georgia Institute of Technology (2000)

12. Chen, G.: SOLAR: Building a context fusion network for pervasive computing. PhD thesis, Dartmouth College, Hanover, New Hampshire, USA (2004) Technical Report TR2004-514.

13. Cohen, N.H., Purakayastha, A., Wong, L., Yeh, D.L.: iqueue: a pervasive data-composition framework. In: 3rd International Conference on Mobile Data Management, Singapore (2002) 146–153

14. Cohen, N.H., Lei, H., Castro, P., II, J.S.D., Purakayastha, A.: Composing pervasive data using iql. In: 4th IEEE Workshop on Mobile Computing Systems and Applications (WMCSA 2002), Callicoon, New York (2002) 94–104

15. Heer, J., Newberger, A., Beckmann, C., Hong, J.I.: liquid: Context-aware distributed queries. In: Proceedings of Fifth International Conference on Ubiquitous Computing: Ubicomp 2003, Seattle, WA, USA, Springer-Verlag (2003) 140–148

16. Dearle, A., Kirby, G., Morrison, R., McCarthy, A., Mullen, K., Yang, Y., Connor, R., Welen, P., Wilson, A.: Architectural Support for Global Smart Spaces. In: Proceedings of the 4th International Conference on Mobile Data Management. Volume 2574 of Lecture Notes in Computer Science., Springer (2003) 153–164

17. Glassey, R., Stevenson, G., Richmond, M., Nixon, P., Terzis, S., Wang, F., Ferguson, R.I.: Towards a Middleware for Generalised Context Management. In: First International Workshop on Middleware for Pervasive and Ad Hoc Computing, Middleware 2003. (2003)

18. Stevenson, G., Nixon, P.A., Ferguson, R.I.: A general purpose programming framework for ubiquitous computing environments. In: Ubisys: System Support for Ubiquitous Computing Workshop at UbiComp 2003, Seattle, Washington, USA (2003)

19. Thomson, G., Richmond, M., Terzis, S., Nixon, P.: An Approach to Dynamic Context Discovery and Composition. In: Proceedings of UbiSys 03: System Support for Ubiquitous Computing Workshop at the Fifth Annual Conference on Ubiquitous Computing (UbiComp 2003), Seattle, Washington, USA (2003)

20. Gibbons, P.B., Karp, B., Ke, Y., Nath, S., Seshan, S.: IrisNet: An Architecture for a World-Wide Sensor Web. IEEE Pervasive Computing 2 (2003)

21. Deshpande, A., Nath, S., Gibbons, P.B., Seshan, S.: Cache-and-query for wide area sensor databases. In: ACM SIGMOD 2003. (2003)

22. Storz, O., Friday, A., Davies, N.: Towards 'Ubiquitous' Ubiquitous Computing: an alliance with the Grid. In: System Support for Ubiquitous Computing Workshop at the Fifth Annual Conference on Ubiquitous Computing (UbiComp 2003), Seattle (2003)

23. Jean, K., Galis, A., Tan, A.: Context-Aware GRID Services: Issues and Approaches. In: International Conference on Computational Science, Krakow, Poland (2004) 166–173

24. Tuecke et al., S.: Open Grid Services Infrastructure (OGSI), Version 1.0. Global Grid Forum Draft Recommendation (2003)

25. Berman, F., Fox, G., Hey, A.J., eds.: Grid Computing: Making the Global Infrastructure a Reality. Wiley (2003)

26. Lua, E.K., Crowcroft, J., Pias, M., Sharma, R., Lim, S.: A Survey and Comparison of Peer-to-Peer Overlay Network Schemes. IEEE communications survey and tutorial (2004)

27. Rabinovich, M., Spatscheck, O.: Web Caching and Replication. Addison Wesley (2002)
28. Globus Alliance: The Globus Toolkit (2004) http://www-unix.globus.org/toolkit/.
29. Buchholz, T., Krause, M., Linnhoff-Popien, C., Schiffers, M.: CoCo: Dynamic Composition of Context Information. In: The First Annual International Conference on Mobile and Ubiquitous Systems: Networking and Services (MobiQuitous 2004), Boston, Massachusetts, USA (2004)
30. Buchholz, T., Hochstatter, I., Treu, G.: Profile-based Data Diffusion in Mobile Environments. In: Proceedings of the 1st IEEE International Conference on Mobile Ad-hoc and Sensor Systems (MASS 2004), Fort Lauderdale, USA (2004)

Towards LuxTrace: Using Solar Cells to Measure Distance Indoors

Julian Randall, Oliver Amft, and Gerhard Tröster

Wearable Computing Laboratory, IfE,
ETH Zürich, Switzerland
{randall, amft, troester}@ife.ee.ethz.ch

Abstract. Navigation for and tracking of humans within a building usually implies significant infrastructure investment and devices are usually too high in weight and volume to be integrated into garments.

We propose a system that relies on existing infrastructure (so requires little infrastructure investment) and is based on a sensor that is low cost, low weight, low volume and can be manufactured to have similar characteristics to everyday clothing (flexible, range of colours).

This proposed solution is based on solar modules. This paper investigates their theoretical and practical characteristics in a simplified scenario. Two models based on theory and on experimental results (empirical model) are developed and validated.

First distance estimations indicate that an empirical model for a particular scenario achieves an accuracy of 18cm with a confidence of 83%.

1   Introduction

Solar cells and modules are usually applied to the conversion of radiant to electrical energy. However, as we show in this paper, such energy flows may also be considered as data flows, thus extending solar module functionality to a form of receiver. Data can be transmitted for reception by solar modules via IR [1] as well as via fluorescent tubing [2]. In the concept proposed here, the solar modules are used only to track the intensity of indoor radiation (e.g. lights) as a form of context information. By using existing lights infrastructure, investment is minimised. Taking the taxonomy of position estimation approaches of Fox [3], we therefore have a local (e.g. single building) and passive (not transmissive) approach. Conceptually, the solar cells are "outward looking" and measure multiple beacons in the environment.

To the authors' knowledge, such a use of solar modules as a component of a location tracking system has not been previously investigated. Optical location investigations have previously considered various technologies including infra red [4,5], laser [6,7] and video [8,9,10,11]. Further location technologies [12], include inertial [13], ultrasound [14], RF [15] and magnetic [16]. It has also been shown that such technologies can be used in tandem [17,18]. A solar *powered* location system is the MIT Locust [19].

T. Strang and C. Linnhoff-Popien (Eds.): LoCA 2005, LNCS 3479, pp. 40–51, 2005.
© Springer-Verlag Berlin Heidelberg 2005

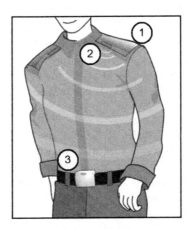

Fig. 1. LuxTrace concept: Wearable location tracking by solar cells

In this paper, a scenario of an office worker walking in a corridor is considered with solar modules integrated into the shoulder area of their clothing. Such modules can have similar characteristics to textiles e.g low cost (2$), low weight (20g), low volume ($2cm^3$) and with a range of colours. A potential concept is shown in Figure 1 in which the flexible solar module system on the shoulder (1) transmits one or more RF pulses only when there is sufficient energy to do so i.e. beneath a light source. This data is collected and processed by the belt worn computer (3) [20]. The environment is assumed to be static [3] as the lights in the corridor are always on during office hours. Whilst it is necessary to process the data from the solar modules, their relatively low bit rate is well adapted to on body processing such as with a body worn computer. Sensors other than solar cells may also be necessary for satisfactory location tracking.

This paper is structured as follows. The investigation of the office worker scenario firstly considers radiant energy received by a solar module from a theoretical perspective. A single fluorescent light tube is modelled from which the radiant energy in a number of interconnected corridors is extrapolated. This model is then validated using a solar module mounted on a wheeled vehicle. The same vehicle is then used to collect training data from which a second model is developed that is specific to similar corridors. The second model is then validated using further data. In the discussion the LuxTrace concept is analysed.

2   Simulation

2.1   Irradiance Modelling

Typical office buildings have windows, varying room architecture, colouring and various ambient light sources, which influence the light intensity and frequency components. These parameters may provide information for location estimation, when included in a radiant energy map and used as reference during indoor navigation. In this first approach, we rely on information extracted from artificial light sources only. More precisely, we consider in this analysis a hallway scenario equipped with regular fluorescent light tubes installed in the ceiling at 2.5 meters from the floor.

Theoretical Model. The source of radiant energy (emitter) creates a field of radiant flux. Many emitters can be modelled as a point source at sufficient distance. The total received flux per area is called *Irradiance* (W/m^2). For the following model, a fluorescent light tube will be approximated as a bounded concatenation of point sources.

As the distance between fluorescent light tube and photovoltaic solar cell (receiver) changes, so does the light intensity received at the cell (irradiance). Irradiance I_{RPS} at a distance r from a point source emitting radiant energy with intensity I_0 is related by the inverse square law [21]:

$$I_{RPS} = \frac{c * I_0}{r^2} \qquad \text{with } c = const. \text{ and } I_0 = const. \qquad (1)$$

For positioning in three-dimensional space the coordinate system x, y, z shown in Figure 2 will be used, with its origin at the centre of the tube $\underline{0}(x_0, y_0, z_0)$.

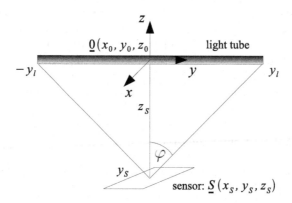

Fig. 2. Schematic for the theoretical model

In the particular case, where a receiver is positioned on a plane with $z_S = const.$ near to the light tube, the irradiance is at maximum, if $x_S = x_0$ and $y_S = y_0$. This can be described as the total planar irradiance I_{TPS}. In the general case, for the irradiance at the solar cell sensor I_{SPS} an arbitrary angle φ must be considered, with $-90° \leq \varphi \leq 90°$ between the point source emitter and the receiving sensor, related by the cosine law: $I_{SPS}(\varphi) = I_{RPS} * cos(\varphi)$.

The distance $\underline{r}$ can be decomposed in the coordinate system by the three coordinates positioning the solar cell $\underline{S}(x_S, y_S, z_S)$ (Figure 2) depending on the position along the light tube (y-coordinate):

$$r(y) = \sqrt{x_S{}^2 + (y_S - y)^2 + z_S{}^2} \tag{2}$$

The cell irradiance from a point source I_{SPS} can be directly related to the coordinate system, by inserting equation 2 into equation 1:

$$I_{SPS}(y) = \frac{c * I_0 * z_S}{(x_S{}^2 + (y_S - y)^2 + z_S)^{3/2}} \tag{3}$$

This relation can be used to simulate the irradiance from a solar cell movement in any direction under light sources at arbitrary heights. The total irradiance I_S for a fluorescent light tube is found by integrating over the chain of point sources (equation 4) in y-direction:

$$I_S = \int_{-y_L}^{y_L} I_{SPS} \; \mathrm{d}y \;\; = \int_{-y_L}^{y_L} \frac{c * I_0 * z_S}{(x_S{}^2 + (y_S - y)^2 + z_S)^{3/2}} \;\; \mathrm{d}y \tag{4}$$

The distance to the light source, e.g. z-component z_S of $\underline{S}$ is assumed as being constant. From the perspective in x-direction, the model assumes the light tube as a point source. Hence, x_S is constant. The total length of the light tube is denoted by l. Hence, the integration limits are described by $y_L = l/2$. I_S is derived in units of W/m.

Practical Data Acquisition. The change in irradiance when varying the distance of a photovoltaic solar cell to the light emitter can be monitored by current or voltage variation. Whilst current is directly proportional to irradiance, voltage is less affected. It varies with the log of intensity:

$$V \propto ln(I_S)$$

For convenience of signal acquisition, solar cell voltage across a 10kΩ resistance was tracked. This resistance is sufficient in our case to ensure that voltage was almost directly proportional to current (and irradiance) [22].

The log correction made does not change the general shape of the waveforms. The resulting simulation graphs are depicted in Figure 3. For the simulation, $c = 1$ and $I_0 = 5W/m^2$ is used. These values are fitted with real measurements to reflect size and type of the solar cells in the model. Since for this simulation example the y-component of the movement is $\Delta y = 0$, the light tubes are approximated as point sources.

Detection of Light Emitter. Figure 3 shows the expected waveform for a straight trajectory equidistant to the walls down a corridor. The light tubes are oriented at right angles to the trajectory and regularly distributed (see Figure 2). Assuming a typical office building height of about 2.5 meters, the distance z_S from an adult shoulder to the ceiling mounted light sources will be about one meter or less. The regular distance d_L between the light tubes is greater than 2.5 meters. Using the theoretical model for the case of $d_L = 2.5m$ and $z_S = 0.8m$ there is a difference of 20% in the log irradiance from the minimum to maximum value. Such a difference is sufficient to detect when the solar module is beneath the tube.

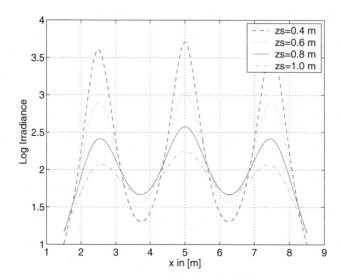

Fig. 3. Log corrected irradiance from three point sources at different ceiling heights

2.2 Environment Effects and Theoretical Model Limitations

The theoretical model does not consider indirect light, such as reflections at walls and cupboards. Fortunately indirect light generally has an order of magnitude less intensity than direct light, so this component of radiant energy has not been included in the theoretical model.

The case of occluded direct light is familiar because it creates distinct shadows hindering radiant energy reaching a sensor. For the intended application using overhead light sources, the possible obstruction area is limited to objects in direct connection with the light source, e.g. the box frame supporting the fluorescent light tube or objects in the line of sight above the solar cell. As this model does not cover human aspects in detail, the head as a possible obstacle is not considered.

2.3 Theory Based Distance Model

In a second modelling step, a 3-dimensional environment for configurable light distributions has been built for the majority of the corridors of our offices. With this approach it is possible to simulate various building scenarios. It is used here exclusively for a section of the corridor scenario.

For the following analysis a distribution of fluorescent light tubes according to the simulated irradiance map in Figure 4 has been chosen. This situation reflects part of a hallway from an existing office building. The scene consists of one long walkway leading to offices and a connecting passage. The section has 3 identical light tubes, equidistant with both walls, oriented in a perpendicular direction to the main corridor access. The distance between the lights is $d_L = 3.7m$. There are no windows or other significant sources of artificial light in the the scenario.

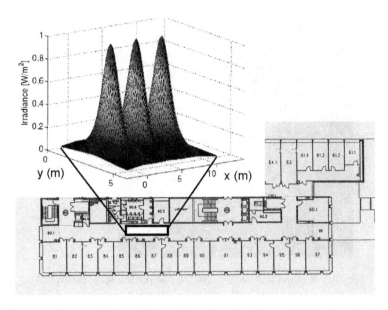

Fig. 4. Example of a simulated irradiance distribution as 3D plot for the scenario

3   Experiments

The initial goal of the experiments was to verify the waveforms calculated in the simulation. At the same time, measurement data were acquired for creating and validating a trained model. This section details the measurement system used and the data sets acquired. Experiments were carried out within the corridor in which radiant energy had been simulated.

3.1   Sensing System

A measurement system based on a trolley (see Figure 5) was built that allows acquisition of the voltage from solar cells. The same set up was used for all experiments. The solar module was positioned in a horizontal plane on top of the trolley. Furthermore, a relatively constant distance z_S between cell and fluorescent light tubes could be maintained, varying by a few millimetres due to ground roughness under the wheels of the trolley for example.

The photovoltaic solar module used for the experiments is an amorphous silicon thin film deposited on glass[1]. The voltage of the solar cells was acquired at 1kHz and 12 bit resolution using a standard data acquisition system.

To associate the acquired waveforms with actual distance down the corridor, two approaches have been used: By using markers on a precisely measured straight trajectory, the waveforms have been tagged. This measurement is con-

---

[1] Manufacturer: RWE SCHOTT Solar, model: ASI 3 Oi 04/057/050.

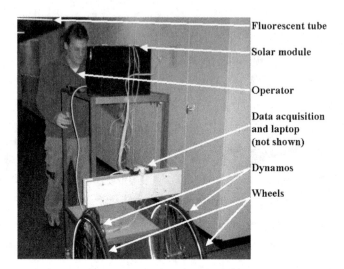

Fig. 5. Measurement trolley used for acquiring solar cell voltage

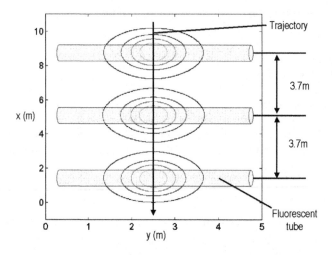

Fig. 6. Experiment setup: Movement on straight line under light tube centre

sidered as being relatively precise but limited in resolution, since data could only be collected practically every 50cm. Therefore a second distance measurement method based on a bicycle dynamo and bicycle wheel appropriate was used. A peak detection of the acquired voltage waveform from the dynamo generator was used to determine the system speed. The accuracy of the wheel measured distance compared with the actual distance was always over 98%. Since this result indicates satisfactory accuracy, the second method was used to acquire ground truth for subsequent experiments.

3.2   Description of Experiments

The experiments were performed pushing the trolley at constant walking speed $(0.55ms^{-1})$. A straight trajectory was taken below the middle of the light tubes as shown in Figure 6. The distance between the horizontal solar cell and the fluorescent light tubes was $z_S = 73cm$, the distance between the light tubes was $d_L = 3.7m$.

Irradiance at the solar module was measured over trajectories with distances in the range of 2m to 10m. Each time the solar module passed under a tube, a waveform peak was measured. A total of 14 such peaks were recorded.

4   Results

4.1   Theoretical Model Validation

The theoretical model used varied over the range of 0.4V to 3.3V whilst the average of the measured values was in the range 0.1 to 2.7V. The error for ten peak waveforms (see Figure 7) was less than 0.3V with a confidence level of 81%. Part of the error can be attributed to the theoretical model being for a bare fluorescent light tube rather than the measured data which was for an installed light tube including a reflective housing.

4.2   Distance Estimation

Distance information can be extracted from the amplitude of the waveform by using a mapping between voltage and known distance from training data. To

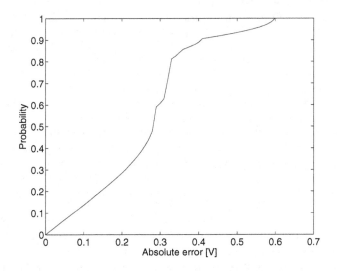

Fig. 7. Cumulated error of the simulation model compared to the average measured voltage

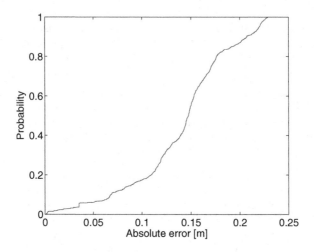

Fig. 8. Cumulated error of estimated distance by the empirical model

build the model, training data sets were segmented into single peak waveforms and scaled to the known light to light distance of $d_L = 3.7m$. This empirical model was used in a simple mapping function which relates observed voltages from the solar cell and the average distance under the light source.

Since the minimum and maximum amplitude of the test data varied in the range of $\pm10\%$, the empirical model did not provide information if the amplitude was greater or less than the average curve. To improve the estimation, the last available averaged speed information obtained with the same model was used in these periods, without information from mapping, to support the distance estimation. The average distance estimation error obtained with this method is less than 18cm with a confidence of 83% over a distance of 7.4m (see Figure 8).

5   Discussion

The assessment of whether solar modules could contribute to a location tracking system can be based on relatively standardised topics. A location system assessment taxonomy proposed by Hightower [23] includes scalability, cost, recognition and limitations.

Scalability of the solar module system will depend on a number of factors including amount of area with no distinct light source and superposition of radiant energy from different sources. Another aspect that we consider as part of *scalability* is the number of sensors. Given the relatively low cost of the solar cells, it can be anticipated that a number of sensors could be used for each user. It would then be possible with non co-planar solar modules positioned on the user to support trilateration.

The *cost* of the solar module system is a function of installation and maintenance. Indoors, lights and on body computers (e.g. mobile phone) are generally available; therefore the incremental hardware will only be the sensor node(s). Installation software costs might include a location tracking program and a simplified map of the building including light source locations. Incremental maintenance costs of the system would be zero assuming that lighting infrastructure is not changed and bulb replacement service(s) exist.

Recognition of the user context may be enhanced both by collecting further data from the solar modules (e.g. light sources can be distinguished by frequency and spectra) as well as including complementary sensors. A simple example of a complementary sensor is an accelerometer or pedometer that would provide information about user movement when he or she is not under sufficient incident radiant energy.

Limitations of using solar modules for optical measurements compared with using a charge coupled device (CCD) camera are much lower pixel rates. These rates may be partly mitigated by the use of lateral effect photo-diodes [24, 25]. Another mitigating factor with solar modules is their lower response times compared with CCDs. Sensor response time can be important in virtual reality applications for avoiding user nausea [26]. Further limitations are mentioned in [22].

6   Conclusion and Future Work

In this paper, the LuxTrace concept of using solar cells as sensors has been presented and one kind of solar cell characterised. For a scenario representative of an office worker walking down a corridor, distance moved has been determined within 18cm with a confidence level of 83%. The results provide evidence that distance travelled and therefore instantaneous speed of a moving object can be estimated satisfactorily using only the output of solar cells and a model based on theory or acquired waveforms (empirical model). Indirectly these results also support the case that a garment integrated location tracking system will be achievable.

Based on these encouraging results, we intend to investigate a number of further avenues. Both models are rudimentary and could be improved or replaced by models based on probabilistic algorithms for example. Also, as solar modules are low cost, a number of them could be used simultaneously in future experiments to allow the estimation of orientation for example. Experiments with alternative (flexible) solar cells, such as manufactured by VHF technologies [27], integrated into clothing would enable location systems embedded into garments as well as allow the influence of gait to be investigated. Finally, whilst the physical limits of what can be achieved with solar modules are an intrinsically valuable result, solar modules could also be combined with further (location tracking) technologies.

Acknowledgement

Thank you to Martin Burri and Christian Morf for performing experiments and preparing a number of figures. We are also grateful to Stijn Ossevoort for proposing the name *LuxTrace* in particular.

References

1. Nishimura, T., Itoh, H., Nakamura, Y., Yamamoto, Y., Nakashima, H.: A compact battery-less information terminal for real world interaction. In: Pervasive Computing: Proceedings of the International Conference. (2004) 124–139
2. Talking Lights: Talking Lights, Boston MA USA. (http://www.talking-lights.com/how.htm)
3. Fox, D.: Markov Localization: A Probabilistic Framework for Mobile Robot Localization and Navigation. PhD thesis, Institute of Computer Science TU Dreseden Germany (1998)
4. Want, R.; Hopper, A.: Active badges and personal interactive computing objects. Consumer Electronics, IEEE Transactions on, Vol.38, Iss.1 (1992) 10–20
5. Ward, M., Azuma, R., Bennett, R., Gottschalk, S., Fuchs, H.: A demonstrated optical tracker with scalable work area for head-hounted display systems. In: Proceedings of the symposium on Interactive 3D Graphics. (1992) 43–52
6. Evans, J., Chang, T., Hong, T., Bostelman, R., Bunch, W.: Three dimensional data capture in indoor environments for autonomous navigation, NIST Internal Report 6912. Technical report, (www.isd.mel.nist.gov)
7. Sorensen, B., Donath, M., Yang, G.B., Starr, R.: The Minnesota Scanner: a prototype sensor for three-dimensional tracking of moving body segments. (Robotics and Automation, IEEE Transactions on, Vol.5, Iss.4) 499–509
8. Piscataway, N.: Multi-camera multi-person tracking for Easy Living. In: 3rd IEEE Int'l Workshop on Visual Surveillance. (2000) 3–10
9. Thrun, S., Bennewitz, M., Burgard, W., Cremers, A., Dellaert, F., Fox, D., Haehnel, D., Rosenberg, C., Roy, N., Schulte, J., Schulz, D.: Minerva: A second generation mobile tour-guide robot. In: Proc. of the IEEE International Conference on Robotics and Automation (ICRA'99). (1999)
10. Starner, T., Maguire, Y.: A heat dissipation tutorial for wearable computers. In: Digest of Papers. (1998)
11. Zobel, M., Denzler, J., Heigl, B., Nöth, E., Paulus, D., Schmidt, J., Stemmer, G.: Mobsy: Integration of vision and dialogue in service robots. In: International Conference on Computer Vision Systems (ICVS). (2001) 50–62
12. Tauber, J.A.: Indoor location systems for pervasive computing. Technical report, Theory of Computation Group Massachusetts Institute of Technology (2002)
13. Vildjiounaite, E., Malm, E.J., Kaartinen, J., Alahuhta, P.: Location estimation indoors by means of small computing power devices, accelerometers, magnetic sensors, and map knowledge. In: Pervasive '02: Proceedings of the First International Conference on Pervasive Computing, Springer-Verlag (2002) 211–224
14. Ward, A., Jones, A., Hopper, A.: A new location technique for the active office. IEEE Personnel Communications, 4(5) (1997) 42–47
15. Bahl, P., Padmanabhan, V.N.: Radar: An in-building RF-based user location and tracking system. In: INFOCOM. (2000) 775–784

16. Rabb, F.H., Blood, E., Steiner, T.O., Jones, H.R.: Magnetic position and orientation tracking system. IEEE Transaction on Aerospace and Electronic Systems, AES-15(5) (1979) 709–718
17. Lee, S.W., Mase, K.: A personal indoor navigation system using wearable sensors. In: Proceedings of The Second International Symposium of Mixed Reality (ISMR01), Yokohama. (2001)
18. Randell, C., Muller, H.: Low cost indoor positioning system. In Abowd, G.D., ed.: Ubicomp 2001: Ubiquitous Computing, Springer-Verlag (2001) 42–48
19. Starner, T., Kirsch, D., Assefa, S.: The Locust Swarm: An environmentally-powered, networkless location and messaging system. In: 1st International Symposium on Wearable Computers. (1997) 169–170
20. Amft, O., Lauffer, M., Ossevoort, S., Macaluso, F., Lukowicz, P., Tröster, G.: Design of the QBIC wearable computing platform. In: Proceedings of the 15th IEEE International Conference on Application-specific Systems, Architectures and Processors. (2004)
21. Randall, J.: On the use of photovoltaic ambient energy sources for powering indoor electronic devices. PhD thesis, EPFL Lausanne Switzerland (2003)
22. Amft, O., Randall, J., Tröster, G.: Towards LuxTrace: Using solar cells to support human position tracking. In: Proceedings of Second International Forum on Applied Wearable Computing, Zrich, CH, 17-18th March 2005. (2005 to appear)
23. Hightower, J., Borriello, G.: A survey and taxonomy of location systems for ubiquitous computing. IEEE Computer, 34(8) (2001) 57–66
24. Pears, N.: Active triangulation rangefinder design for mobile robots. In: Proceedings of the 1992 IEEE/RSJ International Conference on Intelligent Robots and Systems. (1992)
25. Pears, N.: An intelligent active range sensor for vehicle guidance: system overview. In: Proceedings of the 1996 IEEE/RSJ International Conference on Intelligent Robots and Systems, Vol.1. (1996)
26. Janin, A.L., Zikan, K., Mizell, D., Banner, M., Sowizral, H.A.: Videometric head tracker for augmented reality applications. In: Proc. SPIE Vol. 2351, p. 308-315, Telemanipulator and Telepresence Technologies, Hari Das; Ed. (1995) 308–315
27. VHF Technologies: Flexcell from VHF Technologies, Yverdon-les-Bains Switzerland. (http://www.flexcell.ch)

Three Step Bluetooth Positioning

Alessandro Genco

University of Palermo, Department of Computer Engineering,
Viale delle Scienze, edificio 6, 90128 Palermo, Italy
genco@unipa.it

Abstract. This paper discusses a three step procedure to perform high definition positioning by the use of low cost Bluetooth devices. The three steps are: Sampling, Deployment, and Real Time Positioning. A genetic algorithm is discussed for deployment optimization and a neural network for real time positioning. A case study, along with experiments and results, are finally discussed dealing with a castle in Sicily where many trials were carried out to the end of arranging a positioning system for context aware service provision to visitors.

1 Introduction

Many pervasive computing applications rely on real time location to start and manage interaction with people in a detected area. Time and space information are therefore basic elements in arranging mobile context aware services which take into account context factors such as who, why, where, when. Dealing with wide areas, multiple interaction devices, such as remote multi-displays, could be available in one service hall. In such cases a pervasive system can start interaction with who explicitly addresses a selected device by means of some manual action on a touch screen or a mouse, or by means of some voice sound. Nevertheless, there are some kinds of application, as for instance advertising messages, which require interaction to start autonomously. People who are around should be attracted by some customized message exactly arranged on his personal profile and current position in a display neighborhood.

We may feel some worry in looking at a pervasive system as a big brother; however there is some convenience for us in customized services and furthermore, such an interaction modality could be the one preferred by people, because it does not require any manual action to be performed. Once preserved the not invasive requirement of pervasive applications, it is undoubted that system proactive behavior could be a general suitable approach to mobile human computer interaction.

Besides the problem of selecting the nearest interaction device, position aware services may need to rely on position data which must be more accurate than simple location. There are several pervasive applications indeed, which require a maximum error in position coordinates to be kept very low, less than one meter for instance. This is the case of a security system, which is arranged to protect an area around a

T. Strang and C. Linnhoff-Popien (Eds.): LoCA 2005, LNCS 3479, pp. 52–62, 2005.

precious artifact. There are also some cases which require additional position data, like the human body compass angle. People who are looking at an object, as for instance visitors who are looking at an artifact, or factory operators who are checking some manufacturing process could be provided with context aware information which take into account who is looking at what.

The above two basic positioning elements are to be used in conjunction with a higher level point of view to allow a system to arrange those services someone may expect in a given reality [1], [2], [3].

2 Why Bluetooth

We used Bluetooth (IEEE802.15.1) in our positioning experiments because of two main reasons. One is that Bluetooth technology is widely implemented in cellular/smart phones, thus being something quite chip and wearable, and therefore very easy to own. The other reason is that the Bluetooth (BT) technology embedded in cellular phones, allows distances to be estimated by link quality values within a BT covered area which we can suppose to be a 30~40 m. circle approximately. We have also to mention some problems encountered in using cellular embedded Bluetooth devices which mostly deal with BT service implementation by different brand factories. In many cases we had to deal with compatibility problems or service restrictions.

However, given the Bluetooth amazing commercial success, we can hope in near future to deal with standardized Bluetooth services.

Actually, WiFi (IEEE802.11x) can also be used for positioning, as well as any other RF communication technology which provides link quality values. Nevertheless, most of positioning problems which come from link quality measure unreliability, can be discussed with similar considerations for a class of technologies. Therefore, apart some different featuring specifications, discussions on Bluetooth can be considered as representative of a group of communication technologies which are capable of providing positioning information.

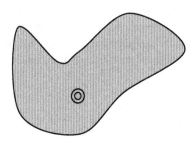

Fig. 1. *An Iso-LQ curve*

An actual problem of positioning by RF communication technologies comes from estimating distances on link quality measurements, which are affected by a high degree of uncertainty. Measured RF link quality equal values actually draw a region which is very unlikely to be a circle because of obstacles and noises. Therefore, position estimation by triangulation, even performed on more than three reference nodes, cannot be accurate. An irregular-shaped region around a RF terminal (Fig. 1) is a more realistic case to be tackled by means of methods which are capable of dealing with uncertainty and site depending solutions.

Distance estimation between a mobile device and a number of reference devices whose location is known is a research topic of several approaches [4]. Some contributions can be found in literature with the end of arranging solutions to be free from site noises.

Among these, ActiveBats [5] and Cricket [6] are based on ultrasound *time-of-flight* lateration, with an accuracy of few cm or less. The *time-of-flight* method estimates distance between a moving object and a fixed point by measuring the time a signal takes to travel from the object and the fixed point at a known speed. This method could be a good one because time of flight and distance have a reliable relationship. The actual problem is in clock accuracy requirement. A 1 µs error in timing leads to a 300 m error in distance estimation.

The Ascension Technology MotionStar system [7] is based on magnetic sensors moving in a magnetic field around their source. This system provides a very high accuracy but needs very expensive hardware.

RX power level positioning method is quite similar to TOA positioning. Both methods locate mobile devices on the intersection of three (or more) circles. The circles radius is evaluated on the measured strength of received signals, thus assuming a direct relationship between signal strength and distance which unfortunately, as said above, can be affected by obstacles and noises.

The Angle Of Arrival (AOA) method processes the direction of a received signal. Position is estimated by triangulation when two reference devices at least measure the signal angle of arrival from a mobile device [8]. This method obviously requires some expensive hardware to evaluate angles of arrival.

The Cell Identity (CI) method looks at the network as divided into cells, each cell being the radio coverage area of a single reference device. A mobile device connected to a given reference device is assumed to be inside its cell. Cells overlapping and connectivity-induced geometric constraints can improve accuracy [8]. One more time, as mentioned above, radio coverage cannot be assumed to be a circle and therefore accuracy cannot be high.

3 Bluetooth Positioning

Hallberg et al. [9] developed two different methods based on BT Received Signal Strength Indicator (RSSI) values: the direct method, which requires a BT device to be programmed, and the indirect method, without any programming being needed. The first one gives a good accuracy by programmable hardware. The second one is cheaper, but its accuracy is very poor, with a worst-case error of 10 meters.

SpotOn [10] and MSR RADAR [11] are based on RF signal power level measurement. They process the RSSI (Received Signal Strength Information) value to give an accuracy of 3-4 meters or more.

The BT Local Positioning Application (BLPA) [12] uses RSSI values to feed an extended Kalman filter for distance estimation. A good accuracy is achieved only by theoretical RSSI values, while unreliability of actual values gives unreliable distance estimation. The BT Indoor Positioning System (BIPS) [13] is designed for tracking mobile devices in motion inside a building. The BIPS main task is real-time tracking of visitors in a building. This led researchers to deal mainly with timing and device discovering, thus achieving an accuracy of 10 meters.

Finally, Michael Spratt [14] proposed the Positioning by Diffusion method based on information transferred across short-range wireless links. Distance estimation is achieved by geometric or numeric calculations.

4   Three Step BT Positioning

BT devices measure RX power level by using both RSSI and *Link Quality* (LQ) parameters. These are implemented in the BT module and can be read through HCI (*Host Controller Interface*) commands [15]. LQ is a quite reliable parameter for distance estimation, differently RSSI only allows to know whether a device is in a given base station power range or not [16]. The use of LQ is recommended by the BT standard specifications, so it is available on most commercial devices. LQ represents the quality of a link in a range from 0 to 255, and a correlation can be assumed between distances and LQ values. We know LQ values are not reliable in measuring distances. Therefore, we need to avoid geometrical concerns and let a BT positioning system to take advantage from LQ values to be processed according to their site depending specificity. Here we discuss a three step procedure which turned out to be capable of providing high definition positioning by the use of low cost BT devices. The three steps are: site LQ sampling, BT base station deployment, and finally, real time positioning.

4.1   Positioning Step 1: Site Sampling

The end of this step is to collected a first sample of LQ measures to allow us to attach a set of LQ ranges to each cell. A range is a set of three values: the lowest, the highest and the mean value of all LQ values measured in a cell from a given BT base station in one point.

The site we investigated is the Manfredi's Castle in Mussomeli - Italy (Fig. 2a), whose map is sketched in Fig. 2b. We split the tourist area in a number of cells, which are rooms, roads, and areas around artifacts. There are two cell types: cells that represent rooms, parts of large rooms, or parts of roads; and cells which represent sub-areas around columns, portals, or other artifacts. We assumed an irregular

Fig. 2a. *Manfredi's Castle in Mussomeli (Italy)*

1. External wall door
2. Stable
3. Castle entrance
4. Courtyard
5. Big arc
6. Vestibule
7. Gothic portal
8. Barons Hall
9. Destroyed bodies
10. Three women room
11. Chimney hall1
12. Cross vault halls
13. Semicircular Turret
14. Maid lodging
15. Male
16. Chapel
17. Polygonal external wall
18. External wall stable
19. Hayloft
20. Defense fencing
21. Double lancet windows

Fig. 2b. *Castle map*

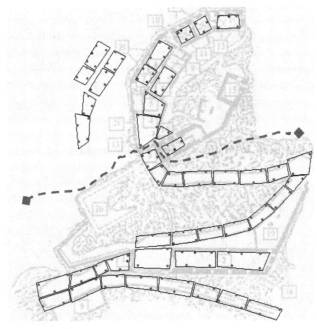

Fig. 3. *Cells layout*

quadrilateral shape for the first kind of cell, a width (typically a diagonal), and a maximum error of 1 meter. Differently, we assumed a circular shape for the second type, a diameter of 2 meters, and a maximum error of 0.5 meters.

Fig. 3 sketches the cells layout. Blue points are centers of circular cells; red points are places where Bluetooth base stations (BT-BS) are allowed to be put.

The analysis of this first set of measures suggested us some considerations. One is that is very hard to find any correlation between obstacles and link quality. Some walls turned out to stop BT coverage; other walls seemed to be glass or air. Only very deep walls or floors turned out to completely stop BT coverage. For instance, BT-BS's which were placed in lower floor areas, did not read any LQ value from mobile terminals (BT-MT) moving on higher floor areas, and vice-versa.

Mainly due to the end of dealing with indoor and outdoor areas separately, we decided to split the site area in two sub-areas (green dotted line in Fig. 3). The results of these measurements are in a matrix whose generic (i,j) element contains a LQ values range measured from the BT-BS at the ith position to the BT-MT moving within the jth cell. A range can also be read as an estimation of the maximum theoretical accuracy in a cell (8 LQ units in a 2 m. cell cannot give an accuracy greater than 2/8 m.). A generic (i,j) range set to [0,0] tells us that the BT-BS placed at the ith position cannot detect any BT-MT in the jth cell. Table 1 shows part of the output file, where rows are for N_S possible stations, and columns are for N_A areas.

Table 1. LQ Ranges

	1	2	...	N_A-1	N_A
1	0,0	160,165	...	161,185	163,175
2	0,0	180,188	...	195,230	201,255
:	...	...	...	...	...
N_S-1	210,212	190,205	...	0,0	145,156
N_S	240,250	181,185	...	0,0	0,0

4.2 Positioning Step 2: BT Base Station Deployment

Several BT positioning experiments were carried out according to different
positioning methods, namely triangulation [16], fuzzy logic [17] and neural network.
A common result of these experiments is that positioning accuracy can be heavily
affected by erroneous arrangements of the available base stations. Actually, we are
unlikely to be allowed to put a base station in the middle of a room, for instance, and
further constraints may come when dealing with a heritage site; base stations should
be invisible and only selected places are available. Therefore, a relevant step in
arranging a BT positioning system should be to optimize BT-BS deployment in a
subset of places which are the only ones permitted by site specific constraints. The
problem can be enounced in the following terms: given the total number of places
where a base station can be put, select a minimal subset which allows the system to
evaluate the position of a BT mobile terminal in any part of the site, with the highest
accuracy degree.

Many optimization methods can be used to this end; we used a genetic algorithm
because of its easy scalability. Each possible deployment is represented by an
individual chromosome of a population. A chromosome has as many genes as places
where BT-BS can be put. Each gene represents a possible BT-BS position, which can
be set either to *true* if a BT-BS is placed in that position, or to *false*.

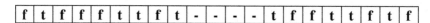

Fig. 4. *Deployment Chromosome*

An acceptable solution has to return a deployment layout whose coverage is unique
for all areas, also achieving a required accuracy.

An optimal solution maximizes coverage quality, and minimizes the number of
BT-BS required. The quality index of each station-area couple (*s,a*) is defined by the
ratio (1) with n_a being the number of sub-areas to be singled out by positioning.

$$q_{s,a} = \frac{[LQ_{\text{sup}} - LQ_{\text{inf}}]_{s,a} + 1}{n_a} \tag{1}$$

Each deployment chromosome includes a number of BT-BS, along with their $q_{s,a}$ value. The chromosome quality takes into account all $q_{s,a}$ values, which represent the contribution of each BT-BS s to the whole chromosome quality.

4.2.1 Some Genetic Algorithm Details

We start generating a population of 50 chromosomes and assigning each gene a probability p to be included in a chromosome. Each chromosome is checked for acceptability and fitness value.

Once the initial population is generated, evolution starts. A maximum of 10 chromosomes are killed each step (20% of population) depending on age. Each chromosome has a percentage probability to die which is equal to its age. So, if a chromosome is 20, it has a 20% probability to die. A constant number of surviving chromosomes are then coupled to generate new chromosomes thus replacing the killed ones. Coupling is performed according to a one-point-crossover and alternating gene exchange (one time the initial part and one time the final part). The evolution steps are repeated 1000 times.

A best solution is detected at each evolution step, and eventually it replaces the previous one if better. At the end of the process an absolute best solution is singled out.

4.2.2 Experiment Results

Here we discuss the experiments carried out in the upper area of the Mussomeli's castle. We split the area in 16 quadrilateral cells and 6 circular cells (Fig. 5). Results

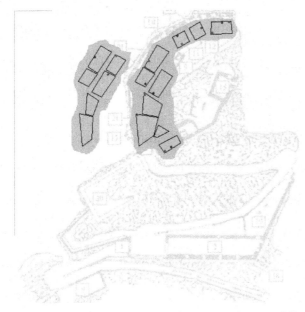

Fig. 5. *Castle's upper area layout and coverage. Areas in different levels are sketched side by side*

Table 2. Upper Area BT-BS Deployment Selection

Initial probability (%)	Maximum allowed error		Fitness	BT-BS
	Quadrilateral areas	Circular areas		
10	1	0,5	0,14	10
	0,75	0,375	------	------
	0,5	0,25	------	------
	0,25	0,125	------	------
15	1	0,5	0,13	12
	0,75	0,375	------	------
	0,5	0,25	------	------
	0,25	0,125	------	------
20	1	0,5	0,17	10
	0,75	0,375	0,16	11
	0,5	0,25	------	------
	0,25	0,125	------	------
25	1	0,5	0,12	12
	0,75	0,375	0,11	15
	0,5	0,25	------	------
	0,25	0,125	------	------
30	1	0,5	0,11	14
	0,75	0,375	0,11	16
	0,5	0,25	------	------
	0,25	0,125	------	------

are shown in Table 2. First column lists the probability for a gene in a chromosome to be "t"; second column lists the maximum allowed error (in meters) for each cell type, third column lists fitness values, and fourth column lists the number of BT-BS's deployed by a solution.

Table 2 tells us that a minimal number of ten BT base stations could be effective for the investigated area, thus achieving an accuracy of 1 m. for rooms and 0.5 m. for cells around artifacts. Some better accuracy can be achieved at the cost of deploying a greater number of base stations.

4.3 Positioning Step 3: Real Time Positioning by Neural Network

The high degree of uncertainty entailed by LQ values leads to the high complexity of their relationships with mobile device position. As previously remarked, positioning needs most to relay on site dependent solutions and not to deal with geometric laws. Each site has its own obstacles and environment noises which affect LQ measures; therefore, we need to assume each LQ value as specific for a given place. A positioning system need to learn LQ values as they are, without any concern with distances between BT-MT and BT-BS.

A Neural network is a solution in such a direction. A neural network can learn the LQ distribution, tune its weights, and then, be ready to provide real time position fast estimates. We also carried out some experiments by means of fuzzy logic which gave good results in terms of accuracy. Unfortunately, fuzzy algorithm computational complexity turned out to be high for real time positioning. Differently the neural network turned out to be very fast.

Experiments were carried out on a single-layer neural network with n inputs, one output, a linear activation function and no hidden layer. This simple network gives its output as (2), where x_j are inputs and w_j are weights. Input is a m-dimensional array of LQ raw values, and no preliminary processing is required. We started our experiments with 10 base stations which we placed within the castle upper area according to a deployment given by the genetic algorithm.

$$y = \sum_j w_j x_j + \theta \qquad (2)$$

The training set is given by the LQ values read by mobile devices placed in 5 known positions. Once trained, the network gives a good accuracy with a maximum relative error of 3% of the theoretical accuracy produced by the deployment optimization step.

5 Conclusions

This paper demonstrates the relevance of arranging a Bluetooth positioning system according to a three step procedure: sampling, deployment optimization, and neural network real time positioning. The first step suggested us to avoid geometrical strategies because of the unpredictability of how obstacles and noises can affect link quality. The second step demonstrates that a theoretical accuracy can be set according to an optimal base station deployment. The third step proved the effectiveness of a neural network especially as far as real time positioning is concerned.

Our best result, in a case study on positioning in a castle, was 10 base stations and accuracy better than 0.5 meters. Even better accuracy can be achieved according to different problem setup, for instance, by increasing the number of base stations to be deployed. The used approach gave solutions which were specific for the castle problem; nevertheless, the same approach can be used for detecting optimal arrangements of Bluetooth base stations for positioning in any area, as well as for using any other RF communication technology which is capable of providing link quality values.

Layout optimization should be considered an unavoidable step for positioning. Once base station deployment has been optimized, various methods can be adopted for actual positioning.

Acknowledgements

This work has been partly supported by the CNR (Council of National Research of Italy) and partly by the MIUR (Ministry of the Instruction, University and Research of Italy).

References

1. A. Genco, *HAREM: The Hybrid Augmented Reality Exhibition Model*, WSEAS Trans. On Computers, Issue 1, vol. 3, 2004, ISSN 1109-2750
2. A. Genco, *Augmented Reality for Factory Monitoring*, Proc. of int. conf. ISC'2004 - Industrial Simulation Conference, June 7-9 Malaga, Spain, 2004
3. F. Agostaro, A. Genco, S. Sorce, *A Collaborative Environment for Service Providing in Cultural Heritage Sites,* Proc. of International Conference on Embedded and Ubiquitous Computing (EUC-04) August 26-28, Aizu Wakamatsu, Japan 2004
4. J. Hightower, G. Borriello, *Location Sensing Techniques*, technical report, IEEE Computer Magazine, August 2001, pp. 57-66
5. A. Harter, A. Hopper, P. Steggles, A. Ward, P. Webster, *The Anatomy of a Context Aware Application*, Proceedings of 5th Annual Int. Conference on Mobile Computing and Networking, ACM Press, New York, 1999, pp. 5968.
6. N. B. Priyantha, A.K.L. Miu, H. Balakrishnan, S. Teller, *The cricket compass for context-aware mobile applications*, Proceedings of the 7th annual international conference on Mobile computing and networking, July 2001.
7. Technical Description of DC Magnetic Trackers, Ascension Technology Corp., Burlington, Vt., http://www.ascension-tech.com/products/motionstarwireless.php
8. I. Jami, M. Ali, R. F. Ormondroyd, *Comparison of Methods of Locating and Tracking Cellular Mobiles*, IEE Colloquium on Novel Methods of Location and Tracking of Cellular Mobiles and Their System Applications (Ref. No. 1999/046), 1999
9. Hallberg, J.; Nilsson, M.; Synnes, K.; *Positioning with Bluetooth*, Proceedings of 10th IEEE International Conference on Telecommunications ICT 2003, Volume: 2, Pages: 954 – 958
10. J. Hightower, C. Vakili, G. Borriello, R. Want, *Design and Calibration of the SpotOn Ad-Hoc Location Sensing System*, University of Washington, Seattle, August 2001, http://portolano.cs.washington.edu/projects/spoton/
11. P. Bahl and V. Padmanabhan, *RADAR: An InBuilding RFBased User Location and Tracking System*, Proc. IEEE Infocom 2000, IEEE CS Press, Los Alamitos, California., 2000
12. Kotanen, A.; Hannikainen, M.; Leppakoski, H.; Hamalainen, *T.D.; Experiments on Local Positioning with Bluetooth*, Proceedings of IEEE International Conference on Information Technology: Coding and Computing [Computers and Communications] (ITCC) 2003, Pages:297 – 303
13. Anastasi, G.; Bandelloni, R.; Conti, M.; Delmastro, F.; Gregori, E.; Mainetto, G; *Experimenting an indoor bluetooth-based positioning service*, Proceedings of 23rd International Conference on Distributed Computing Systems Workshops, 2003, Pages: 480- 483
14. M. Spratt, *An Overview of Positioning by Diffusion*, Wireless Networks, Volume 9, Issue 6 Kluwer Academic Publishers, November 2003
15. Specification of the Bluetooth Core System 1.1, http://www.bluetooth.org
16. F. Agostaro, F. Collura, A. Genco, S. Sorce, *Problems and solutions in setting up a low-cost Bluetooth positioning system*, WSEAS Trans. On Computer Science, Issue 4, vol. 3, 2004, ISSN 1109-2750
17. F. Agostaro, A. Genco, S. Sorce, A *Fuzzy Approach to Bluetooth Positioning*, WSEAS Transactions on Information Science and Applications, ISSN 1790-0832, issue 1, vol. 1, July. 2004, pag. 393

MoteTrack: A Robust, Decentralized Approach to RF-Based Location Tracking

Konrad Lorincz and Matt Welsh

Harvard University,
Division of Engineering and Applied Sciences,
Cambridge MA 02138, USA
{konrad, mdw}@eecs.harvard.edu
http://www.eecs.harvard.edu/~konrad/projects/motetrack

Abstract. In this paper, we present a robust, decentralized approach to RF-based location tracking. Our system, called MoteTrack, is based on low-power radio transceivers coupled with a modest amount of computation and storage capabilities. MoteTrack does not rely upon any back-end server or network infrastructure: the location of each mobile node is computed using a received radio signal strength signature from numerous beacon nodes to a database of signatures that is replicated across the beacon nodes themselves. This design allows the system to function despite significant failures of the radio beacon infrastructure. In our deployment of MoteTrack, consisting of 20 beacon nodes distributed across our Computer Science building, we achieve a 50th percentile and 80th percentile location-tracking accuracy of 2 meters and 3 meters respectively. In addition, MoteTrack can tolerate the failure of up to 60% of the beacon nodes without severely degrading accuracy, making the system suitable for deployment in highly volatile conditions. We present a detailed analysis of MoteTrack's performance under a wide range of conditions, including variance in the number of obstructions, beacon node failure, radio signature perturbations, receiver sensitivity, and beacon node density.

1 Introduction

Using radio signal information from wireless transmitters, such as 802.11 base stations or sensor network nodes, it is possible to determine the location of a roaming node with close to meter-level accuracy [1, 2]. Such *RF-based location tracking systems* have a wide range of potential applications. We are particularly concerned with applications in which the *robustness* of the location-tracking infrastructure is at stake. For example, firefighters entering a large building often cannot see due to heavy smoke coverage and have no *a priori* notion of building layout. An RF-based location tracking system would allow firefighters and rescuers to use a heads-up display to track their location and monitor safe exit routes [3]. Likewise, an incident commander could track the location of multiple rescuers in the building from the command post. Such capabilities would have greatly improved FDNY rescue operations on September 11, 2001, according to the McKinsey reports [4].

T. Strang and C. Linnhoff-Popien (Eds.): LoCA 2005, LNCS 3479, pp. 63–82, 2005.

We note that our system needs to be installed and calibrated before it can be used. We consider this part of bringing a building "up to code", similar to installing smoke detectors, fire and police radio repeaters in high-rise buildings, and other such safety devices. For scenarios where an offline calibration is infeasible (e.g. because the emergency is in a remote location such as a field, highway, etc.), our scheme as described in the paper is not appropriate. It remains an open research question how to address this issue, and we provide some suggestions in the future work section.

RF-based location tracking is a well-studied problem, and a number of systems have been proposed based on 802.11 [1, 5, 6, 2, 7] or other wireless technologies [8]. To date, however, existing approaches to RF-based localization are *centralized* (i.e., they require either a central server or the user's roaming node, such as PDA or laptop, to compute the user's location) and/or use a *powered infrastructure*. In a fire, earthquake, or other disaster, electrical power, networking, and other services may be disabled, rendering such a tracking system useless. Even if the infrastructure can operate on emergency generator power, requiring wireless *connectivity* is impractical when a potentially large number of wireless access points may themselves have failed (e.g., due to physical damage from fire).

In addition, most previous approaches are *brittle* in that they do not account for lost information, such as the failure of one or more transmitters, or perturbations in RF signal propagation. As such, existing approaches are inappropriate for safety-critical applications, such as disaster response, in which the system must continue to operate (perhaps in a degraded state) after the failure of one or more nodes in the tracking infrastructure.

In this paper, we present a *robust, decentralized* approach to RF-based localization, called *MoteTrack*. MoteTrack uses a network of battery-operated wireless nodes to measure, store, and compute location information. Location tracking is based on empirical measurements of radio signals from multiple transmitters, using an algorithm similar to RADAR [1]. To achieve robustness, MoteTrack extends this approach in three significant ways:

- First, MoteTrack uses a decentralized approach to computing locations that runs on the programmable beacon nodes, rather than a back-end server.
- Second, the location signature database is replicated across the beacon nodes themselves in a fashion that minimizes per-node storage overhead and achieves high robustness to failure.
- Third, MoteTrack employs a dynamic radio signature distance metric that adapts to loss of information, partial failures of the beacon infrastructure, and perturbations in the RF signal.

In our deployment of MoteTrack, consisting of 20 beacon nodes distributed over one floor of our Computer Science building, we achieve a 50th percentile and 80th percentile location-tracking accuracy of 2 meters and 3 meters respectively, which is similar to or better than other RF-based location tracking systems. Our approach to decentralization allows MoteTrack to tolerate the failure of up to 60% of the beacon nodes without severely degrading accuracy, making the system suitable for deployment in highly volatile conditions. We present a detailed analysis of MoteTrack's performance under a wide range of conditions, including variance in the number of obstructions, beacon node failure, radio signature perturbations, receiver sensitivity, and beacon node density.

2   Background and Related Work

A number of indoor location tracking systems have been proposed in the literature, based on RF signals, ultrasound, infrared, or some combination of modalities. Our goal is to develop a system that operates in a decentralized, robust fashion, despite the failure of individual beacon nodes. This robustness is essential in order for the system to be used in disaster response, firefighting, or other critical applications in which a centralized approach is inappropriate.

As mentioned previously, RF-based location tracking has been widely studied [1, 9, 10, 11, 8, 12, 5, 6, 2]. Given a model of radio signal propagation in a building or other environment, received signal strength can be used to estimate the distance from a transmitter to a receiver, and thereby triangulate the position of a mobile node [13]. However, this approach requires detailed models of RF propagation and does not account for variations in receiver sensitivity and orientation.

An alternative approach is to use empirical measurements of received radio signals to estimate location. By recording a database of radio "signatures" along with their known locations, a mobile node can estimate its position by acquiring a signature and comparing it to the known signatures in the database. A weighting scheme can be used to estimate location when multiple signatures are close to the acquired signature. All of these systems require that the signature database be collected manually prior to system installation, and rely on a central server (or the user's mobile node) to perform the location calculation.

Several systems have demonstrated the viability of this approach. RADAR [1] obtains a 75th percentile location error of just under 5 meters, while DALS [12] obtains an 87th percentile location error of about 9 meters. These basic schemes have also been extended to improve accuracy for tracking moving targets [9]. MoteTrack's basic location estimation uses a signature-based approach that is largely similar to RADAR. Our goal is not to improve upon the accuracy of the basic signature-based localization scheme, but rather to improve the robustness of the system through a decentralized approach.

Ultrasound-based systems, such as Cricket [14, 15] and the Active Bat [16], can achieve much higher accuracies using time-of-flight ranging. However, these systems require line-of-sight exposure of receiver to ultrasound beacons in the infrastructure, and may require careful orientation of the receiver. Such an approach is acceptable for infrequent use by unencumbered users in an office environment, but less practical for rescue workers. A multimodal system would be able to achieve high accuracy when ultrasound is available and well-positioned, and fall back on less-accurate RF signal strength otherwise. Infrared-based systems, including the Active Badge [17], can localize a user to a specific area with direct line-of-sight exposure to the IR beacon, but suffer errors in the presence of obstructions and differing light and ambient IR levels (as in a fire).

2.1   MoteTrack Goals and Challenges

We first define what we mean by *robustness* with respect to location tracking. Signature-based localization schemes require a set of base stations, generally at fixed locations, to either transmit periodic beacon messages or receive signals from mobile nodes. One form of robustness, then, is graceful degradation in location accuracy as base stations fail (say, due to fire, electrical outage, or other causes).

Another form of robustness is resiliency to information loss. For example, a mobile node may be unable to communicate with an otherwise active base station, due to poor antenna orientation, multipath fading, interference, or other (perhaps transient) effects. If the tracking system assumes complete information when comparing RF signatures, this partial information loss may lead to large errors.

A third type of robustness has to do with perturbations in RF signals between the time that the signature database was collected and the time that the mobile node is using this information to estimate location. Due to the movement of base stations, furniture, opening or closing of doors, and other environmental conditions, an RF signature may no longer be valid after it has been initially acquired. The tracking system should work well even in the presence of this kind of variance in the received RF signals.

The final type of robustness has to do with the location estimation computation itself. As mentioned before, most of the previous work in this area has employed a central server to collect RF signatures and compute a mobile node's location. This approach is clearly undesirable since this server is a single point of failure. Traditional fault-tolerance schemes, such as server failover, are still susceptible to large-scale outages of electrical power or the wired network infrastructure.

Given these goals, a number of challenges arise that we wish to address through MoteTrack. First, the collection of RF signatures and location calculation must be resilient to loss of information and signal perturbation. This requires a signature distance metric that takes loss into account, avoiding explosion of error when one or more base stations cannot be contacted.

Another set of challenges has to do with decentralizing the location tracking system. One approach is to allow the base station nodes themselves to perform location estimation, rather than relying on a central server. This leads to questions about the required resources and cost of the base stations, and whether they can be readily programmed to provide this functionality. An alternative is to allow the mobile device to perform location estimation directly. In its simplest form, the entire RF signature database could be stored on the mobile node. In cases where a mobile user only carries a small RF beacon or listener (e.g., embedded into a firefighter's equipment), this may not be feasible.

3   MoteTrack Overview

In this section we give an overview of the MoteTrack system, shown in Figure 1. Mote-Track is based on low-power, embedded wireless devices, such as the Berkeley Mica2 sensor "mote." The advantages of this platform over traditional 802.11 base stations are that Mica2 motes are inexpensive, small, low-power, and (most importantly) *programmable* — we can easily push new programs and data to each device via their radio. However, the MoteTrack approach could be readily applied to other wireless networks

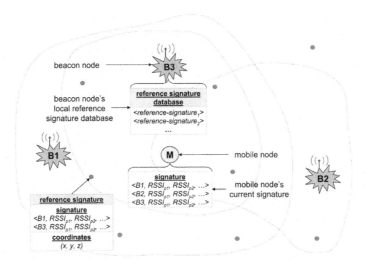

Fig. 1. The MoteTrack location system. *$B1$, $B2$, and $B3$ are beacon nodes, which broadcast beacon messages at various transmission powers ($p1$, $p2$, etc.). Each beacon node stores a subset of all reference signatures. M is a mobile node that can hear from all three beacon nodes. It aggregates beacon messages received over some time period into a signature. The areas marked by red perimeters indicate the reachability of beacon messages from the corresponding beacon node*

based on 802.11, Bluetooth, or 802.15.4, given the ability to program base stations appropriately.

In MoteTrack, a building or other area is populated with a number of Mica2 motes acting as *beacon nodes*. Beacon nodes broadcast periodic *beacon messages*, which consist of a tuple of the format {*sourceID, powerLevel*}. *sourceID* is the unique identifier of the beacon node, and *powerLevel* is the transmission power level used to broadcast the message. Each mobile node that wishes to use MoteTrack to determine its location listens for some period of time to acquire a *signature*, consisting of the set of beacon messages received over some time interval. Finally, we define a *reference signature* as a signature combined with a known three-dimensional location (x, y, z).

The location estimation problem consists of a two-phase process: an *offline* collection of reference signatures followed by *online* location estimation. As in other signature-based systems, the reference signature database is acquired manually by a user with a laptop and a radio receiver. Each reference signature, shown as gray dots in Figure 1, consists of a set of *signature tuples* of the form {*sourceID, powerLevel, meanRSSI*}. *sourceID* is the beacon node ID, *powerLevel* is the transmit power level of the beacon message, and *meanRSSI* is the mean received signal strength indication (RSSI) of a set of beacon messages received over some time interval. Each signature is mapped to a known location by the user acquiring the signature database.

In MoteTrack, beacon nodes broadcast beacon messages at a range of transmission power levels. Using multiple transmission power levels will cause a signal to propagate at various levels in its medium and therefore exhibit different characteristics at the receiver. In the most extreme case, a slight increase in the transmission power may make the difference between whether or not a signal is heard by a receiver. Varying transmission power therefore diversifies the set of measurements obtained by receiving nodes and in fact increases the accuracy of tracking by several meters in our experiments (see Section 6.4).

3.1   Location Estimation

Given a mobile node's received signature s and the reference signature set R, the mobile node's location can be estimated as follows. (In this section, we discuss the approach as though it were centralized; in Section 4 we present our decentralized design.) The first step is to compute the *signature distances*, from s to each reference signature $r_i \in R$. We employ the Manhattan distance metric,

$$M(r, s) = \sum_{t \in T} |meanRSSI(t)_r - meanRSSI(t)_s|$$

where T is the set of signature tuples represented in both signatures, and $meanRSSI(t)_r$ is the mean RSSI value in the signature tuple t appearing in signature r. Other distance metrics, such as Euclidean distance, can be used as well. In our experiments, the Manhattan and Euclidean distance metrics both produced very similar results, and the Manhattan distance is very efficient to compute on nodes with low computational capabilities.

Given the set of signature distances, the location of a mobile node can be calculated in several ways. The simplest approach is to take the centroid of the geographic location of the k nearest (in terms of signature space) reference signatures. By weighting each reference signature's location with the signature distance, we bias the location estimate towards "nearby" reference signatures. While this method is simple, using a fixed value for k does not account for cases where the density of reference signatures is not uniform. For example, in a physical location where few reference signatures have been taken, using the k nearest reference signatures may lead to comparison with signatures that are very distant.

Instead, we consider the centroid of the set of signatures within some ratio of the nearest reference signature. Given a signature s, a set of reference signatures R, and the nearest signature $r^\star = \arg\min_{r \in R} M(r, s)$, we select all reference signatures $r \in R$ that satisfy

$$\frac{M(r, s)}{M(r^\star, s)} < c$$

for some constant c. The geographic centroid of the locations of this subset of reference signatures is then taken as the mobile node's position. We find that small values of c work well, generally between 1.1 to 1.2 (see Section 6.9). In this paper, we choose a specific, empirically-determined value for c. An interesting future research question is how this parameter can be determined automatically.

4   Making RF-Based Localization Robust

In this section, we describe our approach to making RF location tracking robust to beacon node failure and signal perturbations. MoteTrack must ensure that there are *no single points of failure* and that the location estimation algorithm can *gracefully handle incomplete data and failed nodes*.

We address the first requirement by making our system completely decentralized. The location estimation protocol relies only on local data, local communication between nodes, and involves only currently operational nodes. The reference signature database is carefully replicated across beacon nodes, such that each beacon node stores a subset of the reference signatures that is carefully chosen to maximize location tracking accuracy.

We address the second requirement by using an adaptive algorithm for the signature distance metric that accounts for partial failures of the beacon node infrastructure. Each beacon node dynamically estimates the current fraction of locally failed beacon nodes and switches to a different distance metric to mitigate location errors caused by these failures.

4.1   Decentralized Location Estimation Protocol

Given a mobile node's signature s and a set of nearby beacon nodes contained in s, the first question is how to compute the mobile node's location in a way that only relies upon local communication. We assume that each beacon node stores a *slice* of the reference signature database (which may be partially or wholly replicated on other nodes). Using Mica2 motes as the beacons, the limited storage capacity (128KB ROM and 4KB of RAM) implies that the entire database will not generally be replicated across all beacon nodes.

In MoteTrack, a mobile node first acquires its signature s by listening to beacon messages, and then broadcasts s, requesting that the infrastructure send it information on the mobile node's location. One or more of the beacon nodes then compute the signature distance between s and their slice of the reference signature database, and report either a set of reference signatures to the mobile node, or directly compute the mobile node's location. Each of these designs is discussed in turn below.

k Beacon Nodes Send their Reference Signature Slice. In this first design, the mobile node broadcasts a request for reference signatures and gathers the slices of the reference database from k nearby beacon nodes. The mobile node then computes its location using the received reference signatures. While this approach can be very accurate, it requires a great deal of communication overhead. An alternative is to limit the amount of data that is transferred by contacting only $n < k$ nearby beacon nodes, requesting that each one only send the m reference signatures that are closest (in terms of signature distance) to s. For example, the mobile node can query the n beacon nodes with the largest RSSI value in s.

k Beacon Nodes Send their Location Estimate. An alternative to the previous design allows each of the k beacon nodes to compute its estimate of the mobile node's location using its own slice of the reference signature database. These k location estimates are then reported to the mobile node, which can compute the "centroid of the centroids" according to its RSSI to each beacon. The mobile node simply transmits its signature s and receives k location estimates.

While this version has reasonable communication overheads, our initial evaluations indicated that it does not produce very accurate location estimates. The problem is that for k greater than one or two, some of the beacon nodes are too far from the mobile node and therefore do not store a very relevant set of reference signatures. Since this design does not seem to perform well, we abandoned it for the design described in the next section.

Max-RSSI Beacon Node Sends its Location Estimate. Our third and final design combines the advantages from the first two to obtain both low communication overhead and accurate location estimates. In this design, we assume that the most relevant (closest in signature space) reference signatures are stored on the beacon node with the strongest signal. The mobile node sends a request to the beacon node from which it received the strongest RSSI, and only that beacon node estimates the mobile node's location. As long as this beacon node stores an appropriate slice of the reference signature database, this should produce very accurate results. The communication cost is very low because only one reply is sent to the mobile node containing its location coordinates.

4.2   Distributing the Reference Signature Database to Beacon Nodes

Using the decentralized protocol described above, beacon nodes estimate locations based on a partial slice of the entire reference signature database. Therefore it is crucial that the reference signatures are distributed in an "optimal" fashion. In addition, we wish to ensure that each reference signature is replicated across several beacon nodes in case of beacon node failures. We use two algorithms for database distribution, which we refer to as *greedy* and *balanced*.

Greedy Distribution Algorithm. The *greedy* algorithm has one parameter: `maxRefSigs`, which specifies the maximum number of reference signatures that each beacon node is willing to store locally. The algorithm operates by iteratively assigning reference signatures to beacon nodes as follows. For each signature, a given beacon node accepts and stores the signature if (1) it is currently storing fewer than `maxRefSigs` or if (2) the new reference signature contains a greater RSSI value for the beacon node in question.

The advantages of the greedy approach are simplicity and no requirement for global knowledge or coordination between nodes. For example, beacon nodes can be updated individually without affecting the signatures stored on other beacon nodes.

Balanced Distribution Algorithm. One of the problems with the greedy algorithm is that some reference signatures may never get assigned to a beacon node, while others may be replicated many times. The *balanced* algorithm tries to strike a balance between pairing each beacon node with its closest reference signature, while evenly distributing reference signatures across beacons. This is a variant of a stable marriage algorithm. To ensure that no reference signature is paired with too many beacon nodes, the algorithm prevents the match if either the current reference signature or beacon node have been assigned two more times than any other reference signature or beacon node.

The advantage of the balanced algorithm is that it can ensure balanced distribution of reference signatures while attempting to assign reference signatures to their closest beacon nodes. The disadvantage is that it requires global knowledge of all reference signature and beacon node pairings, and is therefore only appropriate for an offline, centralized initialization phase. If one wishes to update a small set of the beacon nodes, a complete reassignment involving all nodes and reference signatures may have to take place.

The pseudocode for both algorithms can be found in the technical report [18].

4.3   Adaptive Signature Distance Metric

Given that we do not expect the set of signature tuples represented in the reference signature r and mobile node's signature s to be identical, there is a question about how to account for missing data in one signature or the other. If r contains a signature tuple not found in s, this can be due to s being taken at a different location in the building, or the failure of a beacon node. Taking the intersection of the beacon set in r and s is not appropriate, because we wish to capture the low intersection in cases where one signature is largely dissimilar to another.

First, we consider the case with no beacon node failures. In this instance, missing tuples between two signatures indicates that they are at different locations. We define the *bidirectional* signature distance metric as:

$$M_{bidirectional}(r, s) = M(r, s) + \beta \sum_{t \in (s-r)} meanRSSI(t)_s + \beta \sum_{t \in (r-s)} meanRSSI(t)_r$$

That is, each RSSI tuple not found in $(r \cup s)$ adds a penalty to the distance that is proportional to that signature's RSSI value. We determined empirically that value between 0.95 and 1.0 work well for β.

This distance metric is appropriate when few beacon nodes have failed, since it penalizes for all RSSI tuples not found in common between r and s. In case of beacon node failures, however, a larger number of RSSI tuples will appear in the set $(r - s)$, leading to an explosion of error. To minimize the errors introduced from failed nodes, we define the *unidirectional* distance metric:

$$M_{unidirectional}(r, s) = M(r, s) + \beta \sum_{t \in (s-r)} meanRSSI(t)_s$$

which only penalizes tuples found in s (the mobile node's signature) and not in r (a reference signature). Assuming that the reference signatures were acquired while all beacon nodes are operational, the unidirectional metric only compares signatures between operational nodes.

As an example, consider the following signatures:

r	s	
BN 1, RSSI 20	BN 1, RSSI 45	$M_{bidirectional} = \|20 - 45\| + \|90 - 60\|$
	BN 2, RSSI 15	$+15 + 70$
BN 3, RSSI 70		$M_{unidirectional} = \|20 - 45\| + \|90 - 60\|$
BN 4, RSSI 90	BN 4, RSSI 60	$+15$

For simplicity, we do not show multiple power levels in this example. As we will see in Section 6.9, when few beacon nodes have failed, the bidirectional distance metric achieves greater accuracy than the unidirectional metric, because its comparison space is larger. With the unidirectional metric, only operational beacons are considered, but overall accuracy is diminished when few beacon nodes have failed.

Therefore, we employ an *adaptive* scheme that dynamically switches between the unidirectional and bidirectional metrics based on the fraction of local beacon nodes that have failed. Beacon nodes periodically measure their local neighborhood, defined as the set of other beacon nodes that they can hear. This neighborhood is compared to the *original neighborhood* (measured shortly after the system has been installed or recon-figured). If the intersection between the current and original neighborhoods is large, the bidirectional distance metric is used, achieving higher accuracy. If the fraction of failed nodes exceeds some threshold, the unidirectional distance metric is used instead.

This approach makes two assumptions. The first assumption is that the connectivity between beacon nodes does not change substantially over time. To mitigate this prob-lem, we only include a beacon node in the original neighborhood if its RSSI is above some threshold. However, for the current neighborhood we include all beacon nodes regardless of RSSI, and that exist in the original neighborhood. Note that we only in-clude a beacon node if it exists in the original neighborhood. This will eliminate cases when a beacon node's signal temporarily reaches more nodes. The second assumption is that there are no beacon node failures between the time that the reference signature database is collected and the system is deployed for normal operation. We believe this is a valid assumption for most installations and can be readily addressed by reinitializing the original neighborhood set of each node.

5   Implementation and Data Collection

MoteTrack is implemented on the Mica2 mote platform using the TinyOS operating system [19]. We chose this platform because it is designed for low-power operation, is relatively small, and can be deployed unobtrusively in an indoor environment. In addi-tion, the motes incorporate a low-power 433/916 MHz FSK radio, the Chipcon CC1000, which provides both programmable transmission power levels and direct sampling of received signal strength. We expect that MoteTrack could be readily ported to use forth-coming 2.4 GHz 802.15.4 radio chips. We note that MoteTrack runs entirely on the mote devices themselves and does not require a supporting infrastructure, such as back-end servers or PCs, in order to operate. A laptop connected to a mote is used to build the reference signature database, but thereafter the system is self-contained.

The total code size for the beacon and mobile node software is about 3,000 lines of NesC code. In our implementation, the reference signatures for each beacon node are loaded into program memory on the mote storing that segment of the database. This could be readily modified to use a combination of RAM and serial flash or EEPROM. Recall that each beacon node stores a different set of reference signatures depending on the distribution mechanism used.

5.1   Deployment

We have deployed MoteTrack over one floor of our Computer Science building, measuring roughly $1742 \, m^2$, with $412 \, m^2$ of hallway area and $1330 \, m^2$ of in-room area. Our current installation consists of 20 beacon motes (Figure 2).

We collected a total of 482 reference signatures. Each signature was collected for 1 minute, during which time every beacon node transmitted at a rate of 4 Hz, each cycling through 7 transmission power levels (from -20 to 10 dBm in steps of 5 dBM).

We note that in a normal deployment, a *much* smaller dataset is required and the amount of time spent collecting a signature can be on the order of several seconds rather than 1 minute. The large number of reference signatures was gathered in order to evaluate the system under various conditions and parameters. Likewise, we collected many samples for each beacon message transmission power pair, because we suspected the RSSI to vary across samples; however, we discovered that there is *very* little variation between samples and therefore we only need on the order of 2 to 3 samples.

A *beacon message* consists of a three-byte payload: 2 bytes for the source node ID and one byte representing the transmission power level. Therefore, all beacon messages from a source node ID require $2+T$ bytes: 2 bytes for the ID and $T*1$ bytes for each of the T power levels, i.e. $\{sourceID, RSSI_{p=1}, ..., RSSI_{p=T}\}$. A complete *reference signature* consists of 6 bytes for the location size (3 coordinates time 2 bytes per coordinate), 2 bytes for the ID, and up to N beacon nodes with T power levels each. The storage overhead for one reference signature is therefor $6+2+N*(2+T) = 8+2N+TN$ bytes. In our deployment, $N = 20, T = 7$, for a total of 188 bytes per reference signature. The code size for MoteTrack is about 20 KB, leaving 108 KB of read-only SRAM on each beacon node for storing a partition of the reference signature database. Therefore each beacon node can store up to 588 reference signatures. In Section 6.3 we discuss the impact of limiting the amount of per-beacon storage to estimate the effect of much larger reference signature databases.

We divided the collected signatures into two groups: the *training data set* (used to construct the reference signature database) and the *testing data* (used only for testing

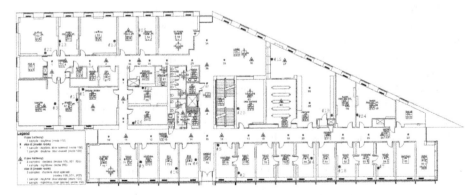

Fig. 2. Signature locations in the testing data set. *The blue dots represent the fixed beacon nodes. The red squares represent acquired signature locations; those with a green triangle were tested with 3 different motes*

	Training data	Testing data
Total signatures	282	200
Daytime	282	170
Nighttime	–	30
Using 3 motes	–	30
Hallway	151	79
In room, door open	67	81
In room, door closed	64	40

Fig. 3. Summary of the number of samples for each scenario of the training and testing data

the accuracy of location tracking). Our analysis investigates effects of a wide range of parameters, including whether signatures are collected in a hallway or in a room, whether the room's door is open or closed, the time of day (to account for solar radiation and building occupancy) and the use of different mobile nodes (to account for manufacturing differences). We collected at least 30 signatures for each of the various parameters to ensure that results are statistically significant. Figure 2 shows a map of the testing data sets, and Figure 3 summarizes the data.

6   Evaluation

In this section we present a detailed evaluation of the performance of MoteTrack along a number of axes. First, we look at the overall accuracy of MoteTrack. Although accuracy is not our focus, we do need to understand how the system performs under various parameters. We evaluate the accuracy on our entire floor which includes hallways and rooms, the location estimation protocols, algorithms for selecting reference signatures, type of database distribution, number of transmission powers used, and the density of beacon nodes and reference signatures.

Second, we look at robustness with no beacon node failures. Here we investigate the effects of radio signature perturbations, using different motes, time of day, and obstacles such as doors.

Finally, we look at robustness with beacon node failures. Here we examine how MoteTrack performs under extreme failures of the beacon infrastructure and evaluate our adaptive signature distance metric.

These results were obtained using an offline simulation of the MoteTrack protocol in order to give us the maximum flexibility in varying experimental parameters. In all cases the real reference signature database acquired in our building was used to drive the simulation. The simulator captures the effect of beacon node failure, RF perturbations, distribution of the reference signature database, and the different algorithms for signal distance and centroid calculation. The system is fully implemented on real motes and we have demonstrated a full deployment of MoteTrack in our building along with a real-time display of multiple user locations superimposed on a map.

6.1   Location Estimation Protocols

We first evaluate the accuracy of the system over the entire floor in the context of three location estimation protocols. Two decentralized location estimation protocols and a centralized one: having a closest (in terms of RSSI) beacon node compute the location, receiving reference signatures from several ($k = 3$) nearby beacon nodes, and computing the location based on all of the received signatures. The centralized version is used as a benchmark for comparison purposes.

Figure 4 shows the cumulative distribution function (CDF) for the protocols. As we can see, the accuracy of the 3 versions is nearly identical suggesting that the closest beacon node does in fact store most of the relevant reference signatures for accurately estimating the mobile node's location. Likewise, the additional overhead of receiving reference signatures from k beacon nodes is unjustified.

Our deployment uses the first decentralized protocol (i.e., closest or maxRSSI beacon node sends the location estimate), and it's the accuracy a user of the system should expect to get. As we can see, 50% and 80% of the location estimates are within 2 m and 3 m respectively from their true location. This is more than adequate for applications that require locating persons, such as tracking the location of rescue personnel or locating patients.

For the rest of this section we consider only the decentralized version where the closest beacon node computes the location.

6.2   Selection of Reference Signatures

The next parameter of interest is the algorithm used to select reference signatures that are close (in terms of signal space) to the mobile node's signature. Figure 5 compares the k-nearest selection approach to the relative signature distance threshold technique. The k-nearest algorithm computes the centroid location of the k closest reference signatures. The relative threshold scheme limits the set of reference signatures based on a threshold that is proportional to the signature distance to the nearest reference signature. For k-nearest, small values of k are appropriate for computing the location centroid, but values above this introduce significant errors. The relative thresholding scheme is more accurate as it limits the set of locations considered according to the signature distance metric. The optimal distance threshold is around 15-20% of the closest reference signature.

6.3   Distribution of the Reference Signature Database

Next we look at the different techniques for replicating reference signatures across beacon nodes. This aspect of the design is crucial because each beacon node stores only a subset of the full signature database. We look at two algorithms: *greedy* and *balance* reference signature distribution. For each of these we also vary whether a given signature is stored only on the beacon node that is closest to the signature (*closest BN*), or replicated across $k = 3$ beacon nodes (*k=3 BN*). To estimate the effect of growing the reference signature database beyond its current size (282 signatures), we artificially limited the maximum number of reference signatures that each beacon node could store.

Figure 6 shows the results of this experiment. As the maximum storage capacity of each beacon node is decreased, the balanced distribution achieves the best results. In most cases, replicating each signature across k beacon nodes achieves better results than storing it only on the closest beacon node. When the memory capacity of the beacon nodes is not limited, there is less noticeable difference between the approaches as it is more likely that any given beacon node has the relevant set of signatures.

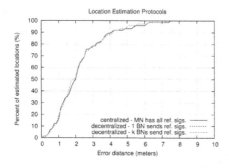
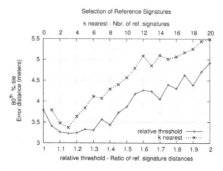

Fig. 4. Location Estimation Protocols. *Under normal circumstances both decentralized protocols perform nearly identical to the centralized one*

Fig. 5. Two reference signature selection algorithms. *The relative threshold algorithm performs better than the k-nearest one*

6.4   Transmission of Beacons at Multiple Power Levels

Recall that beacon nodes cycle through transmitting beacons at different power levels ranging from -20 dBm to 10 dBm. Initially it was not clear if transmitting at multiple power levels would noticeably improve accuracy. Figure 7 shows the 80th percentile error distance as the number of beacon transmission power levels is varied. The error is averaged across all combinations of N power levels, with N ranging from 1 to 7, i.e. $\binom{7}{N}$. As the figure shows, increasing the diversity of power levels increases the 80th percentile accuracy by nearly 2 m. However, increasing the number of transmission power levels involves a trade-off in terms of higher storage for reference signatures.

6.5   Density of Beacon Nodes and Reference Signatures

Of particular interest to someone deploying MoteTrack is the number of beacon nodes and reference signatures needed to achieve a certain accuracy. For this experiment we artificially restricted the set of beacon nodes represented in the reference signature database. For each number of beacon nodes we hand-selected the appropriate number of nodes that were approximately uniformly distributed throughout the building, avoiding any "clusters" of beacon nodes.

Figure 8 shows how location error varies with the number of beacon nodes deployed in the building, which also represents the overall density of nodes. It appears that there is a critical number of beacon nodes required after which the accuracy of the system increases marginally. In this case the critical density is around 6 to 7 nodes which is about $0.004 \frac{beacon\ nodes}{m^2}$.

Likewise, varying the number of reference signatures has a strong effect on location tracking accuracy. Figure 9 shows that the error distance decreases quickly up to the first 25 reference signatures and begins to stabilize after 75 reference signatures, representing a signature density of $0.043 \frac{reference\ signatures}{m^2}$.

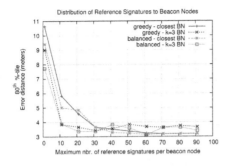

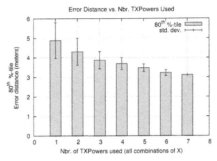

Fig. 6. Greedy vs. balanced distribution of **Fig. 7. The effect of varying the number of** **the reference signature database.** *When the* **transmission power levels used to transmit** *memory size of beacon nodes is limited, the* **beacons.** *Increasing the diversity of beacon* *balanced algorithm outperforms the greedy* *power levels increases accuracy considerably* *one*

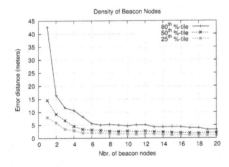

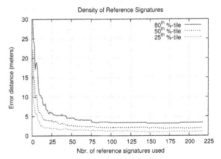

Fig. 8. Density of beacon nodes. *After 6 to 7* **Fig. 9. Density of reference signatures.** *The* *beacon nodes (0.004 $\frac{beacon\ nodes}{m^2}$) additional* *accuracy begins to stabilize after 75 reference* *beacon nodes provide diminishing returns* *signatures (0.043 $\frac{reference\ signatures}{m^2}$)*

6.6   Robustness to Perturbed Signatures

We now turn our attention to the robustness of the system under no beacon failures. We begin by looking at the effects of radio signature perturbations.

The RF propagation in a building may change slightly over time or more drastically in a disaster, when the building's characteristics may alter from events such as walls collapsing. To understand these implications, we evaluate how the accuracy of MoteTrack changes for various perturbation levels of a signature's RSSI measurements.

For each percentage, we perturbed the RSSI measurements of all signatures (i.e. the testing data) by up to a *maximum percentage* of the entire RSSI range. The perturbation amount for each RSSI is taken from a uniform distribution between zero and maximum percentage. As we can see in Figure 10, MoteTrack is quite robust to RSSI perturbations. For a maximum perturbation of 40%, the 80th percentile has an accuracy of under 5 m and, for the 50th and 25th percentiles it has an accuracy of under 3 m and under 2 m respectively.

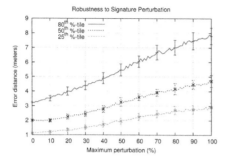
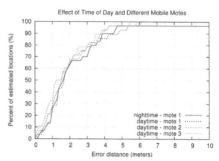

Fig. 10. Robustness to the perturbation of signatures' RSSI measurements. *The accuracy of MoteTrack degrades linearly with increased perturbation levels. All results are averaged over 30 trials*

Fig. 11. Effects of time of day and manufacturing differences between motes. *No consistent differences are found when varying the time of day or the mobile node*

6.7   Time of Day and Different Motes

Next we look at the effects of two other parameters: the time of day and manufacturing differences between motes. Time of day examines how the system reacts to changes in building occupancy and movement; the use of different motes accounts for the overall effect on the system from variation between motes.

For this experiment we collected a daytime data set between the hours of 9:00am and 4:00pm on a weekday, using 3 different motes. We also collected a nighttime data set at 1:00am, when few occupants are in the building, using only a single mote. Mote 1 was used to collect a larger number of data points. To ensure a fair comparison, only the locations that are common to all four data sets (three motes during the day and one mote at night) were used here.

Figure 11 shows that the accuracy is largely unaffected by these parameters, so we expect it to work well even for different mobile nodes and times of day.

6.8   Effect of Hallways, Rooms, and Door Position

Hallways tend to act as waveguides while walls and doors contribute to signal attenuation. We first investigate how the accuracy of a mobile user is affected by its location in the building. Figure 12 shows the cumulative distribution function (CDF) of the location error for signatures obtained in the hallway and inside rooms, with doors opened and closed. In the hallway, nearly 80% of location estimates are within 2 m of their true location, while in rooms the 80th percentile is slightly under 4 m.

As we can see from the table in Figure 3, the density of reference signatures in the hallway is higher then inside rooms. In the hallway the density is $0.36 \frac{\text{reference signatures}}{m^2}$ and inside rooms it is $0.05 \frac{\text{reference signatures}}{m^2}$. In order to make a fair comparison, we pruned the hallway data set to have the same density as the rooms data set, and plotted the pruned hallway data set (labeled *hallway pruned*). As we can see, the error distance for the 80th percentile increased to just under 2.9 m. For *hallway and rooms*, after pruning the hallway part of the data set, we found a slight increase in error from 3.2 m

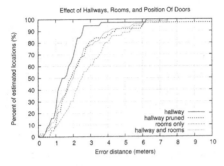

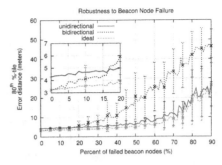

Fig. 12. The effects of hallways, rooms, and position of doors on location tracking accuracy. *This data represents a limited data set in which the density of reference signatures in the hallway was pruned to match that of the in-room reference signatures*

Fig. 13. Robustness to beacon node failure. *The unidirectional algorithm is more robust to large beacon node failures, but yields poorer accuracy when there are fewer failures. All results are averaged over 30 trials. The vertical bars represent the standard deviation*

to 3.5 m. We also looked at the effect of the position of doors and found that they don't make a significant difference.

6.9   Robustness to Beacon Node Failure

Finally, we evaluate MoteTrack's ability to continue providing accurate location estimates even when a large number of beacon nodes have failed. We consider this aspect of MoteTrack to be essential for its potential use in disaster response scenarios. Here, we simulate the effect of failed beacon nodes by selectively eliminating beacon nodes from mobile node signatures, as well as preventing those beacon nodes from participating in the decentralized location calculations.

We evaluated robustness to failure using both the unidirectional and bidirectional algorithms for calculating signature differences. Beyond a certain failure threshold, we expect the unidirectional version to perform better than the bidirectional version, since it only considers RSSI values from beacon nodes that are present at the time the signature is constructed. As we can see in Figure 13, after about 16% of the beacon nodes have failed, the unidirectional version indeed produces more accurate results.

For comparison purposes, we also show the ideal case. The ideal case is when we have perfect knowledge of which nodes failed. This is the best case scenario and although under normal circumstances it's unachievable in a completely decentralized system, it shows the lower bound. In this case the bidirectional algorithm is used but only over RSSI measurements from nodes that did not fail.

Although unidirectional signal distance is more robust, it is less accurate when there are few failed nodes. As mentioned in Section 4.3, MoteTrack decides dynamically which algorithm to use based on the local failure percentage that it last computed. MoteTrack starts out using the bidirectional algorithm and after it estimates that the beacon failure is greater than 16%, it switches to the unidirectional algorithm.

7 Future Work

The current system as described in the paper, requires an offline installation and calibration prior to use. In many cases, such as responding to a mass casualty incidents in an arbitrary area (e.g. a train wreck or a multi-car highway accident), pre-installation and calibration of a beacon node infrastructure is clearly not feasible. For these scenarios, we need an *ad hoc* mechanism for rapidly deploying the location tracking system and populating the beacon nodes with reference signatures.

In an outdoor environment, one approach is to leverage GPS to automatically populate the signature database. For example, medics responding to the scene of a disaster can place beacon nodes at well-spaced (and arbitrary) points around the site. Rather than require every patient or medic to carry a GPS receiver (which are often higher power and bulkier than sensor motes), several medics can carry a PDA equipped with a GPS receiver and MoteTrack transceiver. The PDA can automatically record reference signatures as the medics move around the site, populating the reference signature database on the fly using the greedy distribution, which does not require global knowledge of the beacon nodes and reference signatures. Signature acquisition can be performed rapidly, since each signature requires only a few beacon messages from each node and transmission power, which can be acquired in a very short period of time [20]. In our experiments we obtained good results in about one second. Location tracking accuracy will improve over time as more reference signatures are acquired.

One of the challenges faced is how to deal with the additional error introduced by the GPS location estimate. While in North America, GPS devices using the Wide Area Augmentation System (WAAS) can yield location estimates to within 3 m 95% of the time [21], it is not clear how much this will impact the overall accuracy of the system.

We have already ported our system to the Telos and MicaZ mote platforms, which support the CC2420 802.15.4 radio chip, and observed very similar results in terms of accuracy. For the immediate future, we intend to re-run our measurements with identically-placed beacon nodes to directly compare the performance of MoteTrack with 433 MHz and 2.4 GHz radios.

8 Conclusions

In this paper, we describe how to extend the basic RF approach for localization in order to make it highly robust and decentralized. We achieve this through a decentralized location estimation protocol that relies only on local data, local communication, and operational nodes; by replicating the reference signature database across beacon nodes in a fashion that minimizes per-node storage but achieves high level of robustness to failure; and by using a dynamic signature distance metric that handles incomplete data and adapts to the locally failed beacon nodes.

We implemented, deployed, and extensively evaluated our approach through a system called MoteTrack, based on the Berkeley Mica2 mote. We choose this platform because we believe that many of the applications where robustness is important will want to use small, inexpensive devices that can be embedded in the environment such as walls, in the equipment of rescue personnel, or integrated with vital-sign sensors placed on patients [22].

MoteTrack achieves a 50th and 80th percentile of 2 and 3 meters respectively, and can tolerate a failure of up to 60% of the beacon nodes and signature perturbations of up to 50%, with negligable increase in error.

References

1. Bahl, P., Padmanabhan, V.N.: RADAR: An In-Building RF-Based User Location and Tracking System. In: INFOCOM. (2000) 775–784
2. Youssef, M., Agrawala, A., Shankar, A.U.: WLAN location determination via clustering and probability distributions. In: IEEE PerCom 2003. (2003)
3. Slack, G.: Smart Helmets Could Bring Firefighters Back Alive. FOREFRONT (2003) Engineering Public Affairs Office, Berkeley.
4. McKinsey & Company: Increasing fdny's preparedness.
   http://www.nyc.gov/html/fdny/html/mck_report/index.shtml (2002)
5. Myllymaki, P., Roos, T., Tirri, H., Misikangas, P., Sievanen, J.: A Probabilistic Approach to WLAN User Location Estimation. In: Proc. The Third IEEE Workshop on Wireless LANs. (2001)
6. Smailagic, A., Small, J., Siewiorek, D.P.: Determining User Location For Context Aware Computing Through the Use of a Wireless LAN infrastructure (2000)
7. Ray, S., Starobinski, D., Trachtenberg, A., Ungrangsi, R.: Robust Location Detection with Sensor Networks. IEEE JSAC **22** (2004)
8. Krumm, J., Williams, L., Smith, G.: SmartMoveX on a Graph – An Inexpensive Active Badge Tracker. In: UbiComp 2002. (2002)
9. Bahl, P., Balachandran, A., Padmanabhan, V.: Enhancements to the RADAR User Location and Tracking System. Technical Report 2000-12, MSR (2000)
10. Castro, P., Chiu, P., Kremenek, T., Muntz, R.R.: A Probabilistic Room Location Service for Wireless Networked Environments. In: UbiComp. (2001)
11. Pandya, D., Jain, R., Lupu, E.: Indoor Location Estimation Using Multiple Wireless Technologies. In: IEEE PIMRC. (2003)
12. Christ, T., Godwin, P.: A Prison Guard Duress Alarm Location System. In: IEEE ICCST. (1993)
13. Hightower, J., Want, R., Borriello, G.: SpotON: An Indoor 3D Location Sensing Technology Based on RF Signal Strength. Technical Report UW CSE 00-02-02, University of Washington (2000)
14. Priyantha, N.B., Chakraborty, A., Balakrishnan, H.: The Cricket Location-Support System. In: MobiCom. (2000)
15. Priyantha, N.B., Miu, A., Balakrishnan, H., Teller, S.: The Cricket Compass for Context-Aware Mobile Applications. In: Proc. 7th ACM MobiCom. (2001)
16. Ward, A., Jones, A., Hopper, A.: A New Location Technique for the Active Office. IEEE Personal Comm. **4** (1997)
17. Want, R., Hopper, A., Falcao, V., Gibbons, J.: The Active Badge Location System. Technical Report 92.1, Olivetti Research Ltd. (ORL) (1992)
18. Lorincz, K., Welsh, M.: A Robust, Decentralized Approach to RF-Based Location Tracking. Technical Report TR-19-04, Harvard University (2004)
19. Hill, J., Szewczyk, R., Woo, A., Hollar, S., Culler, D.E., Pister, K.S.J.: System Architecture Directions for Networked Sensors. In: ASPLOS 2000. (2000) 93–104

20. Krumm, J., Platt, J.: Minimizing Calibration Effort for an Indoor 802.11 Device Location Measurement System. Technical Report MSR-TR-2003-82, Microsoft Research (2003)
21. WAAS: What is WAAS? (2005) http://www.garmin.com/aboutGPS/waas.html.
22. Lorincz, K., Malan, D.J., Fulford-Jones, T.R.F., Nawoj, A., Clavel, A., Shnayder, V., Mainland, G., Moulton, S., Welsh, M.: Sensor Networks for Emergency Response: Challenges and Opportunities. IEEE Pervasive Computing (2004)

Correcting GPS Readings from a Tracked Mobile Sensor

Richard Milton and Anthony Steed

Department of Computer Science, University College London,
Gower St., London, WC1E 6BT, United Kingdom
{R.Milton, A.Steed}@cs.ucl.ac.uk

Abstract. We present a series of techniques that we have been using to process GPS readings to increase their accuracy. In a study of urban pollution, we have deployed a number of tracked mobile pollution monitors comprising a PDA, GPS sensor and carbon monoxide (CO) sensor. These pollution monitors are carried by pedestrians and cyclists. Because we are operating in an urban environment where the sky is often occluded, the resulting GPS logs will show periods of low availability of fix and a wide variety of error conditions. From the raw GPS and CO logs we are able to make maps of pollution at a 50m scale. However, because we know the behaviour of the carriers of the devices, and we can relate the GPS behaviour and known effects of CO in the environment, we can correct the GPS logs semi-automatically. This allows us to achieve a roughly 5m scale in our maps, which enables us to observe a new class of expected environmental effects. In this paper we present the techniques we have developed and give a general overview of how other knowledge might be integrated by system integrators to correct their own log files.

1 Introduction

In this paper we describe a study of carbon-monoxide (CO) pollution that uses a small set of mobile CO sensors. Our aim is to make a map of CO pollution in a small area of a city at a scale fine enough to identify per street variation. The sensors are tracked using a GPS sensor. However, because the availability and quality of a GPS signal will vary greatly in an urban area, we have to be careful how we process the GPS log files.

The quality of GPS varies in the urban environment because of the urban canyon effect. A GPS sensor needs to be able to see three or more GPS satellites in order to get a reliable reading. However, buildings around the sensor cause the view of the sky to be obscured, and as the sensor moves the view of the sky will change quite rapidly.

Without processing the GPS logs, we are able to generate maps of CO at a scale of roughly 50m. Although this allows us to see gross variations in CO, such as CO being lower in a relatively traffic-free square, it is not sufficient to see per street variations. Errors in GPS reading will often put the sensor inside buildings, next to the pavements and streets, or will even put the sensor in

T. Strang and C. Linnhoff-Popien (Eds.): LoCA 2005, LNCS 3479, pp. 83–94, 2005.

adjacent streets. Obviously this prevents us identifying pollution variations that exist between streets.

However, in our trials, we know quite a lot about the behaviour of the carriers of the pollution monitors. We can use knowledge about the behaviour to process the GPS logs to increase the accuracy. This has enabled us to identify features at the 5m scale including CO hot spots such as bus stops and road crossings. We discuss five different levels of knowledge of the situation that we can apply to exclude or filter GPS readings: knowledge of geographic extent of trial, knowledge of the building footprints, knowledge of the route undertaken, knowledge of the behaviour of the carrier of the GPS receiver and knowledge about expected sensor readings. We then describe tools that we have built to help us do such corrections semi-automatically.

In the following section we will present some related work on environmental sensing and GPS correction. Then in Section 3 we will outline the sensor package and the trials we have undertaken. In Section 4 we give an overview of the type and quality of data that we can capture. Then in Section 5 we discuss how we can refine the GPS readings to give us more focussed data. In Section 6 we will then generalise and discuss tools that should be more applicable to processing tracked data logs. Finally we conclude and present plans for future work.

2   Related Work

The background to the work in this paper is an ongoing study of urban carbon monoxide. The Air Quality Site contains archive data from over 1500 UK monitoring stations going back in some instances to 1972 [1]. Such data sources give a good picture of variation from urban to rural areas. In urban areas some sense of potential variation is conveyed by the difference in readings between kerbside sensors and sensors placed in background areas away from pollutant sources. However they don't capture the detail of per street variation.

CO disperses over a matter of hours, but Croxford et al. have shown that this is affected by local street configuration [2]. Croxford's study used a cluster of sensors in fixed placements in a small area around University College London, UK. The Air Quality Strategy for England, Scotland, Wales and Northern Ireland [3], suggests a standard of 10ppm (11.6mg/m^3) running 8-hour mean. In the vicinity of UCL, the Croxford study found a peak CO concentration of 12ppm, but nearby sensors reported much lower values near the background level for CO. Thus, moving pedestrians or vehicles would probably not experience this peak for a long period.

Our approach to bridging this gap between large, sparse and small, dense studies is to use tracked mobile sensors that are carried by pedestrians or mounted on vehicles. There are several related works in the UK that aim to increase understanding of pollution variation. These include the dense sensor network of the Discovery Net e-Science project [4], and the combined sensing and air flow modelling approach of DAPPLE [5].

In this paper we apply knowledge from records of trials and map data to correct GPS readings. GPS is of course a very significant area of research and there are many existing and upcoming enhancements to GPS that can improve its accuracy [6]. Most relevant for this work is fusion of velocity and acceleration data with the GPS position [7]. We wanted to focus on consumer-level GPS because we anticipate sensor package cost being a primary concern when deploying large collections. We also didn't want to weigh down the carriers, nor burden them with any device configuration tasks. Indeed, the user should simply have to power on the devices before going outside. In an urban environment, especially in central London where the buildings are typically close to or higher than the road width, we do not expect to get a GPS fix 100% of the time [8].

3   Trial Description

3.1   Monitor Overview

We will briefly describe the trial; a longer description with more background on the task and scientific justification from an environmental science point of view can be found in [9]. We collected data over two weeks, in collaboration with the 'Dispersion of Air Pollution and Penetration into the Local Environment' (DAPPLE) project. The data collection area was centred around Marylebone Road and Gloucester Place in central London. The local area is densely packed with buildings up to roughly eight stories high, with a wide variety of street widths. We collected data from both pedestrians and walkers.

Fig. 1. The ICOM carbon monoxide sensor on the left, the PDA in the middle and the GPS aerial on the right. The HAICOM GPS unit can just be seen sticking out of the top of the PDA

The equipment pack consisted of an HP Jornada PDA running our own data logging software, an ICOM device from Learian Environmental to measure carbon monoxide and a GPS unit. Two types of GPS units were used: the HAICOM type, or a self-built device based around the GARMIN 15 package. All the equipment was colour coded and divided into separate coloured packs, which were not mixed, to allow for easy tracking of the equipment used for each sample run.

Data is sampled once every second from the ICOM carbon monoxide sensor, combined with the last valid GPS fix and written to a log file on the PDA. Every time the software on the PDA is run, both a GPS log file and a pollution log file are created with a filename that includes the current timestamp. After data collection, these files are uploaded to a database server for analysis.

3.2   System Overview

All the data is stored in a Postgres database, with the reported and corrected GPS positions in separate fields. Correction of the GPS data can take one of two forms: either the data is fitted to the point closest to a route drawn on a map, or an offset is calculated from the position of a known point and applied to a section of the data. With data collection taking place on a limited number of fixed routes, idealised routes have been created to ensure that the corrections are coherent.

3.3   Routes

In these trials, we studied a small number of walkers and cyclists in a small area to prove the concept. To cover larger areas would require more carriers, and eventually we hope to distribute 100s or 1000s of devices to commuters.

Two routes along the length of Marylebone Road were used for most of the walking, in addition to both walking and cycling around the cycle route. On other occasions, different routes around the area were tried (Figure 2).

4   Quality of Captured Data

During the data collection period, up to 3 carriers, either walkers or cyclists were out for up to four 45 minute sessions a day. 87 log files containing GPS positions and carbon monoxide data were collected, resulting in 227,496 separate data points. During the two-week sampling period, a small number of equipment problems were encountered, resulting in the loss of 10 sample runs. Figures 3 and 4 show an example log file.

4.1   Reliability of GPS in an Urban Environment

Overall, the percentage of data points with corresponding GPS positions was 42.8% of the total walking and cycling time, corresponding to 34.6 hours of collection time over two weeks. By comparing daily GPS data for the pedestrians and cyclists, it can be seen that the cyclists have a marginally higher percentage of valid GPS positions. These levels of GPS reliability might sound low, but are not atypical for an urban environment. In [10], a car-mounted consumer GPS unit driven around central London achieved 60% reliability. Given that in this case the aerial placement and vehicle behaviour would probably have afforded a more stable view of the sky, this suggests our availability results are not low.

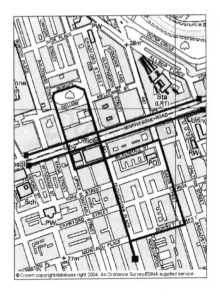

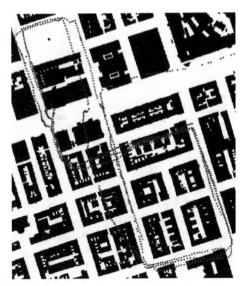

Fig. 2. The two walking routes along Marylebone Road and the figure 8 cycle route

Fig. 3. GPS trace for the cycle route on 19 May 2004 at 11:48

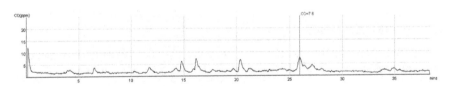

Fig. 4. Carbon monoxide trace for the cycle route on 19 May 2004 at 11:48. The maximum CO reading, 7.6ppm is indicated

When looking at the GPS traces, it is apparent that the GPS is not reliable enough to place a sensor on the correct side of the road. Occasionally, a log file will contain a distinct trace that sticks to the correct side of the road, but the majority meander along the centre of the road, or drift from side to side. For the Marylebone Road walking routes, without looking at the recorded route notes, it is not usually possible to be certain whether the sensor was on the north or south side of the road, unless additional features are present. For the cycle route this is less of a problem as the cyclists are always in the road and the pedestrians always take the same route, although the problem of detecting junctions accurately remains.

The method of standing in a fixed location and averaging the GPS position for a period of time is a standard GPS surveying technique used to increase accuracy, but this is not applicable to mobile CO sensors. Using differential GPS would be more appropriate, but again, this method requires expensive equipment and does

Fig. 5. Reported GPS location while standing at the bottom right corner of the junction for 30 minutes

not lend itself to ad-hoc data collection by people during their normal daily lives. Using inexpensive commercial GPS devices, post processing of the data is not possible due to the pseudo range information not being stored, so the accuracy is limited to about 15m. As was shown previously, the carbon monoxide level can vary significantly from the kerb side to the building, so a more accurate method of positioning samples using standard GPS devices is required.

By analyzing the collected GPS data, there is additional information present in the trace that can be exploited to perform further correction. There is little that can be done about multi-path reflections other than detecting when low elevation satellites are used in the solution or using the GPS unit's own dilution of precision figure to assign a confidence measure to the position.

4.2   ICOM Accuracy

The quoted tolerance of the ICOM CO sensors is +/- 5%, but the nature of the carbon monoxide levels being measured is such that they can change very rapidly. The range of values of most interest is between 0.3ppm and 5.0ppm. Although much larger values do occur, they are very rare and their location is of more interest than their value. A base level of 0.3ppm corresponds to the global background carbon monoxide level.

5   Correction Methodologies

In this section we describe features in the data that can be used to correct GPS. These come from two main sources: knowledge about the behaviour of the carriers and knowledge about expected CO effects. Using such knowledge allows us to correct our GPS logs so that we can discriminate between the readings more accurately, and thus make maps at a finer scale.

In the following figures, raw trace logs (Figures 6, 7, 10 & 11, and also Fig 3) are drawn with buildings in black, background in white and a red-scale for pollution. Summary maps (Figures 8 & 9) are drawn with a similar scale but on a grey background. The method of plotting summary maps was to divide the area into grid squares of either 20 metre or 5 metre spacing and to place data points into the correct square based on GPS position. The mean and variance

Fig. 6. An uncorrected GPS trace on Marylebone Road

Fig. 7. The trace of Figure 6, fitted to a line drawn along the kerbside using the GPS correction tool

Fig. 8. 5 May 2004 09:00-17:00, 5 metre mean boxes without GPS correction

Fig. 9. 5 May 2004 09:00-17:00, 5 metre mean boxes with corrected GPS positions

for each square can then be calculated and plotted. By altering the size of the grid squares, the data can be plotted to within the accuracy limits imposed by the GPS positional error. We use a simple grid-based visualisation because the mean and variance vary at a fine geometrical scale and don't lend themselves well to smooth representations as, say, surfaces.

Fig. 10. Identifying a south side route from a pedestrian crossing

Fig. 11. GPS trace crossing Marylebone Road

5.1   GPS Logfile Manipulation

The Postgres database that stores all of the information in the collected log files is structured so that data for an individual path can be extracted. Each path can be analyzed separately using a tool that plots a colour coded carbon monoxide value over an aerial map using the GPS position logged at the time the sample was taken. Traces can be played back in real-time, or faster, allowing the velocity of the sensor to be tracked. The GPS positions for a path can be corrected by drawing a polyline representing the idealised route on the map and fitting the GPS points to the nearest point on the line. When correcting the data, only points which lie a user-defined distance from the line are corrected, allowing a limit to be set on how far points will be moved. Typically, 20m has been used as this is the GPS accuracy. During the correction process, points that have no valid GPS position, but lie between two valid GPS positions, are linearly interpolated. See Figure 6 and 7 for an example of before and after correction.

Using the viewer tool, GPS tracks that cross into buildings can often be seen, Figure 3 is a good example of this. This type of error is harder to correct, as using the line correction tool will result in a trace with a discontinuity at the corner of the building. The trace needs to be stretched around the building, but this has not been attempted due to the majority of the data fitting the line correction model more closely.

5.2   Fitting GPS Traces to Routes

As an example, data for 5 May 2004 was corrected using the GPS correction tool to see what physical features could be observed. This day was chosen as a good example for wind variation on opposite sides of the road as the wind only varied between 170 degrees and 260 degrees all day, starting at 5 knots and increasing to 16 knots. The difference between the corrected and uncorrected data is clearly visible.

The data in Figure 9 shows a marked difference in carbon monoxide levels between the north and south sides of the road at the east end. Compare to the raw data in Figure 8. On the north side the levels are 1.5 ppm, while on the south side they are 3.5ppm. A factor of 2-3 difference is something that would be expected with the wind coming from the southwest quadrant, dispersing the carbon monoxide on the north side of the road where the wind speed is highest and causing levels to build up on the south side where the air is static.

5.3   Disambiguating Route Choice

In addition to correcting known routes, it is also necessary to be able to work out the route taken from the GPS data. On the Marylebone Road walking routes, which are virtually straight lines, there are points where the walkers move off line and down a side road in order to cross at a crossing away from the junction. The most noticeable instance of this is on the south side of Marylebone Road at the junction with Baker Street, where the pedestrian crossing is a short distance south down Baker Street (Figure 10). By recognising certain features present in the GPS data, it is possible to determine which side of the street the samples were taken and also identify when the sensors reach junctions.

When crossing Marylebone Road at the junctions to the east or west of the council building, there is a characteristic that often appears in the GPS trace due to the position of the crossings relative to the road (Figure 11). The significant deviations to the left are a function of where the crossings are placed in relation to Marylebone Road. Both GPS paths are significantly out of position, placing the pedestrian in the middle of the road while he is still waiting to cross. This fact can be seen when the traces are played back in real-time by noting where the GPS position is stationary while the pedestrian is waiting to cross.

In both these situations, once the behaviour has been identified, appropriate route corrections can be applied. The trace in Figure 10 can be positioned accurately from the shape of the crossing, but the rest of the trace is on the wrong side of the road. This can be corrected by fitting the trace to a line on the correct side of the road. Figure 11 shows a different situation where a strong GPS fix is offset by a constant amount. Matching the shape of the GPS trace to the junction layout and subtracting this from the GPS coordinates can remove the offset.

5.4   Recorded Carbon Monoxide Levels When Crossing Roads

As all the pedestrian routes involve crossing roads at some point, there is a significant amount of data available in the middle of the road. The problem is how to detect when a sensor crosses the road due to the poor quality of GPS positioning information. One route stands out as a good example of road crossings, which was recorded on 10 May 2004 from 15:39, see Figures 12 and 13.

The actual positioning of these points is very hard to do from the GPS position alone and is determined more by when the GPS is stationary or additional satellites suddenly come into view, rather than the actual reported position.

The final peak on Figure 12 is marked as point X as there is no GPS and there is no additional information to identify this point, other than that it is somewhere along Gloucester Place.

It is worth noting that there are eleven crossing points when walking around the cycle route, not all of which produced peaks in this example and where there was a peak at a junction, it might disappear on the next circuit. For these reasons, using CO data to indicate the route taken can only be performed when there is extensive knowledge of the traffic flow patterns. Scientists working in

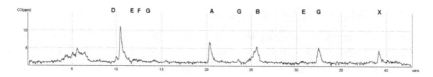

Fig. 12. Carbon monoxide levels while walking around the cycle route

Fig. 13. Map showing positions of carbon monoxide trace

this area can tell from the CO data when one of the pedestrians skips part of their assigned route. This is done using knowledge of the mean CO values for each road, together with where the CO peaks occur.

6   Dealing with Tracking Data

In Section 5 we saw several features in the data that allowed us to determine corrections to GPS log files. To summarise:

- Noting readings that were inside buildings.
- Noting that logs could be constrained to paths.
- Noting the micro-scale behaviour of the carriers when crossing a road allowed us to identify that a crossing was taking place and, in some cases, which side.
- Noting disambiguating paths by noting a particularly high CO reading.

We can generalise these to suggest there are five levels of knowledge that can be exploited when observing GPS logs in order to improve understanding:

- Knowledge of geographic extent of trial. We can discard some GPS readings because they are not in the geographic vicinity. This can be done as a pre-processing stage in the database.

- Knowledge of the building footprints. By overlaying GPS readings over a bitmap showing places where the carrier could not have received a signal we can easily see which locations need correcting. Note it is not appropriate to simply discard the points, as they may only be a few metres inside a building for example. These points can be corrected semi-automatically by moving them to the closest viable position.
- Knowledge of the route undertaken. Once we have filtered based on the impossible location, we can constrain the route based on route knowledge. We have made the distinction from the above, because at this stage we start identifying probable regions where the device was located, not impossible. There are many potential implementations here. Ours uses proximity to a route centre line, but it could be any region or line defined in a Geographical Information System (GIS).
- Knowledge of the behaviour of the carrier of the GPS receiver. At this stage we employ knowledge that is difficult to capture in GIS systems: behaviour of the users in the space. This requires some human knowledge. Although interesting artefacts such as velocity changes can be flagged in the data, a human observer is really needed to interpret this data. Again, a number of tools could be built, ours allows us to watch scaled-time replays of the event logs to get a sense of the behaviour.
- Knowledge of expected correlation with other sensor readings. At this stage we employ knowledge about other readings that can allow us to reason about positions on paths or corrections to paths. In other cases this could be considered implemented in a semi-automatic manner if framed as a sensor fusion problem. However at this time we are interpreting these observations by hand because the observer needs to know the expected effect and the context in which the readings were taken.

7   Conclusions

In this paper we investigated techniques to correct GPS logs. The need for correction arose because GPS readings are only accurate to roughly 20m, and because the urban environment causes quite significant variation in the behaviour of the readings. Several different types of knowledge have been exploited to correct the logs: geographic region, building footprints, routes and user behaviour. We have shown that by semi-automatically correcting the GPS logs, we can start to observe phenomena at a much smaller scale. In the previous section we discussed how these tools might be more generally applicable to other sensor tracking trials.

CO has proved interesting because it shows variation at a scale we can detect without expending effort in correcting GPS traces, but if that effort can be made, we can also find smaller-scale geographic effects. In our case, correcting GPS logs has allowed us to identify features on a 5m scale. This is sufficient for us to note expected properties of CO pollution: differentials across a street on a windy day and increases in CO near bus stops and road junctions.

Acknowledgements

This work was supported by the UK projects Advanced Grid Interfaces for Environmental e-science in the Lab and in the Field (EPSRC Grant GR/R81985/01) and EQUATOR Interdisciplinary Research Collaboration (EPSRC Grant GR/N15986/01). The vector data used was supplied by the UK Ordnance Survey.

References

1. Department for Environment, Food and Rural Affairs (Defra): The air quality archive. http://www.airquality.co.uk/ (verified 2005-02-09) (2005)
2. Croxford, B., Penn, A., Hillier, B.: Spatial distribution of urban pollution: civilizing urban traffic. 5th Symposium on Highway and Urban Pollution (1995)
3. Department for Environment Food and Rural Affairs (Defra): The air quality strategy for england, scotland, wales and northern ireland, 1999 (1999)
4. Discovery Net. Web Resource (2005) http://ex.doc.ic.ac.uk/new/index.php (verified 2005-02-09).
5. DAPPLE (Dispersion of Air Pollutants and their Penetration into the Local Environment). Web Resource (2005) http://www.dapple.org.uk/ (verified 2005-02-09).
6. O'Donnell, M., Watson, T., Fisher, J., Simpson, S., Brodin, G., Bryant, E., Walsh, D.: Galileo performance: Gps interoperability and discriminators for urban and indoor environments. GPS World, June (2003)
7. El-Sheimy, N.: Integrated systems and their impact on the future of positioning, navigation, and mapping applications. Quo Vadis International Conference, International Federation of Surveyors (FIG) Working Week, 21-26 May 2000, Prague. (2000)
8. Steed, A.: Supporting mobile applications with real-time visualisation of gps availability, s. brewster and m. dunlop (eds). MobileHCI 2004, LNCS 3160 (2004) 373–377
9. Steed, A., Milton, R.: Making maps of pollution using tracked mobile sensors. Under submission (2004)
10. Ochieng, W., Noland, R., Polak, J., Park, J.H., Zhao, L., Briggs, D., Crookell, A., Evans, R., Walker, M., Randolph, W.: Integration of gps and dead reckoning for real time vehicle performance and emissions monitoring. The GPS Solutions Journal, 6(4) (2003) 229–241

Web-Enhanced GPS

Ramaswamy Hariharan[1], John Krumm[2], and Eric Horvitz[2]

[1] School of Information and Computer Sciences,
University of California, Irvine, 444 Computer Science Building,
Irvine, CA 92697, USA
rharihar@ics.uci.edu
[2] Microsoft Research, Microsoft Corporation, One Microsoft Way,
Redmond, WA 98052, USA
{jckrumm, horvitz}@microsoft.com

Abstract. Location-based services like reminders, electronic graffiti, and tourist guides normally require a custom, location-sensitive database that must be custom-tailored for the application at hand. This deployment cost reduces the initial appeal of such services. However, there is much location-tagged data already available on the Web which can be easily used to create compelling location-aware applications with almost no deployment cost. Such tagged data can be used directly in applications as well as to provide evidence in models of activity. We describe three applications that take advantage of existing Web data combined with location measurements from a GPS receiver. The first application, "Pinpoint Search", finds web pages of nearby places based on GPS coordinates, queries from a Web mapping service, and general Web searches. The second application, "XRay", uses the mapping service to find businesses in a building by pointing a GPS-equipped electronic compass at the building. The third application is called "Travelogue", and it builds a map and clickable points of interest to help automatically annotate a trip based on GPS coordinates. Finally, we discuss the use of Web-based data as rich sources of evidence for probabilistic models of a user's activity, including a means for interpreting the explanation for the loss of Web signals as users enter structures.

1 Introduction

Location-based services use knowledge of a user's location to index into services and data that are likely useful at that location. For instance, a reminder application like comMotion[1] can give the user relevant information at a given location, like, "You're near a grocery store, and you need milk at home." A so-called "electronic graffiti" system, such as Stick-e Notes[2], supports users who want to leave electronic notes for themselves or others that are associated with a particular location, like "There is a better Thai restaurant one block north of here." Location-based tour guides such as Cyberguide[3] offer relevant information about the exhibit or site at which the user is standing. These and most other location-based services share a need for a custom database dedicated to storing and serving data for specified locations. Reminder systems must have reminders, electronic graffiti needs digital tags, and tour guides need site information.

T. Strang and C. Linnhoff-Popien (Eds.): LoCA 2005, LNCS 3479, pp. 95 – 104, 2005.
© Springer-Verlag Berlin Heidelberg 2005

While a custom store of location-indexed data can lead to interesting applications, there is already a wealth of location data available on the Web that can be exploited for location services without the data deployment costs of traditional applications of this type. This paper demonstrates three applications that use existing location data on the Web in conjunction with position information from a GPS receiver. In this way, we avoid the deployment costs of creating a database of location information, relying instead on what is already available. These applications show that it is possible to create useful location-based services using existing location data.

The three applications are:

- Pinpoint Search – Convert a measured (latitude, longitude) into search terms for Web searches, giving web pages relevant to the user's immediate surroundings.
- XRay – Point a pose-sensitive device at a scene of interest and get a list of what businesses are situated along that direction.
- Travelogue – Help a user recall points of interest from a trip logged with GPS data.

Our maps and point-of-interest data come from the Microsoft MapPoint Web service, which requires a subscription fee. However, using this service is much more economical than building a custom store of location data. Other point-of-interest databases could be used as well. We have implemented these three applications on a desktop computer using real GPS data. Their ultimate target is mobile users, and we foresee few problems modifying the applications for use on a PDA or cell phone. The target device would need a Web connection and a GPS receiver.

2 Pinpoint Search

Pinpoint Search is designed to give information about a mobile user's immediate surroundings. We assume the user is equipped with a GPS receiver and Web-connected mobile computer. Starting with a (latitude, longitude) from the user's GPS, Pinpoint Search uses MapPoint to compute the nearest street address. This is shown in Fig. 1(a) which is a screen shot of a working mockup of Pinpoint Search's client and server parts. In this offline demonstration program, each stored (latitude, longitude) can be resolved, which triggers the street address lookup, the results of which are shown in a popup window on the map.

The conversion from (latitude, longitude) to street address is important, because the street address serves as a good search term for Web searches. This is illustrated in Fig. 1(b), which shows a screen shot giving the result of an MSN® search automatically performed on the resolved street address within the Pinpoint Search program. The search results give links to web pages that contain the nearby street address, resulting in entries for a nearby acupuncturist, chiropractor, and restaurant.

Not all relevant Web pages contain a street address that matches our search term, so we also convert (latitude, longitude) into a list of nearby businesses using the

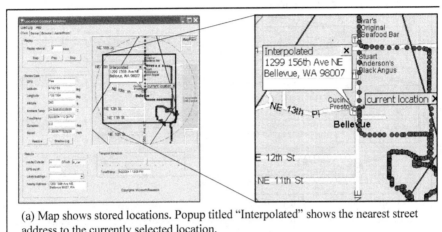

(a) Map shows stored locations. Popup titled "Interpolated" shows the nearest street address to the currently selected location.

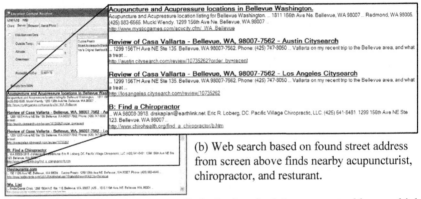

(b) Web search based on found street address from screen above finds nearby acupuncturist, chiropractor, and resturant.

Figure 1: Pinpoint Search converts (latitude, longitude) to a street address, which is used as a search keyword.

MapPoint Web Service. This service uses a database of business locations from Acxiom® that are categorized by type, such as food stores, automobile dealers, and restaurants. As shown in Fig. 2(a), this lookup results in three nearby restaurants, one of which is called "Stuart Anderson's Black Angus". After the user clicks on this result, Pinpoint Search automatically runs another Web search using the restaurant name as a search term. One of the search results is shown in Fig. 2(b), which gives positive and negative reviews of the restaurant. Starting with GPS coordinates, the user could find this review page with only two mouse clicks using Pinpoint Search.

Our implementation of Pinpoint Search runs on a desktop computer and uses stored GPS coordinates for demonstration. It would be straightforward to port it to PDA or cell phone equipped with GPS.

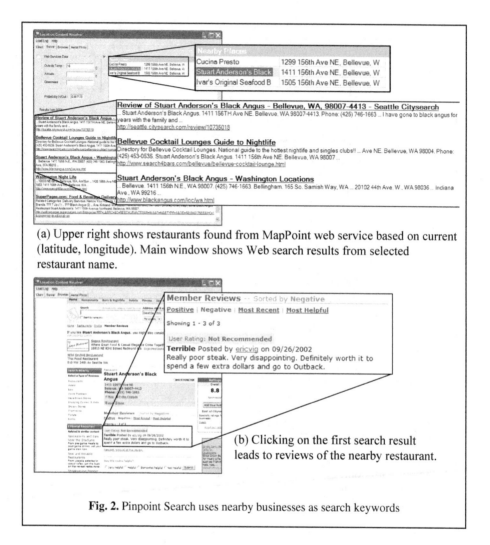

(a) Upper right shows restaurants found from MapPoint web service based on current (latitude, longitude). Main window shows Web search results from selected restaurant name.

(b) Clicking on the first search result leads to reviews of the nearby restaurant.

Fig. 2. Pinpoint Search uses nearby businesses as search keywords

Pinpoint Search shares similarities with other programs. Google Local[4] and Yahoo! Local[5] allows Web searches around a certain geographic area, specified by parts of a street address. Pinpoint Search adds the important preprocessing step of converting a numerical (latitude, longitude) into a street address or business name, which allows the local search to be completely automated based on a GPS receiver. Thus, to use Pinpoint Search, the user does not need to know a street address, postal code, or even the name of the city. Because it starts with a GPS coordinate, the search results are likely to be much more focused on the user's immediate surroundings.

Another similar project is AURA[6], which uses bar codes and other means to create Web search terms, much like we use location data to generate addresses and business names for searches. Like these projects, Pinpoint Search requires no special databases, but instead exploits the extensive amount of data already available on the Web. It is unique in that it uses a measured position to ultimately index into web pages about nearby things.

3 XRay

XRay is a concept for a pose-sensitive query device designed to allow a mobile user to point toward an outside object and discover what is inside or behind it. It is based on a device containing a GPS receiver, electronic compass, and network connection. The user physically points the device at something and issues a query. XRay responds with a list of businesses or other points of interest along the direction of pointing. A working mockup of the program is illustrated in Fig. 3. Here XRay has been pointed toward a street that the user cannot yet see, and its "field of view" has been adjusted

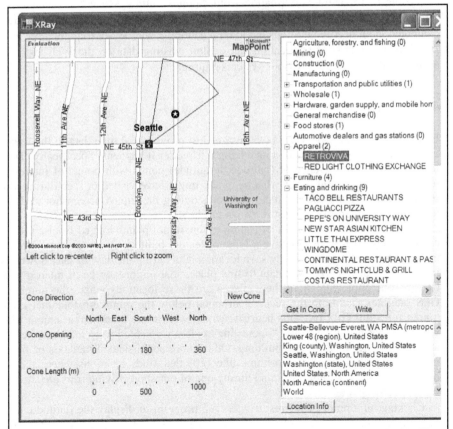

Fig. 3. XRay lets a user point in a direction and query for what is behind and inside. The user can adjust how far XRay "sees" and its field of vision. This data was taken near the University of Washington in Seattle

to query over the length of the unseen block. The list at the right shows different categories of items returned from MapPoint's® Acxiom® database. The "Apparel" and "Eating and drinking" categories have been expanded to show the names of places inside the query cone. Clicking on one of these places puts an icon on the map at its location. Using this program, users can quickly get a sense of what is around them by simply pointing in a direction of interest. As with Pinpoint Search, XRay relies on a rich, existing database of places, meaning that it works "out of the box" with no deployment cost.

Although XRay shows real results, we have not yet found a commercially available combination of devices that would let us exercise it in the field. While there are GPS receivers with a built-in compass, we have not yet found one that can interface to a handheld computer with ubiquitous network access. An ideal platform would be a cell phone with a GPS, an electronic compass, and Web access, which we expect will soon be easy to buy or make. XRay is similar to augmented reality systems, e.g. [7, 8] that project extra information into the user's view. In fact, the "Touring Machine"[7] is capable of projecting information from campus buildings that the user can see. Similarly, UCSD's Active Campus project[9] can show information on a handheld computer about nearby buildings based on positioning with Wi-Fi triangulation. XRay shows that this type of application can exploit existing data on the Web instead of custom-authored data as previous systems required.

4 Travelogue

Pinpoint Search and XRay are both designed to provide real time data to the user. In contrast, Travelogue is aimed at analyzing a trip after it happens. Specifically, Travelogue annotates a sequence of (latitude, longitude) points with points of interest. In this way, it helps a user recall what he or she might have visited or seen during a trip without requiring the user to make any notes during the trip. A screenshot of Travelogue showing points of interest is shown in Fig. 4.

Travelogue starts with a list of (latitude, longitude) points logged from a GPS receiver. Many GPS receivers have a logging feature built in, so a user can simply set up his or her GPS to periodically save location data during the day for later downloading. Travelogue analyzes this data to find places that might later be of interest to the user. In our implementation, these places are those locations where the number of GPS satellites visible to the receiver dropped below the minimum four that are required for a full GPS fix. This heuristic works well in the urban environment where we tested, because most of the "interesting" places we visited were indoors where the satellites are occluded by the building. Other more sophisticated methods of finding interesting places include algorithms like [10] that finds locations where the user dwells. The points of interest found during one of our short driving trips are shown in Fig. 4 as red squares.

Clicking on a point of interest triggers the program to display the (latitude, longitude) of the point, the time of arrival, and the amount of time spent there. Using our heuristic, the time spent is simply the time over which at least four GPS satellites could not be detected. Clicking on a point of interest also populates multiple

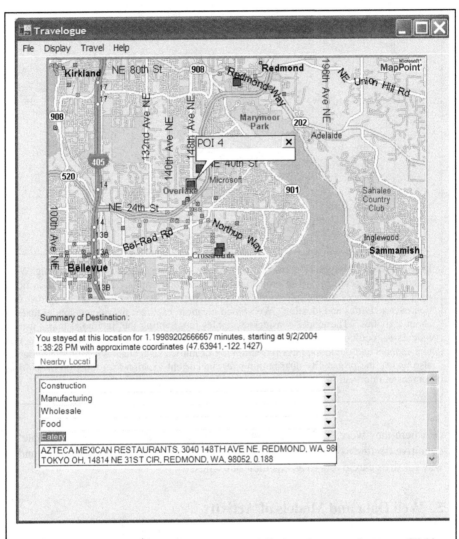

Fig. 4. Travelogue discovers points of interest and displays them as red squares. Clicking on these points triggers the program to display the amount of time spent there and a list of nearby places. This can help a user remember what he or she did during a trip

dropdown lists showing nearby places found in MapPoint's® Acxiom® database. In Fig. 4, the "Eatery" list has been expanded to show nearby restaurants. While the user did not necessarily visit every place on these lists of nearby places, the lists give a simple, fast way of jogging the user's memory for what places he or she actually did visit.

We envision normal users using Travelogue as a way of automatically annotating business and pleasure trips without having to worry about making explicit notes

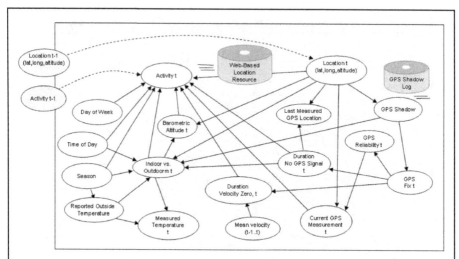

Fig. 5. Dynamic Bayesian network for inferring location and activity, employing Web-based location resources. The probabilistic dependencies among random variables highlight the influences of multiple sources of contextual evidence on the probability distribution over activities and location. Web-based location resources provide evidential updates about activities. The model considers variables representing the likelihood that a user is indoors vs. outdoors as a function of multiple variables including differences in temperature indoors and outdoors, GPS fix, and a log of known GPS shadows. Multiple variables also update the probability distribution over the current location as a function of multiple sources of information

on where they were. Travelogue could also be used as a memory aid for people with cognitive decline as a way to help them remember what they have been doing and as a way for their caregivers to assess their activities.

5   Web Data and Models of Activity

Semantic content associated with locations on the web can provide rich sources of evidence about users' activities over time. We have been exploring general probabilistic models with the ability to fuse multiple sources of information. Such models can be used to perform inferences about a user's activities from the historical and short-term GPS data, as well as extended sensing with such information as temperature[11], barometric pressure, ambient light and sound, and Web data. Web content can be used to update, in an automated manner, a set of key resources and venues available at different locations, providing Bayesian dependency models with sets of resources that are coupled to an ontology of activities (*e.g.,* shopping, restaurants, recreation, government offices, schools, entertainment etc.). Such information can be used as a rich source of evidence in a probabilistic model that computes the likelihood of different plausible activities. Inferences can further take into consideration the dwelling of a user at a location with zero or small velocities and the complete loss of GPS signals at particular locations for varying periods of time, indicating that a user has entered a structure the blocks receipt of GPS signals. The timing, velocity, and frank loss of signal after a slowing of velocity provide rich evidence about a user's interests or entries into different proximal buildings and structures, as characterized by the content drawn from the Web about resources in the region of the last seen GPS coordinates. Such reasoning can be enhanced by a tagged log of prior activities noted by a user. Reasoning about losses of GPS signal can take into consideration a log of known "GPS shadows," that are not associated with being inside buildings, such as those occurring inside "urban valleys," as GPS access is blocked by tall structures.

We are pursuing rich probabilistic models of activity and location based on multiple sources of information, including information available from the Web and from logs of prior activities and GPS availability. Figure 5 displays a time slice of a more general dynamic Bayesian network model, showing probabilistic dependencies among key measurements and inferences. Server icons signify access of information from the Web about local resources as well as access from a store of known GPS shadows. The model is designed to make inferences about the probability distribution over a user's activities and over the location of a user, even when GPS signals are unreliable or lost temporarily. Sub-inferences include computation about whether a user is indoors or outdoors, employing information about the loss of GPS signals, a log of GPS shadows, information about local resources to the current location, and sensed temperature. We call out two key variables from an adjacent, earlier time slice to highlight the potential value of including dependencies among variables in adjacent time slices.

6 Conclusion

Knowledge of a user's raw (latitude, longitude) is not normally very useful. However, with publicly available databases, location measurements can be converted into useful information. We first reviewed three simple applications. Pinpoint Search takes (latitude, longitude) and finds web pages that are pertinent to the user's immediate surroundings. XRay lets a user point in a direction of interest and retrieve listings of places in that direction. Travelogue helps a user remember interesting places encountered during a trip. These applications share the trait of starting with raw GPS readings and using publicly available Web data to produce useful information. Although each of these applications has the potential for more development, together they show how the Web can be used in a simple way to enhance GPS. Finally, we reviewed the prospect of using the Web, in addition to other sources of information, to support rich probabilistic inferences about a user's activities and location. Such inferences can provide a window into a user's activities as well as access to location information even when GPS fixes become erroneous or are lost completely. Indeed, such models can take losses of GPS signal as valuable evidence for making inferences about activities and location.

References

1. Marmasse, N. comMotion: A Context-Aware Communication System. in CHI 99. 1999.
2. Pascoe, J. The Stick-e Note Architecture: Extending the Interface Beyond the User. in IUI 97. 1997. Orlando, Florida, USA.
3. Abowd, G.D., et al., Cyberguide: A Mobile Context-Aware Tour Guide. ACM Wireless Networks, 1997(3): p. 421-433.
4. http://local.google.com/.
5. http://local.yahoo.com/.
6. Smith, M., D. Davenport, and H. Hwa. AURA: A mobile platform for object and location annotation. in Fifth International Conference on Ubiquitous Computing - UbiComp 2003. 2003. Seattle, WA, USA.
7. Feiner, S., et al. A touring machine: Prototyping 3D mobile augmented reality systems for exploring the urban environment. in First IEEE Int. Symp. on Wearable Computers (ISWC '97). 1997. Cambridge, MA, USA.
8. Kutulakos, K.N. and J.R. Vallino, Calibration-Free Augmented Reality. IEEE Transactions on Visualization and Computer Graphics, 1998. 4(4): p. 1-20.
9. Griswold, W.G., et al., ActiveCampus - Sustaining Educational Communities through Mobile Technology. 2002, University of California at San Diego: San Diego, CA, USA.
10. Hariharan, R. and K. Toyama. Project Lachesis: Parsing and Modeling Location Histories. in Third International Conference on GIScience. 2004. Adelphi, MD, USA.
11. Krumm, J. and R. Hariharan. TempIO: Inside/Outside Classification with Temperature. in Second International Workshop on Man-Machine Symbiotic Systems. 2004. Kyoto, Japan.

The COMPASS Location System

Frank Kargl and Alexander Bernauer

University of Ulm, Dep. of Multimedia Computing, Ulm, Germany

Abstract. The aim of *COMPASS* (short for *COM*mon *P*ositioning *A*rchitecture for *S*everal *S*ensors) is to realize a location infrastructure which can make use of a multitude of different sensors and combine their output in a meaningful way to produce a so called *Probability Distribution Function (PDF)* that describes the location of a user or device as coordinates and corresponding location probabilities. Furthermore, COMPASS includes a so called translator service, i.e. a build-in component that translates PDFs (or coordinates) to meaningful location identifiers like building names and/or room numbers. This paper gives a short overview on the goals and abilities of COMPASS.

1 Motivation

There are a lot of situations in mobile computing where mobile nodes need to determine their current position. Ubiquitous computing applications derive context information from this position, e.g. in order to determine whether a user is currently at home, at work or on the way in between. Location-aided routing protocols for ad-hoc networks need position information to support their routing decisions. Navigation systems naturally rely on precise position information to plan the further route of a car or pedestrian. To support this large demand that applications have for precise location information, a number of commercial and research projects are working on this subject. Section 2 gives an overview on some of these activities.

We have identified two major challenges that are not completely resolved yet:

1. Location information from multiple sensors needs to be combined effectively in order to present one and only one position to the application. Any single location sensor has drawbacks, e.g. is usually not available inside buildings, RFID sensors or WLAN/Bluetooth APs are only available where installed etc. So in order to provide reliable and pervasive location support, an architecture must use multiple sensors, combine their results and present this to the application. The application should not need to worry about what sensor(s) were used for the current position information. Additionally combining the results from multiple sensors may improve the precision of overall results.

2. Raw coordinates may not really be useful to an application that needs to know the position in terms of buildings, rooms, street names etc. So a location system should include an infrastructure to resolve the raw position information to some kind of *symbolic position*.

T. Strang and C. Linnhoff-Popien (Eds.): LoCA 2005, LNCS 3479, pp. 105–112, 2005.

The primary focus of COMPASS will be to address these two issues by both including many different sensors into the system using a plugin interface and by providing a translator that is able to derive symbolic location information from the raw coordinates received from the locator. COMPASS is a software framework that can be used by arbitrary applications for location retrieval.

2   Related Work

The need of location systems is almost as old as mobile computing itself. Many of them use satellite navigation systems like GPS [G93] or the future Galileo system [G05]. A major problem of satellite navigation systems is the fact that the antenna of the receiver usually needs a direct line-of-sight towards a number of different satellites. So they are only useful for outdoor navigation.

As many ubiquitous computing projects include mostly indoor scenarios, researchers started to develop specialized indoor location systems. Prominent examples include the Cricket Location-Support System [P00] or the Bat [H97]. These systems make use of different kinds of sensors, like scanning for ultrasound or radio beacons or observing nearby WLAN or Bluetooth access points.

Unfortunately most of these location systems do not work together and many can use only one single kind of sensor. So there is a clear need for a framework that can combine the output of different kinds of location sensors into one single and consistent result.

Such a system has been proposed as part of the HeyWow project [H03]. In [A01, W02] the authors suggest the use of so-called probability density functions (PDFs) to represent the location measurement of one sensor or the combined measurement of multiple sensors. Other similar projects include [B03, H02].

As our architecture is based in part on these ideas, we first give some details on how position is represented in COMPASS before describing the architecture itself.

3   Position Representation

A major issue in positioning systems is how to express the position. COMPASS knows two kind of position representations: a geocoordinate based representation and one that delivers a semantical description of the current position, like the current room number or a street address. The functionality of COMPASS includes a mechanism to translate a geocoordinate to a semantic position description automatically, as sensors often deliver the first representation whereas applications often need the semantic representation.

No matter what kinds of sensors are in use to determine geocoordinates, most of them will inevitably introduces some kind of error. E.g. GPS has a typical error of a few meters, estimating the position based on available WLAN access points will deliver results with a precision of a few dozens to a few hundreds of meters, depending on local conditions. So delivering a single point as position information will never be accurate.

Therefor COMPASS expresses all position information as *Probability Distribution Functions (PDFs)* like introduced in [A01, W02]. PDFs represent a two or three dimensional area in which they express the probability of being at a certain position. PDFs use a Cartesian coordinate system with a north-south (y), east-west (x) and up-down (z) axis. Additionally a PDF contains the origin expressed as WGS 84 coordinates. This way, multiple PDFs can be correlated and combined.

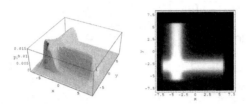

Fig. 1. Example of Probability Distribution Function (from [A01])

Figure 1 shows an example PDF which might represent a user that is inside a building with crossing corridors. This information can e.g. be the result of a radio sensors which detected that the mobile user entered this corridor and has not left it since. Combined with PDFs from other sensors, the position within the corridor might be narrowed down further.

In parallel to coordinates, positions may also be provided in a symbolic representation. At the moment we use a hierarchical string of the form "'country.city.streetname.streetnumber.roomnumber"' for simplicity. But we are now switching to a more powerful RDF-based representation that offers a very flexible description of locations. See the final section for an outlook on this mechanism.

4   Architecture and Components

4.1   Principles

COMPASS is designed to run on mobile devices. Therefore memory capacity, CPU speed and power consumption have to be taken into account. Depending on the application's needs the desired accuracy of position determination can reach from some centimeters to several hundred meters. COMPASS is designed to work with different degrees of accuracy to be usable for a wide spectrum of applications.

To gain a maximum of flexibility any dynamic content is separated from the COMPASS system and displaced to remote databases. A database is accessed using Web Services technology and is generally called service within the context of COMPASS. The application can optionally influence the selection of the services. It is possible for both the application and the COMPASS system to replace services at runtime without deep impact.

4.2   Overview

Figure 2 shows the overall architecture of COMPASS.

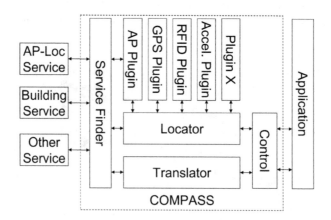

Fig. 2. COMPASS architecture

COMPASS has a plugin based design. For any source of position information exists a corresponding plugin. The plugins are connected to the so called Locator and deliver a PDF to it on demand. A plugin may use a service for accessing additional information. The task of the Locator is to determine the compound PDF of all PDFs supplied by the plugins. Additionally the Locator computes the position of the highest probability.

A plugin may register itself at the Locator to always get the latest compound PDF. This is useful for sensors which provide only relative position informations. To provide a human readable representation of the position with the highest probability there is a Translator component.

It will use webservices, that are able to convert PDFs to symbolic location informations. These symbolic location informations may be represented as hierarchical strings like "'germany.ulm.university.main_building.o27.3303'". It is possible to specify the hierarchical depth of the response string. In the future we plan to use a more flexible, XML-based format instead of simple strings.

The Service Finder is responsible for finding proper RPC services and to assign them to the plugins and the Translator. Using standard Web Service technologies like WSDL, UDDI and SOAP, it will find local services that provide information to plugins or the Translator.

Locator and Translator are called by the Control unit which provides the API. It is also responsible for initialization of all components. The API provides either the compound PDF or the WGS84 coordinates of the most likely position. Additionally the application can retrieve the symbolic position information.

4.3   Plugins

Currently four plugin types are supposed to be used:

AP plugin: This module uses WLAN access points as source of position information. It needs a service to resolve the access point's MAC to a geographic position. The service provides the WGS84 coordinates of the AP and a power density spectrum. From this information the plugin can compute a PDF with respect to properties of the sensor's antenna. If multiple access points are within reach one PDF for each access point can be computed.

GPS plugin: A NMEA capable GPS sensor is used to retrieve the current position. This can immediately be transformed into a PDF with Gaussian distribution. The error depends on receiving conditions, number of available satellites, etc.

RFID plugin: RFID tags have only a short range. But if they are scattered throughout a building at doors and gateways they can provide position information with a very good accuracy. The RFID plugin is statefull and for instance logs if one enters a room. A service is needed for resolving the tag's IDs and to retrieve information about the building's structure. The RFID plugin typically provides a PDF with equal distribution for the whole room which was entered last.

Acceleration plugin: This plugin uses a gyro sensor to gain relative position information. With the help of the last know position the plugin can compute a PDF. The distribution function usually is a sphere around the last known position. If a compass is additionally used the sphere can be clipped.

5   Implementation

5.1   Probability Distribution Functions

A PDF maps from Cartesian coordinates to probabilities. Therefore every PDF has an origin, which is given in WGS84 coordinates. A PDF can be accessed by supplying coordinates and retrieving the corresponding probability. To implement a PDF on a hardware we need a finite and discrete representation. So every PDF has a resolution and a maximum expansion for each dimension. A PDF always covers a cuboid. When iterating over the cuboid the sum of the probabilities must always be one. It is expedient to agree on a maximum resolution for all PDFs. Our implementation uses a maximum resolution of 10 centimeters. This should be adjusted depending on the precision of the existing sensors and the desired accuracy.

The naive approach for an implementation of a PDF is a three dimensional array. This is easy to implement and has minimum time penalty for accessing. But this approach is not practical for all possible PDFs, as depending on the resolution and the expansion of the PDF it quickly blasts the memory capabilities of any mobile device. To save memory the intern resolution can be reduced by using interpolation. Of course accuracy suffers from this.

A second possibility is to have a function representing a mathematical formula, which calculates the probability on demand. This approach has minimum memory requirements but is expensive at runtime. A mathematical description of some physical behavior often differs from reality or the best known formula is too complex for computation at runtime. Furthermore reality may differ from the mathematical description locally because of some irregularities such as obstacles which are not considered by the formula. So this approach is not practical for all possible PDFs, either.

To combine the advantages of both approaches we define the PDF container. A PDF container is either a PDF or a list of PDF containers. In the second case requests are delegated to the proper containers. So a container looks like a PDF but can cover a hierarchy of sub-PDFs, each being optimized for access on the set of coordinates they cover. When accessing the container it has to determine which container is responsible.

If the number of sub-PDFs is small, three dimensional polygon intersections are a good way to determine the responsible sub-PDF. If the number of sub-PDFs is large, Z curves [B99] are used to map the three dimensional coordinates to a linear search index of a binary tree.

5.2   Modules

Plugins: Every plugin creates its own thread on initialization. The Locator can trigger the determination of a PDF. When finished the PDF is returned using a callback to the Locator. The plugin is allowed to deliver a list of PDFs when there are multiple sources of information. But is is also allowed to compute a compound PDF on its own and deliver only this one. The plugin is allowed to return no PDF, if it is unable to determine one. The stub of the webservice is supplied by the Service Finder. A plugin is allowed to cache results from a service. But if the Service Finder assigns a new service the cache has to be invalidated. A plugin is allowed to be statefull. If the Service Finder assigns a new service the state has to be reset if it depends on the service. Otherwise the state is allowed to be reset.

Locator: The task of the Locator is to poll all plugins on demand and to deliver the compound PDF of all returned PDFs. The Locator creates its own thread on initialization. When triggered by the control unit the locator triggers every plugin to determine the PDFs. As soon as every plugin delivered one ore more PDF or after a timeout the compound PDF is built and returned to the control unit via callback. The compound PDF is a special PDF container. The difference to normal PDF containers is that there can be multiple responsible PDFs for a set of coordinates. For mathematical details on how to combine different PDFs, see [A01, W02].

Translator: The translator uses the point of maximum likelihood from the combined PDF and searches via the service finder for a suitable service that can determine a symbolic representation for that position.

Service Finder: The service finder is responsible to find webservices and to supply the plugins and the translator with a proper stub. The service finder is allowed to assign any service at any time to any module. It is also allowed to remove a service when the service is not reachable. In this case the module is unable to fulfill its task. The Service Finder can use several techniques to find an webservice. This includes Jini, UDDI or SLP.

6   Summary and Outlook

COMPASS provides an architecture that allows the concurrent use of multiple location sensors that can assist each other in finding positions and reduce potential errors. In addition the Translator provides symbolic position information that can be retrieved from local web services.

We are currently finishing a prototype implementation that includes two plugins (GPS and AP). As soon as this is finished, we will do some real-world analysis in order to verify, that this approach is practical and really decreases errors.

On the conceptual layer, we are investigating how to create a more flexible description of symbolic position information. Depending on the current context, an application may understand a location as "being in a certain city", "being inside a specific building", "being near some important monument", at a certain postal address, etc. In order to express this information about a given geolocation, COMPASS represents such facts as RDF/XML documents using the Resource Description Framework developed by the W3C [W05] as part of their Semantic Web initiative. The following RDF document shows an example, where the object "myself" is located at some coordinates, at an certain address and inside a specific building.

```
<?xml version="1.0" encoding="iso-8859-1" ?>
<rdf:RDF xmlns:rdf="http://www.w3.org/1999/02/22-rdf-syntax-ns\#">
  <Object rdf:nodeID="myself">
    <locatedAt> <Coordinates rdf:nodeID="myCoords" lc:type="WGS84">
        <longitude parseType="Resource">
          <rdf:value>010°01.538E</rdf:value>
          <coordUnit>degree<coordUnit>
        <longitude>
        <latitude rdf:parseType="Resource">
          <rdf:value>48°27.182N</rdf:value>
          <coordUnit>degree<coordUnit>
        <latitude>
    <Coordinates> </locatedAt>
    <locatedAt>
      <Address>
        <street>Albert-Einstein-Allee</street> <number>11</number>
        <city rdf:resource="http://ulm.de/" />
      </Address>
    </locatedAt>
```

```
<locatedAt>
  <Building rdf:about="http://uni-ulm.de/campus/UniOst">
    <inPart><BuildingPart>O27</BuildingPart></inPart>
  </Building>
</locatedAt>
</Object>
</rdf:RDF>
```

This way applications may be enabled to combine the location information with other documents from the Semantic Web like route information. Then e.g. navigation systems may automatically infer that you are on an Autostrada in Italy and that there the general speed limit is 130 km/h.

References

[A01]   M. Angermann & J. Kammann & P. Robertson & A. Steingaß & T. Strang, *Software representation for heterogeneous location data sources within a probabilistic framework*, International Symposium on Location Based Services for Cellular Users, pp. 107–118, Locellus 2001.

[B99]   Christian Boehm & Gerald Klump & Hans-Peter Kriegel, *XZ-Ordering: A space-filling curve for objects with spatial extension*, Proceedings of Advances in Spatial Databases, 6th International Symposium, SSD'99, Hong Kong, China, pp. 75–90, July 20–23, 1999.

[B03]   Jürgen Bohn & Harald Vogt, *Robust Probabilistic Positioning Based on High-Level Sensor-Fusion and Map Knowledge*, Technical Report nr. 421, ETH Zurich, Apr. 2003.

[G93]   I. Getting, *The Global Positioning System*, IEEE Spectrum 30, 12, pp. 36–47, December 1993.

[G05]   *Galileo Project Webpage*, http://europa.eu.int/comm/dgs/energy_transport/galileo/index_en.htm

[H97]   A. Harter & A. Hopper *A New Location Technique for the Active Office*, IEEE Personal Communications 4, 5, pp. 43–47, October 1997.

[H03]   *HeyWow Project* http://www.heywow.com/

[H02]   J. Hightower & B. Brumitt & G. Borriello *The location stack: A layered model for location in ubiquitous computing*, In Proceedings of the 4th IEEE Workshop on mobile Computing Systems & Applications (WMCSA 2002), IEEE Computer Society, pp. 22–28, Callicoon, NY, USA, June 2002.

[P00]   Nissanka B. Priyantha & Anit Chakraborty & Hari Balakrishnan, *The Cricket location-support system*, Mobile Computing and Networking, pp. 32-43, 2000.

[W05]   W3C, *The Resource Description Framework (RDF)*, http://www.w3.org/RDF/.

[W02]   K. Wendlandt & A. Ouhmich & M. Angermann & P. Robertson, *Implementation of Soft Location on mobile devices*, International Symposium on Indoor Localisation and Position Finding, InLoc 2002, Bonn, Germany, 2002.

The xPOI Concept

Jens Krösche[1] and Susanne Boll[2]

[1] Oldenburg Research and Development Institute for
Computer Science Tools and Systems (OFFIS), Escherweg 2,
26 121 Oldenburg, Germany
jens.kroesche@offis.de

[2] Carl von Ossietzky University Oldenburg, Faculty II,
Department of Computing Science, Escherweg 2,
26 121 Oldenburg, Germany
boll@informatik.uni-oldenburg.de

Abstract. Today, most mobile applications use geo-referenced points of
interest (POIs) on location-based maps to call the user's attention to interesting spots in the surroundings. The presentation of both, maps and
POIs, is commonly location-based but not yet adapted to the individual
user's needs and situation. To foster the user's information perception by
emphasising the location-based information that is most relevant to the
individual user, we propose the xPOI concept – the modeling, processing, and visualisation of *context-aware* POIs. We introduce a data model
for xPOIs, supporting the exchange of xPOIs and define an architecture
to process and present xPOIs in cooperation with a mobile information system. With the integration of context-awareness into POIs, we
contribute to the development of innovative location- and context-aware
mobile applications.

1 Introduction

Developments in areas like mobile computing, localisation, and wireless networks
have recently increased the emergence of a wide range of mobile applications.
One observation that can be made here, is that distinct tasks, such as navigation and orientation support through maps are part of nearly every mobile
application. Typically, these applications provide not only a map but also a visualisation of geo-referenced information like points of interest (POIs). Although
many applications provide similar features with regard to map and POI presentation, nearly all of them build their own proprietary mechanism to realise them.
Furthermore, as the situation of a mobile user constantly changes, so should the
visualisation of the POIs; behind every corner new impressions and situations
– the context – could come up that might influence the user and should therefore be reflected by the applications. Today, factors like role, location, and time
play an increasingly important role when selecting and displaying information.
However, today's applications provide only static POIs that do not adapt to the
user's context.

T. Strang and C. Linnhoff-Popien (Eds.): LoCA 2005, LNCS 3479, pp. 113–119, 2005.

In this paper, we present a concept for conte**x**t-aware points of interest (*x*POIs) that make POIs aware to the user's context. The *x*POI concept allows to adapt the presentation of POIs according to a current user situation. This helps the user to focus on those POIs that are most "important" in a given situation, guiding the attention of the user to the needed information. Another benefit is the emergence of new interaction methods: Since *x*POIs are sensitive to the current context situation, they can adapt their visualisation, and in addition trigger interaction. Hence, the user is not restricted to a direct interaction with the system; rather an additional indirect interaction by changing his or her context situation is possible.

In the following, we propose a data model for *x*POIs that reflects context-aware visualisation and interaction aspects. In addition to the *x*POI data model, we present a processing engine (*x*POI engine) which is able to analyse and process instances of this *x*POI data model and generate the appropriate *x*POI presentations. We illustrate the usage of the *x*POI-concept in prototypes developed on our mobile Niccimon platform [1].

2   Related Work

Only a few attempts can be found to establish a standard for POIs, like for instance POIX [2] or NVML [3] and recently OpenLS [4]. POIX and NVML both define POIs and/or route information usable in a city information system, without taking into account context-awareness. However, these "standards" provide first insights and stimuli towards a structured POI data model. OpenLS on the contrary defines services and data types for location services. On this behalf it provides a data type for POIs, which only reflects basic features, e. g., position and time. With regard to context management, the project Nexus [5] works in the field of federating context and context reasoning, whereas we aim at utilising context and modeling context-aware behaviour for POIs. An interesting example of context-aware interactive objects has been proposed by the Stick-e Notes [6] project. In which a single or compound context condition causes the invocation of an electronic note. However, this approach addresses a context based invocation of those notes, but does not support more complex context situations and *dynamic* context-aware presentation of POIs. Today, POIs are in the most cases more or less hard-wired within the application, so that they are all known beforehand, defined in an application specific data model, and presented in a uniform manner; dynamic POI exchange or context-aware visualisation is integrated in the POI concepts, can not be found.

3   The General Requirements and Goals

It is the goal of the *x*POI concept to provide self-describing "all-in-one" *x*POIs that carry sufficient information to be processed, managed, and presented according to the users context. This requires a suitable data model to specify and structure the needed information. With regard to the design of the data model, the two central aspects are: visualisation and context-awareness. Consequently, a suitable representation of the xPOI's visualisation is needed. To achieve as much platform-independence and standard conformity as possible, the visualisation representation should follow widely accepted presentation standards. In order to reflect context-awareness, the xPOI data model needs to contain the information about the context situation to which the xPOI is "sensitive"/"aware". The goal to dynamically adapt the visualisation of a xPOI necessitates a mechanism that correlates a context situation with the visualisation presentation, which defines a resulting adaption. Since we also aim at exchanging xPOIs between applications, we further need a suitable transport/exchange format, preferably platform-independent. Finally, a mechanism that is able to process and use xPOIs together with context information is needed. Our approach to meet these requirements in the xPOI concept is presented in the following.

4   Design of xPOIs, Context, and xPOI Engine

Based on the general concepts and requirements, we present in this section the design of the main components of the xPOI concept in more detail: the design of the xPOI data model, the context data model, and our xPOI engine which analyses and processes the xPOIs and the context to generate a context-based xPOI visualisation and interaction.

4.1   The xPOI Data Model

Our xPOI data model comprises five main sections, each responsible for one distinct aspect: identification, management, visualisation, context-awareness, and messaging.

Identification: The *identification* section contains the information needed to identify an xPOI. It comprises an unique xPOI id, one or more types or categories of the xPOI, an action entry to distinguish an initial xPOI distribution from an update or delete call, and last, but most important, the actual position of the xPOI.

Management: This section contains information for the later *management* of xPOIs. Issues addressed are: information concerning security aspects like access rights or the xPOI creator, together with the temporal validity, update version, and history of the xPOI.

Visualisation: In the *visualisation* section, the possible visualisation forms of the xPOI are defined. This visualisation can range from simple text over 2D graphics to auditory and haptic representations. To restrain complexity, we concentrate our work on a 2D representation of the xPOIs. Generally, there are two ways to present an xPOI: By a bitmap or by a vector object. For the context-aware presentation we apply suitable transformations on both the vector definition and the bitmap definition. For example, in the case of a 2D presentation, the result of this transformation is a SVG object which is later used by the surrounding application to present the xPOI on a map.

Context-Awareness: The context-aware presentation of a xPOIs is achieved by a transformation of a visualisation information according to one of the context-aware rules, defined in the *context-awareness* section. We use Event-Condition-Action (ECA) rules known from, e. g., active databases [7], to define the situation in which a rule is to be executed. Based on the overall rule characteristics, a user may cause an *event*, e. g., by selecting an xPOI on the map or by changing the context situation, and/or a specific context situation/*condition* is given, e. g., the user reached a distinct position, and an *action* is performed. Driven by the requirements of our application domain, we restrict the set of supported user **events** to define these rules at this stage of the project to on the one hand "POI-Pressed, POI-Released, and POI-Clicked" for direct user interaction with the xPOI and on the other hand indirect updates on the supported context represented by variables. To specify our **conditions** the supported variables for context like position, position quality, time, date, velocity, role and stress factor are combined with operators and elements of the context domains. The following operators are supported:

conditional:	AND, OR, and NOT
relational:	EQUAL, LESS, and GREATER
spatial:	DISTANCE and INSIDE
constructors:	POINT, RECT, TIME, and DATE

In later versions, we will add more operators/context information and examine the usage of different formalisms to specify the context-aware rules. Nevertheless the use of ECA rules is sufficient to demonstrate the usage and advantages of context-aware POIs. Today, we support two types of **actions**. First of all, an XSLT transformation can be applied on an appropriate visualisation information. The result of this transformation is the mentioned SVG presentation of the xPOI. Another action could be the usage of the messaging mechanism to send a message to the xPOI creator realising context-aware interaction. The following example illustrates an ECA rule for an xPOI, in which a xPOI changes its presentation, if the user move's into a defined area at a specific day, showing only a meta statement for the action part.

Event:	UPDATE($position_{now}$) OR UPDATE($date_{now}$)
Condition:	INSIDE($position_{now}$, RECT(POINT(0,0), POINT(9,9)))
	AND
	$date_{now}$ EQUALS DATE(2005,6,21)
Action:	*use visualisation form 3*

Messaging: In the section *messaging*, we define information that is needed if a context-aware rule results in a messaging action instead of a visualisation transformation, enabling interprocess communication. Due to the required platform-independence, information regarding the messaging channels like TCP/IP sockets or other mechanisms are provided in this section.

The goal to integrate context-awareness in the POI presentation and foster the exchange and reusability of POIs requires a uniform but flexible xPOI definition. Therefore, we utilise XML and XML-Schema to specify, express and exchange our xPOIs.

4.2   Context

In order to use context information information such as the position of the user, the position quality, time, date, velocity, role, and stress factor of the user, we developed a data model for context. It incorporates additional meta data like position information, time related factors, creator, trustworthiness etc. and is not described any further. Nevertheless, due to the modular design of the xPOI engine (see Section 4.3), any other context data model could be used. In addition to the aforementioned context information, which can be used in the definition of the ECA rules, we use one additional context element, the *event horizon*. The event horizon describes the area in the real world, which the user is able to comprehend at the moment. This special context element is used only during the processing of xPOIs.

4.3   Processing of xPOIs

In order to utilise xPOIs, a mechanism to analyse and process these xPOIs is one of the key factors. We propose a so called "xPOI engine" depicted in Figure 1, containing all the necessary functionality.

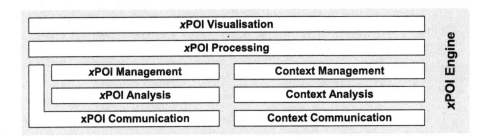

Fig. 1. The xPOI engine for xPOI analysis and processing

This component is designed to be integrated as a module into the later mobile information system, to receive the xPOIs and context information, process both, and either calculate the xPOIs SVG representations or trigger the context-aware interaction through informing the xPOI originator. Figure 1 shows the layers of the xPOI engine in which xPOIs and context information are integrated, pre-processed, and managed. Both context information and xPOIs are integrated into the xPOI-engine via the communication layer. Afterwards xPOIs as well as the context are analysed and transformed into an internal object representation. Context information and xPOIs are then transferred to their respective management layer, where they are stored "persistently". Depending on the event (i. e., a new xPOI has been integrated or an old one has been updated, an user interaction occurred, or the context situation has changed) the affected xPOIs and context information are enquired by the processing layer from both management layers. Here, the *event horizon* is used to extract only those xPOIs from the management layer that are of interest in the given situation. After processing the involved xPOIs, either their new SVG representation is transferred to the visualisation layer and from there to the surrounding system or in case of a communication action the communication layer is used to propagate the information to the xPOI originator, where it may initialise a context-aware interaction.

5   Using xPOIs in Mobile Information Systems

To show the practical usage of and evaluate our xPOI concept, we integrated the concept into our mobile, multimedia, and location-aware Niccimon platform [1]. Through this, applications that use the Niccimon platform as system foundation are able to utilise the xPOI concept. The proposed xPOI engine is implemented and integrated in the Niccimon mediator as an additional horizontal module.

The first project to integrate the new xPOI concept was our mobiDENK demonstrator [8]. MobiDENK represents a typical guide system, displaying the position of the user and POIs in his or her surrounding on a map, providing location-based information. We are currently testing the new xPOIs by addressing proximity awareness, user interests, and time, to adapt their visualisation. Furthermore, we will integrate the xPOI concept in our project, Sightseeing4U [9], which provides personalised sightseeing tours on mobile devices. Here, the xPOIs will be adapted according to the user's interests regarding architecture or landscaping.

6   Conclusion

In this paper, we introduced a "standard" xPOI data model that enables the creation, exchange, and processing of context-aware points of interest in mobile, location- and context-aware applications. These xPOIs each carry the information to be identified, managed, and, most important, be visualised dependent on the user's current context situation. We not only presented the data model for xPOIs but also the processing architecture in which they are embedded. The concept invites new interaction models in mobile computing. Based on this "context sensitive interaction" users can interact with the system by changing their context, e. g., walking around, changing their role from tourist to business person, and the like. Due to the platform-independent exchange format of xPOIs, applications can easily exchange and integrate foreign context-aware POIs. When thinking of teenagers and their struggle for individuality the exchange of individualised xPOIs among themselves in an instant messenger or "friend finder" application could prove to be an interesting and promising scenario. Nevertheless there are some open issues like for instance the diffuse definition of context or the definition of "useful" rules together with a sound visualisation transformation. But, the implementation and application of the xPOI concept so far is promising. In a world of evolving mobile and context-aware applications, the flexible usage and exchange of context-aware points of interest clearly is an important next step.

References

1. Baldzer, J., Boll, S., Klante, P., Krösche, J., et al.: Location-Aware Mobile Multimedia Applications on the Niccimon Platform. In: Informationssysteme für mobile Anwendungen (IMA'04), Brunswick, Germany (2004)
2. Kanemitsu, H., Kamada, T.: POIX: Point Of Interest eXchange language, version 2.0, document revision 1. W3C NOTE (1999)
3. Sekiguchi, M., Takayama, K., Naito, H., Maeda, Y., Horai, H., Toriumi, M.: NaVigation Markup Language (NVML). W3C NOTE (1999)
4. Bychowski, T., Williams, J., Niedzwiadek, H., Bishr, Y., et al.: OpenGIS Location Services (OpenLS). OpenGIS Implementation Specification OGC 03-006r3 (2004)
5. Hohl, F., Kubach, U., Leonhardi, A., Rothermel, K., Schwehm, M.: Next Century Challenges: Nexus - An Open Global Infrastructure for Spatial-Aware Applications. In: 5th Conf. on Mobile Computing and Networking, Seattle, USA (1999) 249–255
6. Pascoe, J.: The Stick-e Note Architecture: Extending the Interface Beyond the User. In: IUI '97: Intelligent User Interfaces, Orlando, Florida, USA (1997) 261–264
7. Gatziu, S., Geppert, A., Dittrich, K.R.: The SAMOS Active DBMS Prototype. Technical Report 94.16, IFI (1994)
8. Krösche, J., Boll, S., Baldzer, J.: MobiDENK – Mobile Multimedia in Monument Conservation. IEEE MultiMedia **11** (2004) 72–77
9. Boll, S., Krösche, J., Scherp, A.: Personalized Mobile Multimedia meets Location-Based Services. In: INFORMATIK 2004 - Informatik verbindet, Volume 2 "Multimedia-Informationssysteme". Volume 51 of LNI., Ulm, Germany (2004)

The GETA Sandals: A Footprint Location Tracking System

Kenji Okuda, Shun-yuan Yeh, Chon-in Wu, Keng-hao Chang, and Hao-hua Chu

Department of Computer Science and Information Engineering,
Institute of Networking and Multimedia,
National Taiwan University,
#1 Roosevelt Road, Section 4, Taipei, Taiwan 106
{r90092, r93124, r92079, r93018, hchu}@csie.ntu.edu.tw

Abstract. This paper presents the design, implementation, and evaluation of a footprint-based indoor location system on traditional Japanese GETA sandals. Our footprint location system can significantly reduce the amount of infrastructure required in the deployed environment. In its simplest form, a user simply has to put on the GETA sandals to track his/her locations without any setup or calibration efforts. This makes our footprint method easy for everywhere deployment. The footprint location system is based on the dead-reckoning method. It works by measuring and tracking the displacement vectors along a trial of footprints (each displacement vector is formed by drawing a line between each pair of footprints). The position of a user can be calculated by summing up the current and all previous displacement vectors. Additional benefits of the footprint based method are that it does not have problems found in existing indoor location systems, such as obstacles, multi-path effects, signal noises, signal interferences, and dead spots. However, the footprint based method has a problem of accumulative error over distance traveled. To address this issue, it is combined with a light RFID infrastructure to correct its positioning error over some long distance traveled.

1 Introduction

Physical locations of people and objects have been one of the most widely used context information in context-aware applications. To enable such location-aware applications in the indoor environment, many indoor location systems have been proposed in the past decade, such as Active Badge [1], Active Bat [2], Cricket [3], smart floor [4], RADAR [5], and Ekahau [6]. However, we have seen very limited market success of these indoor location systems outside of academic and industrial research labs. We believe that the main obstacle that prevents their widespread adoption is that they require certain level of *system infrastructural support* (including hardware, installation, calibration, maintenance, etc.) inside the deployed environments. For example, Active Badge [1], Active Bat [2], and Cricket location systems [3] require the installation of infrared/ultrasonic transmitters (or receivers) at fixed locations (e.g., ceilings or high walls) in the environments. In order to attain high location accuracy and good coverage, the system infrastructure requires large number of transmitters (or

T. Strang and C. Linnhoff-Popien (Eds.): LoCA 2005, LNCS 3479, pp. 120–131, 2005.

receivers) installed in the deployed environments. This is beyond the reach of ordinary people to afford, operate, and maintain the infrastructure. WiFi based location systems such as RADAR [5] and Ekahau [6] require an existing WiFi network in the deployed environment. For example, the Ekahau location system recommends a WiFi client to be able to receive signals from 3~4 WiFi access points in order to attain the specified location accuracy of 3 meters. This high density of access points is unlikely in our everyday home and small office environments. In addition, most WiFi based location systems require users' calibration efforts to construct a radio map by taking measurements of WiFi signal strength at various points in the environment. This forms another barrier for users. Smart floor [4] can track the location of a user by using pressure or presence sensors underneath the floor tiles to detect the user's gait. This infrastructure cost is expensive because it requires custom-made floor tiles and flooring re-construction.

Significantly reducing the needed system infrastructure serves as our main motivation to design and prototype a new *footprint location system* on traditional Japanese GETA (pronounced "gue-ta") sandals. This footprint location system can compute a user's physical location solely by using sensors installed on the GETA sandals. To enable location tracking, a user simply has to wear the GETA sandals with no extra user setup & calibration effort. This system works by attaching location sensors, including two ultrasonic-infrared-combo readers and one ultrasonic-infrared-combo transmitter, on the GETA sandals. The basic idea can be described by looking at a person walking from location A to location B on a beach. He/she will leave a trial of footprints. To track a person's physical location, the system continuously measures a *displacement vector* formed between two advancing footprints (advancing in the temporal sense). To track a user's current location relative to a starting point, the system simply sums up all previous *footprint displacement vectors* leading to his/her current footprint location. This idea is similar to the so-called (deduced) *dead-reckoning navigation* dated back to the medieval time when the sailor/navigator would locate himself/herself by measuring the course and distance sailed from a starting point. In our system, this dead reckoning idea is adapted in tracking human footprints. We believe that having a *wearable location tracker* is an important advantage in our footprint location system over infrastructure-based indoor location systems. Users simply need to wear our GETA-like shoes, and our location system can work anywhere they want to go.

In addition to the benefit of low infrastructure cost, the *footprint location system does not have problems commonly found in existing indoor location systems*. For example, existing wireless based solutions (e.g., using radio, ultrasonic, or infrared) can experience poor position accuracy when encountering obstacles between transmitters and receivers, multi-path effects, signal noises, signal interferences, and dead spots. On the other hand, our footprint location system avoids almost all of these problems. The reason is that the location sensors (ultrasonic-infrared transmitters and readers) in our footprint method only need to cover a small sensing range, which is the short distance between two sandals in a maximum length of a walking step (< 1.5 meters). Assume walking on a relatively smooth surface, the footprint location sensors are unlikely to encounter any obstacles or experience multi-path effects, signal noises, and signal interferences over this small sensing range. This is in contrast to existing wireless (radio, ultrasonic, or infrared) based location systems where the sensing range must be large enough to cover the distance between fixed location sensors in the environment and a mobile location sensor on a user. This short sensing range in our footprint method also brings two additional advantages: (1) location sensors can significantly reduce its power consumption due to short sensing range, and (2) location sensors (ultrasonic-infrared) have high accuracy under such short sensing range (e.g., 0.2 *mm* in static setting).

There is one important shortcoming in our footprint location system called the *error accumulation* problem. It is inevitable that a small amount of error is introduced each time we take measurements to calculate a displacement vector. Consider a user has walked *n* steps away from a starting point. His/her current location is calculated as a sum of these *n* displacement vectors. This means that the current location error is also the sum of all errors from these *n* previous displacement vectors. In other words, the error in the current footprint measurement will be a percentage of the total distance traveled. To address this error accumulation problem, we utilize a small number of passive RFID tags with known location coordinates in the environment. A small RFID reader is also placed under a GETA sandal to read these RFID tags. When a user walks on top of a location-aware RFID tag, the known location coordinate of that RFID tag is used instead of the calculated footprint location. Encountering a RFID tag has the same effect as resetting the accumulated error to zero. Although these location-aware RFID tags are considered system infrastructure, they constitute very light infrastructure because (1) RFID tags are relatively inexpensive in cost (< \$1 each) and easy to install, and (2) only a very small number of RFID tags are needed. Based on our measurements in Section 4, the average error per footstep is only about 4.6*mm*. If we want to limit the average error to 46*cm*, we only need to install enough RFID tags in the environment such that a user is likely to walk over a RFID tag approximately every 100 steps.

There are several pervious systems that are also based on incremental motion and dead reckoning. Lee et al [11] proposed a method to estimate the user's current location by recognizing a sequence of incremental motions (e.g., 2 steps north followed by 40 steps east, etc.) from wearable sensors such as accelerometers, digital compass, etc. Lee's proposed method differs from our footprint tracking system in that it can only recognize a few selected locations (e.g., bathroom, toilet, etc.) rather than track location coordinates. Point research [12] provides a vehicle self-tracking system that provides high location accuracy by combining the dead-reckoning method (wheel motions) and GPS. The solution from Point research differs from our method which is based on footprint tracking in normal human walking motion rather than mechanical wheel movements.

At the time of this paper writing, we have gone through three design-and-evaluation iterations. Rather than presenting only the last (3rd) design and evaluation, we think that readers may also be interested to know these intermediate designs as well as mistakes we made on them. The remainder of this paper is organized as follows. In Sections 2 to 4, we describe our three design-and-evaluation iterations, including performance evaluations and discussions about design mistakes. Section 5 draws our summary and future work.

2   Initial Design: Design Version I

The human walking motion can be modeled by stance-phase kinematics shown in Fig. 1. A forwarding walking motion is consisted of a sequence of three stances – *heel-strike*, *mid-stance*, and *toe-off*. In the heel-strike stance, the body weight pushes down from the upper body to the lower body, resulting in both feet in firm contact with the ground. This generates a footprint on the ground. In the mid-stance, the body raises one (left) foot forward and above the ground. In the toe-off stance, the body weight again pushes down on the forwarded (left) foot, again resulting in both feet in contact with the ground. This creates another footprint on the ground.

The basic idea behind our footprint location tracking system is to (1) detect heel-strike and toe-off stances, and then (2) take measurement of two feet's *displacement vector* v_d (i.e., the footprint vector) on the ground. As shown in Fig. 2, given a starting point in a location tracking region (x_{start}, y_{start}), e.g., the entrance of home or a building, we can compute the current position of a user, who has walked n number of steps away from the starting point, by summing up all displacement vectors Σv_{di}, for $i=1..n$, corresponding to these n footsteps.

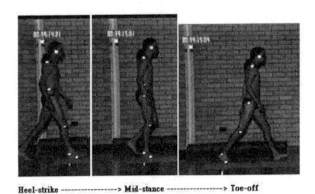

Heel-strike -------------> Mid-stance -------------> Toe-off

Fig. 1. Three stances in a normal human walking motion

2.1   Footprint Positioning Algorithm

To measure the displacement vector v_d for each footprint, we place two ultrasonic-infrared-combo receivers on the left sandal and two ultrasonic-infrared-combo transmitters on right sandal shown in Fig. 3. The components for ultrasonic-infrared transmitters and receivers are obtained by disassembling the NAVInote's [8] electronic pen and base unit. In order to make both the receivers and transmitters face

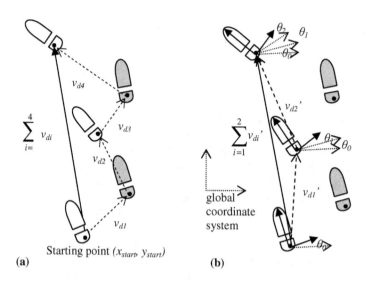

Fig. 2. The user has walked four footsteps 1-4. Fig. 2(a) shows these displacement vectors (v_{di}) corresponding to these displacement vectors. Fig. 2(b) shows θ_i as the rotational angel between the current local coordinate system and the previous local coordinate system in the previous footstep

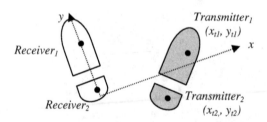

Fig. 3. It shows the locations of ultrasonic-infrared receivers and transmitters on the sandals. The coordinates of the transmitters on the right sandal is relative to the local coordinate system on the left sandal

directly toward each other during normal walking motion[1], they are placed on the inner sides of the sandals. The prototype of the GETA sandals is shown in Fig. 7. Through NAVInote APIs, we obtain the (x, y) coordinates of these two transmitters located on the right sandal. Denote them as (x_{t1}, y_{t1}) and (x_{t2}, y_{t2}) as shown in Fig. 3. Note that the ultrasonic-infrared-combo technology can achieve very fine position accuracy and resolution at the short sensing range between two sandals. Under static setting, the measured average positioning error is $< 0.2mm$ and the resolution is $< 0.2mm$.

---

[1] We assume that people don't intentionally walk cross-legged.

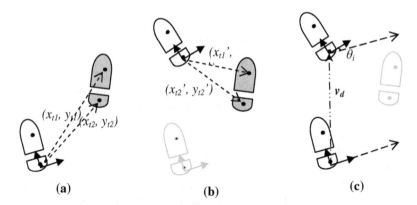

(a)   (b)   (c)

Fig. 4. Before moving the left foot, the coordinates of the transmitters on the right sandal, (x_{t1}, y_{t1}) and (x_{t2}, y_{t2}), are recorded as shown in (a). After walking the left foot, (x_{t1}', y_{t1}') and (x_{t2}', y_{t2}') are recorded as shown in (b). To calculate v_d, we have to consider the rotation angel θ_i, translate (dx, dy) to the coordinate system of the left foot, and then transform them into the global coordinate system to get the displacement vector v_d, as shown in (c)

The coordinates of these two transmitters are measured relative to the *local coordinate system* of the left sandal, where the origin of this local coordinate system is at the heel position and the *y-axis* forms a straight line from the heel to the toes. Since moving left foot also changes the local coordinate system, it is necessary to *re-orientate* the displacement vector from its local coordinate system to a global coordinate system. The global coordinate system is set to be the coordinate of the starting point. To perform this orientation translation, we need to compute the *orientation angle* θ of local coordinate system relative to the global coordinate system.

Denote the current step as the *i*-th left footstep. The orientation angle θ can be calculated as $\Sigma\ \theta_i$, where θ_i is the rotational angle between the *i*-th left footstep's coordinate system and the *(i-1)*-th left footstep's coordinate system. This means that to compute the orientation angle θ, we need to compute θ_i for each new left footstep as illustrated in Fig. 2(b).

Fig. 4 shows (x_{t1}, y_{t1}) and (x_{t2}, y_{t2}) as the recorded coordinates of two transmitters on the right foot before moving the left foot, and (x_{t1}', y_{t1}') and (x_{t2}', y_{t2}') as their recorded coordinates after moving the left foot. As the left foot moves, the coordinate system on the left foot rotates θ_i and then translates into (dx, dy). This gives us the following four sets of equations, which are sufficient to solve for three unknowns: θ_i and (dx, dy).

$$\begin{bmatrix} \cos\theta_i & \sin\theta_i \\ -\sin\theta_i & \cos\theta_i \end{bmatrix} \begin{bmatrix} x_{t1} \\ y_{t1} \end{bmatrix} - \begin{bmatrix} dx \\ dy \end{bmatrix} = \begin{bmatrix} x_{t1}' \\ y_{t1}' \end{bmatrix} \qquad (1)$$
$$(2)$$
$$\begin{bmatrix} \cos\theta_i & \sin\theta_i \\ -\sin\theta_i & \cos\theta_i \end{bmatrix} \begin{bmatrix} x_{t2} \\ y_{t2} \end{bmatrix} - \begin{bmatrix} dx \\ dy \end{bmatrix} = \begin{bmatrix} x_{t2}' \\ y_{t2}' \end{bmatrix} \qquad (3)$$
$$(4)$$

We can then compute v_d using summed θ, dx, and dy.

$$\begin{bmatrix} \cos\theta & \sin\theta \\ -\sin\theta & \cos\theta \end{bmatrix} \begin{bmatrix} dx \\ dy \end{bmatrix} = v_d$$

Some readers might wonder why we use two transmitters instead of one transmitter. The reason is that one transmitter only gives two equations, which are insufficient to solve three unknowns. With the additional transmitter, it can give two additional equations needed to solve three unknowns.

Prior to the above-mentioned geometry calculation, we need to detect the heel-strike and toe-off stances to measure (x_{t1}, y_{t1}) and (x_{t2}, y_{t2}). We call these two stances the *steady state* because when both feet are in contact with the ground, the measured coordinates on two transmitters are *stable* (do not change much) for some small period of time. When we detect the steady state, we record the coordinates of two transmitters and then calculate the displacement vector.

Assume that the user moves the right foot and the left foot in an interleaving manner. We can track the position of the left foot by first computing two displacement vectors from left footprint to the right footprint and right footprint back to the left footprint.

2.2 Performance Evaluation

We have evaluated the performance of our initial design. The results have shown poor positioning accuracy. The main cause of poor accuracy is due to the interference of the signals from the two transmitters. Since the receivers can not distinguish two distinct signals from two transmitters, it can calculate incorrect coordinates on two transmitters. This leads to miss-detection of the steady state and incorrect calculation on the displacement vectors. Although we tried to filter out these incorrect coordinates, our results still showed high 49% rate of steady state miss-detections. When a miss-detection occurs, dx, dy, θ, and displacement vector will also be calculated incorrectly. This leads to rapid error accumulation. Note that even a small error in the rotation angle θ, which is used to re-orient the displacement vector, can significantly reduce the position accuracy.

An additional problem is that we have not found a working method to distinguish if a person is moving forward or backward and which (left or right) foot is moving. This problem can be explained as follows. Consider the 1st case that a person is moving forward: if the right foot is moving forward, the x-coordinates of both transmitters will increase; on the other hand, if the left foot is moving forward, the x-coordinates of both transmitters will decrease. Consider the 2nd case that a person is moving backward, the situation is reverse, i.e., the x-coordinate will decrease(increase) when right(left) foot is moving backward. Given increasing x-coordinates on transmitters, it can be either right foot moving forward or the left foot moving backward. As a result, it is impossible to distinguish if a person is moving forward/background & left/right (foot movement).

3 Revised Design: Design Version II

Design II tries to fix the following three problems from design I: (1) interferences from two transmitters, (2) incorrect detections of heel-strike and toe-off stances, and (3) indetermination of forward/backward & left/right movements. Design II solves these problems by incorporating additional sensors into the GETA sandals. To accurately detect the heel-strike and toe-off stances, we have added two pressure sensors at the bottom of both sandals to sense when both feet are in contact with the ground. These pressure sensors are also used to distinguish the forward/backward & left/right movements. To eliminate interferences from two transmitters, we remove one transmitter from the right sandal and incorporated an orientation sensor by InterSense InterTrax2[9] on the front of left sandal. Fig. 7 shows the GETA sandal prototype of the revised design (version II).

3.1 Revised Footprint Positioning Algorithm

Since the orientation sensor can provide θ value for the global coordinate system, it removes one unknown in our calculation. This leads to a simpler algorithm than in version I. By measuring (x_t, y_t) and θ at the time of the heal-strike and toe-off stances, the displacement vector in the current footstep can be calculated by performing a simple rotational transformation. The displacement vector to the starting point is the sum of all the displacement vectors corresponding to the all previous footsteps.

3.2 Performance Evaluation

Fig. 5 shows the measured positioning error over different traveling distances and walking speeds. It has shown that two problems in design I have been addressed. The positioning accuracy is very good at short walking distances: the average error after walking a little more than $5m$ is only $0.36m$, or approximately 6.8%. It also shows that our new design can accurately detect the heel-strike & toe-off stands, and then take measurements to compute the displacement vector. It can be seen that the error increases only slightly with increasing walking speed. However, we can clearly observe the problem of *error accumulation* in our footprint-only method, as the positioning error increases *super-linearly* with increasing walking distance.

The error is contributed from two main sources: (1) the displacement error vector from the ultrasonic-infrared-combo device, and (2) the orientation error from the orientation sensor. The displacement error is relatively small and stable due to the high accuracy in the ultrasonic-infrared-combo device. However, the displacement error is *accumulative* in future location calculation, so the error distance follows a linear growth pattern. Note that orientation error is more destructive than displacement error, i.e., even a one-time orientation error can make the positioning error grow linearly over walking distance. This can be explained by looking at Fig. 6. After the one-time orientation error of θ_{error} occurs, the calculated path will forever deviate from the real path, leading to linear grow in error displacement. In addition, we have found that our orientation sensor becomes inaccurate after rotating over 90 degrees. In order to get more accurate rotation angle θ, we reset the orientation sensor after each left step, and then sum up each rotation θ_i to get the orientation θ. Due to this extra calculation, the orientation error of θ_i also becomes accumulative.

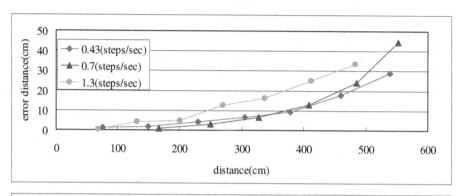

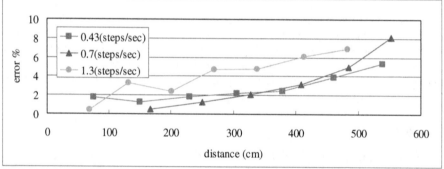

Fig. 5. The positioning accuracy (error) under different walking speeds over the walking distance

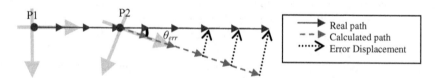

Fig. 6. Illustration of the accumulation of the error of

4   Final Design: Design Version III

Design III tries to fix the error accumulation problems in design II. Design III incorporates location-aware passive RFID tags & readers that can reset the accumulated error whenever the user steps on top of a RFID tag with a pre-determined location coordinate. These location-aware passive RFID tags forms a passive RFID grid that can be used to bound the accumulated error in design II. Since a higher RFID grid density means higher probability that a user will step on top of a passive RFID tag (therefore resetting the positioning error), the ideal density of the RFID grid can be chosen to achieve the needed positioning accuracy in the deployed environment.

The RFID solution has two parts: (1) a Skyetek M1[10] RFID reader is installed at the bottom of the left sandal, and (2) a set of passive RFID tags with the read range of 4.5 *cm* are placed in the grid fashion. We only attach one additional RFID reader to the left sandal, and the other device configuration is the same as in design II (Fig. 7.).

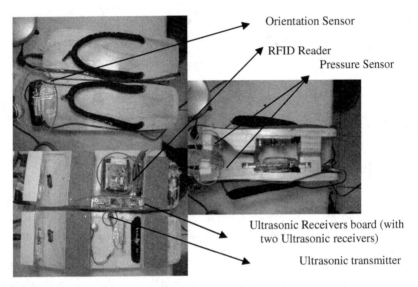

Orientation Sensor

RFID Reader
Pressure Sensor

Ultrasonic Receivers board (with
two Ultrasonic receivers)

Ultrasonic transmitter

Fig. 7. It shows the prototype of final design (version III) of the GETA sandals. Prototype of design (version II) does not have the RFID reader. Prototype of design (version I) does not have the orientation sensor but has an additional transmitter

In the target environment, a server is used to maintain the table mappings between RFID tag IDs and corresponding location coordinates. When a user enters the target environment, the GETA sandal downloads its mapping table. The positioning algorithm is revised as follows. When the GETA sandal steps on top of a RFID tag, it looks up the cached mapping table to find the location coordinate of this RFID tag. Then, the current location of the user is set to the location coordinate of this RFID tag rather than from the footprint tracking method.

4.1 Performance Evaluation

We have evaluated the performance of GETA sandal (version III) in a 15x15 square meters testing environment. We have two different configurations of passive RFID grids. The first configuration places one tag every 3*m*, and the second configuration places one tag every 5*m*. Fig. 8 shows the measured positioning error over walking distance for these two configurations. The error is reset to zero when a user steps on top of a RFID tag. Fig. 8 also shows that under a random walk, there is a probability that a user may not step on a RFID tag every 3*m* or 5*m*. As a result, the errors continue to accumulate past 3*m* or 5*m* until a user eventually steps over a RFID tag.

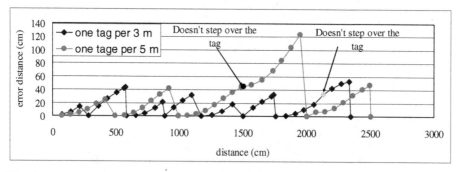

Fig. 8. The positioning accuracy (error) under different walking speeds over the walking distance

5 Conclusion and Future Work

This paper describes the design, implementation, and evaluation of our footprint-based indoor location system on traditional Japanese GETA sandals. Our footprint location system can significantly reduce the amount of infrastructure needed in the deployed indoor environments. In its simplest form, the footprint location system is contained within the mobile GETA sandals, making it easy for everywhere deployment. The user simply has to wear the GETA sandals to enable his/her location tracking with no efforts in calibration and setup. In addition to the benefit of being low infrastructure cost, the footprint based method does not have problems in infrastructure-based indoor location systems such as noises, obstacles, interferences, and dead spots. Although the footprint based method can achieve high accuracy per moving footstep, it has a problem that positioning error can be accumulated over distance traveled. As a result, it may need to be combined with a light RFID infrastructure to correct its positioning error over some long distance traveled.

There are two yet-to-be-addressed problems in our current prototype of GETA sandals: wear-ability, RFID tag placement, and stair climbing. The current wear-ability is unsatisfactory due to interconnecting all sensor components to a Notebook PC through hardwiring. In our next prototype, we would like to replace all hardwiring with wireless networking (e.g., Bluetooth), and replace processing on the Notebook with a small embedded processor. To further reduce the RFID infrastructure, we are interested to locate strategic frequently visited spots in an environment and to place these RFID tags. Stair climbing is a serious problem because the stair becomes the obstacle blocking the sensors between two sandals. To address this problem, we use the strategy of putting RFID tags at the entrances of the stairs. We can treat a stair as a transition path from one floor space to another. Then we can use the RFID to know when we move into or out of a stair and change the position to the new floor space.

Acknowledgements

This work was supported by NSC grants #93-2213-E-002-088, #93-2218-E-002-146, and #93-2622-E-002-033. We would like to thank anonymous reviewers for their valuable comments.

References

1. R. Want, A. Hopper, V. Falcao, and J. Gibbons. The Active Badge Location System. ACM Transaction on Information Systems, 10:1, pp. 91-102, 1992.
2. A. Harter, A. Hopper, P. Steggles, A. Ward, and P. Webster. The anatomy of a context-aware application. In Proc. of 5th MOBICOM, pages 59–68, 1999.
3. N. B. Priyantha, A. Chakraborty, and H. Balakrishnan. The Cricket Location-Support System. In Proc. of the 6th MOBICOM, Boston, MA, USA, August 2000.
4. R. J. Orr and G. D. Abowd. The Smart Floor: A Mechanism for Natural User Identification and Tracking. GVU Technical Report GIT-GVU-00-02, 2000.
5. P. Bahl and V. Padmanabhan. RADAR: An In-Building RF-based User Location and Tracking System. In Proc. of the IEEE INFOCOM 2000, pages 775-784, March, 2000.
6. Ekahau, http://www.ekahau.com/
7. T. Amemiya, J. Yamashita, K. Hirota, and M. Hirose. Virtual Leading Blocks for the Deaf-Blind: A Real-Time Way-Finder by Verbal-Nonverbal Hybrid Interface and High-Density RFID Tag Space. In Proc. of the 2004 Virtual Reality (VR'04).
8. NaviNote technology, http://www.navinote.com
9. InterSense, http://www.isense.com
10. Skyetek RFID engineering, http://www.skyetek.com/index.php
11. E. Foxlin and L. Naimark. Miniaturization, Calibration, Accuracy Evaluation of a Hybrid Self-Tracker, IEEE/ACM Internationl Symposium on Mixed and Augmented Reality (ISMAR 2003), Oct. 2003.
12. Point Research Corp., http://www.pointresearch.com/

Improving the Accuracy of Ultrasound–Based Localisation Systems

Hubert Piontek[1], Matthias Seyffer[1], and Jörg Kaiser[2]

[1]Universität Ulm
{piontek, matthias.seyffer}@informatik.uni-ulm.de
[2]Otto–von–Guericke–Universität Magdeburg
kaiser@ivs.cs.uni-magdeburg.de

Abstract. We present an improvement to ultrasound–based indoor location systems like Cricket [1]. By encoding and modulating the ultrasound pulses, we are able to achieve greater accuracy in distance measurements. Besides improving the distance measurements, we improve the position update rate by synchronizing the active beacons. We also propose a method that could further improve the update rate by superimposing encoded ultrasound pulses. Further, an experimental evaluation of our improvements is presented.

1 Introduction

Localisation is still one of the challenges in mobile ubiquitous application scenarios. Because knowledge of the position is the basis for location–based services, navigation and ad–hoc cooperation, considerable research has been expended and many approaches can be found already as working systems and in the literature. However, most of them only allow a rather coarse grained determination of the position. Consider the coordination of autonomous vehicles which use accurate location to coordinate access to junctions or intersections on a factory floor or in a large stock building. It would be highly beneficial if this could be based on an accurate and reliable positioning system. Another example which we aim at is to use robots in a "mixed reality" scenario where they move in a simulated virtual scene and interact with virtual robots. This application requires a very accurate localization of the robots in the real world. Applications like this add a new dimension to techniques of simulating real world settings. We consider in our application mobile robots with a physical size of about 0.4m in length which move with a moderate speed of about 0.7m/sec (see figure 1). From this, we can derive some primary requirements for the location system:

The accuracy of the system must at least be in the order of the size of the robots or better. Our design goal here is a position accuracy of 0.30m. The position update rate must be at least 1/sec. Our robots are capable of moving up to 0.7m/sec. This would correspond to about two robot lengths at full speed. The system must be scalable in space and in the number of clients.

T. Strang and C. Linnhoff-Popien (Eds.): LoCA 2005, LNCS 3479, pp. 132–143, 2005.

Fig. 1. Cooperating robots

A crucial point in a localization system often is the question of the division of labour between the infrastructure provided in the environment and the respective components on the mobile vehicles. This also affects problems of energy consumption and required computational resources. In a mobile robot, a substantial amount of energy is needed for mobility of vehicles. Compared to this, the energy need for the moderate computational requirements of a location system may be relatively low. However, when considering the infrastructure, it should be simple, easy to deploy and, because it often needs to be operated with a local power source, the energy demand should be low. Thus, we aimed at relatively simple infrastructure composed from simple beacons with low computational requirements and low energy consumption.

The principle of operation is based on distance measurements to at least three beacons and subsequent trilateration. The distance is determined by the differences between the time which a radio signal and an ultrasound signal need to travel from a beacon to the respective receivers on the mobile entity. This is very similar to the principle of the cricket location system [1]. Actually, we first built up the hardware of a cricket system and tried to use it in our mobile environment. However, we discovered two major drawbacks of this system:

1. it was rather difficult to obtain a precise edge of an ultrasound signal. This is mainly because of the limited bandwidth of the ceramic ultrasound transducer. Even when we used a high energy ultrasound pulse, which is not desirable because of energy constraint, the results were not satisfactory. It limits the localization accuracy substantially.
2. the update rate of the cricket system is not sufficient to accommodate high accuracy localization of mobile systems.

The contribution of this paper therefore is to investigate ways of how to improve such a location system. Firstly, we will present and evaluate an impulse compression technique to overcome the first drawback. Then we will describe ways to improve the update rate. The rest of the paper is organized as follows: the next section discusses related work. Section 3 will introduce the general architecture of the location system and will describe the hardware. It also will briefly address the synchronization of the beacons to avoid collisions on the radio channel. Section 4 will sketch the pulse compression techniques and present results. Section 5 will give a short overview over the activities to improve the update rate of the location system. Finally, a we will provide a conclusion.

2   Related Work

There are three basic methods to determine one's location:

1. sensing the location by means of a *sensor grid*: a sensor (e. g. magnetic or pressure sensitive) can detect an object in its close vicinity. Distributing those sensors in a regular grid allows to determine one's position within this grid. The most common example for such a system is a touch screen as found on most PDAs. For our application scenarios of autonomous robots, such (tight) sensor grids would be too costly to deploy. The original Cricket [1], and the Active Badge Location System [10] are of this kind. Both provide location information at the granularity of a room, which does not meet our requirements.
2. sensing the direction of at least two landmarks (of known location) to a common reference (for 2D positioning), and using *triangulation* to determine one's position.
3. sensing the distance to at least three landmarks, and using *trilateration* to determine one's position. A popular example for this method is GPS, which is proven to work reliably at a accuracy down to several meters in an outdoor environment. However, it does not work inside buildings. Laser technology is known for its accuracy, but we considered it to be too expensive, and too complex for our goals. The Cricket Compass [2], Cricket v2 [3], The Bat [8], [9] and RADAR [6], [7] are examples for this type of positioning systems that work indoors. Because we are aiming at a distributed system that is composed of fixed, active beacons and passive mobile clients, The Bat and RADAR are ruled out. The Cricket system comes closest to our goals: it uses easily deployable components. The autonomous system architecture fits nicely into our view of the world where we aim at autonomous components that interact without any central coordination.

RADAR [6], [7] is an RF–based system for locating (and tracking) users inside buildings. It uses the signal strength information from wireless networking equipment. The system is capable of locating users to within a few meters of their actual location. The system uses a combination of empirically determined and theoretically computed signal strength information, as the propagation of RF inside buildings is hard to cope with. One of the advantages of RADAR is, that the means to provide location information also provides traditional data network services. However, the trackable entities are laptop computers. A major drawback is the need for empirical data, which must be collected before the system can go to *real–time mode*, i. e. into operation.

Cricket [1] originally aimed at supporting a user to find his location within one or two square feet. The system architecture is similar to our own: Cricket uses *beacons* that basically advertise their position. *Listeners* that move through instrumented areas use the time of flight of ultrasound signals to estimate the distance to all beacons in (ultrasound) range. The current position is the area advertised by the beacon which is closest to the listener. Cricket's *granularity* is *portions of a room*. As in any other system utilizing both radio and ultrasound signals, corresponding radio und ultrasound signals must be correlated at the receiver. Cricket does not modulate the ultrasound signal, so it needed a different mechanism: Typically, radio signals can be received at much greater distances than ultrasound signals. This ensures, that whenever an ultrasound signal is received, so is the radio signal. Using a small bandwith radio link, and having long enough radio messages, it is assured that the ultrasound signals arrive while the radio message is still being transmitted. In the absence of interference, this ensures that the correct correlation of radio and ultrasound signal is done. Errors in measurement due to changes in the speed of sound — e. g. due to temperature — are irrelevant because only the closest beacon is used to determine the current position.

The Cricket Compass [2] improves the original Cricket system on the listener side only. Besides mere distance measurement, the receiver orientation towards the beacon is determined. This can be done using five ultra soundreceivers in a V–shape, and measuring phase differences in the incoming utra sound signal. With Cricket Compass, positioning in terms of absolute coordinates within a room was introduced. This requires at least four beacons. The method described in [2] overcomes the problem of not knowing the speed of sound, so no further sensory equipment is needed.

Cricket v2 [3] is based on improved and simplified hardware components. Cricket hardware units can be configured to either be a beacon or a listener. The API has been extended, and a software distribution allowing to develop Cricket enabled applications e. g. in Java is available.

3   Positioning System

The basis for our location system is distance measurement. A *client* measures the distance to several (3 or more) *beacons*. Taking these distance measurements, it calculates its position using multilateration. Beacons are attached to the ceiling. They are equipped with a radio module and an ultrasound transmitter. Within each room, all beacons should be mounted at the same height (see figure 2a). A client (e. g. on a mobile robot) can determine its position in a three dimensional space by measuring the distance to at least three beacons: Two measurements limit one's position to somewhere on a circle (denoted by the thick vertical black line between the intersections of the two circles in figure 2b). The third measurement reduces this to two points on this circle (The intersection of both circles in figure 2c). One of them can be discarded, as it is above the beacons, which is impossible because of the directional characteristics of the ultrasound transmit-

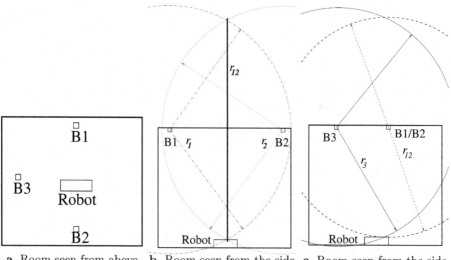

a. Room seen from above   **b.** Room seen from the side   **c.** Room seen from the side
      (X–Y plane)   (Y–Z plane)   (X–Z plane)

Fig. 2. System Architecture

ters. The measurements ideally should be done simultaneously, especially if the robot is moving.

3.1   Distance Measurement

Distance is measured using the difference in time–of–flight of RF signals and ultrasound signals.

The time difference for travelling a distance d between the ultrasound signal and the radio signal is

$$t = t_{us} - t_{rf} = \frac{d}{v_{\text{ultrasound}}} - \frac{d}{v_{\text{radio}}}$$

For a distance d of 10m, the radio signal needs about $t_{rf} \approx 30$nsec. The ultrasound signal, however, will need about $t_{us} \approx 30$msec. As $t_{rf} \ll t_{us}$, t_{rf} can safely be omitted from the above term.

Unfortunately, the speed of sound is not constant. It varies e. g. with temperature. Between $-20°$C and $40°$C, it can be approximated in a linear fashion: $v_{\text{ultrasound}} = (331.6 + 0.6 \cdot T)$ m/sec where T is in $°$C. Not dealing with temperature would introduce rather large errors, e. g. 3.4% or 0.34m when measuring a distance of 10m and going from $10°$C to $30°$C.

There are two possibilites for dealing with unknown speed of sound: (1) try to approximate the speed of sound using sensors, e. g. temperature sensors, and (2) using one more beacon, and introduce it as another unknown variable in the positioning calculations. Of course, the latter assumes, that the speed of sound remains constant within a room for the duration of a positioning operation. This enhancement is based on the work of the Cricket Compass [2].

3.2   Synchronisation of Beacons

The beacons become active in a time–triggered fashion, both for simplicity, and for avoiding collisions (see section 5). The timely properties of the radio channel is well known in our setup, as we use low–power RF modems [11] that show a well–defined behaviour.

3.3   Hardware

Our initial approach was to build a Cricket v1 clone using the same analogue tone decoder for detecting the ultrasound pulse that was originally used with Cricket. The beacons simply sent a constant ultrasound "tone" for a small period of time. The receiver should ideally detect this tone right at the first "edge". In practice, it took the tone decoder several milliseconds for the incoming carrier tone to be detected. With this setup, the distance measurements had errors in the range of several tens of centimeters for perfectly aligned ultrasound transmitters and receivers. When the ultrasound parts were only slightly misaligned, we had even worse readings (see also [5]). We were able to improve the system by discarding the analogue tone decoder. Instead, we fed the amplified input signal to a comparator circuit. The output is a binary signal that was directly fed into a microcontroller's capture unit. Tone detection was done in software [13]. We recently became aware that Cricket changed to the same technique [4], [5]. The results were promising for aligned ultrasound transmitters/receivers: all measurement were within ±2cm of the actual distance. Measurement errors grew with the misalignment of the transceivers. when misaligned by more than 35°, the measurements became completely unreliable. Our current approach is to use pulse compression on the ultrasound channel to get accurate distance measurements. The theoretical background is briefly described in section 4, and in more detail in [12]. The beacons (see figure 3a) use an 8 bit microcontroller [16]. They communicate via radio modules. The modulated ultrasound pulse is created in software, so the hardware is kept as simple as possible. The clients (see figure 3b) are more sophisticated. The incoming ultrasound signal is amplified and fed directly to a low–power DSP (Motorola 56F800, about 30 MIPS). The DSP is in charge of demodulating and correlating the incoming signal. It essentially sends a timestamp to an 8 bit microcontroller (same type as on the beacons) which then does the final calculation of the distance.

4   Pulse Compression

The resolution of distance measurements is directly proportional to the sending of an extremely short pulse (*Dirac–pulse*). Sending a very short, single pulse, however, leads to misdetections, as random noise could be misinterpreted as the expected single pulse. There are two solutions to misdetections: (1) make the pulse a very–high power pulse, so that its signal–noise–ratio is good enough, or (2) generate a longer and weaker pulse with the same amount of energy as the short and very–high power pulse. Due to the limited bandwidth of approx.

a. Beacon   b. Client

Fig. 3. The components of our location system are equipped with a radio module and an ultrasound transducer

2 – 4kHz of ceramic ultrasound capsules and their voltage limit of about 15V, it is impossible to generate a short pulse with high enough energy. The drawback of the longer and weaker pulse is worse distance resolution, but it can be sent with narrow band devices. To achieve proper distance resolution, the ultrasound signal must be encoded in an appropriate way, called *pulse compression*. The receiver can then apply correlation filters, that transform this long, low–power pulse to a short peak, that gives a similar distance resolution as the short, high–power pulse. To encode the signal several methods can be used. But considering the computational power of our beacons, we had to use a simple method. We encode our signal using binary pseudo–noise sequences (PN sequences). These are modulated using binary phase shift keying (BPSK). BPSK matches the computational abilities of our beacons and achieves a good coding efficiency of 1 bps per Hz of bandwidth. For our ultrasound transducers, this yields a maximum data rate of about 2000Bit/s (as we have a usable bandwidth of about 2kHz. The PN–sequences must have the characteristic to provide a good autocorrelation function to get a sharp peak. Barker–Codes are a class of well–known codes that posess the required correlation properties (see figure 4). The disadvantage of Barker–Codes is the limited maximum code length of 13 Bits. The Barker–Code's auto–correlation exhibits a sharper edge than the "triangle" shape of the ping's auto–correlation. This shows that a ping, as used in our first experiments, is not very suitable to achieve good distance resolution. Demodulation and correlation on the receiver side are done in software on the DSP. For best results the signal must be sampled at a rate of 160kHz. The optimal receiver must correlate the incoming signal with the stored reference signal continuously. Such a receiver would need about 166 MIPS at a bitrate of 2000Bit/sec. We use a modified BPSK–demodulator (see figure 5) to limit the demand for computational power. First, the received signal must be transformed from a pass band signal to a base band signal by the quadrature mixer. The signal is splitted into a in–phase and a quadrature component. Before a data reduction stage, the signals are low–pass filtered. The resulting signals are correlated with the stored Barker–Code using the schema in figure 6. These modifications to the

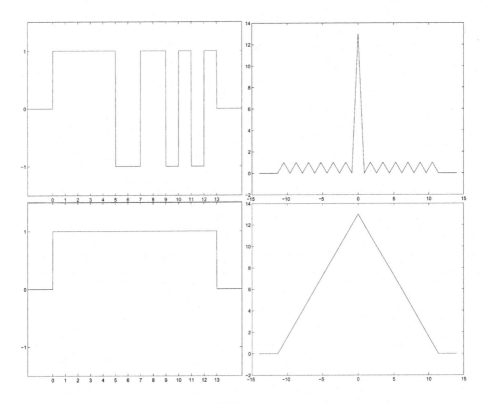

1 and -1 stand for the two symbols usable with BPSK
when transmitting, whereas 0 stands for "transmitter
switched off"

Fig. 4. BPSK modulated Barker–Code (top left, 13 bits), and its auto–correlation (top right) vs. Ping (bottom left, 13 bits), and its auto–correlation (bottom right)

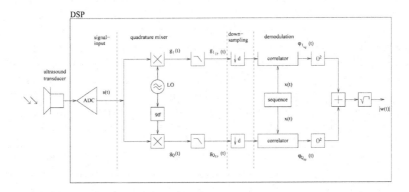

Fig. 5. Demodulation of the incoming signal

BPSK–demodulator reduce the computational requirements to about 10 MIPS. Of course, these modifications degrade accuracy. The overall accuracy we experienced is well within our requirements (see section 4.2). The resulting signal of

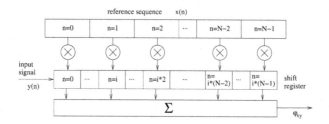

Fig. 6. Correlator

the receiver is an envelope, showing how good the received signal matches the stored signal. The best match and thus the time of arrival of the signal can be easily determined by a search for the global maximum.

4.1   Theoretical Results

Using a sampling rate of 160kHz, a best–case resolution of about 2mm could be expected. Because the signal is phase–coded, the resolution cannot be less than the wavelength of the ultrasound signal, which is about 8.6mm. To achieve this accuracy in practice, very long PN–sequences are needed, which would affect the position update rate. The bitrate of the signal gives an absolute worst–case resolution (upper bound) of 20cm.

4.2   Experimental Results

To get an idea of the behavior in the real world, the test setup in figure 7 was used for first experiments. The data rate was set to approximately 1666Bit/s and the 13 Bit Barker–Code was used. Several measurements were done at distances from 0m to 2m. The sender was mounted below the ceiling 2.13m above the receivers. The first test showed the expected result. It is presented in figure 8a. The upper graph shows the received ultrasound signal. The BPSK encoded Barker–Code is visible starting at a time of about 7msec, corresponding to a distance of 2.35m. An echo is visible at approximately 19msec, corresponding to about 6.5m. The lower graph shows the "detection envelope" with the global maximum corresponding to the beginning of the incoming waveform, and a local maximum for the weak echo. The second test (figure 8b) shows unexpected behavior. The signal is superimposed by several reflections and echoes. In the lower graph, it can clearly be seen (by a human observer) that the beginning of the signal was correctly detected at approximately 6msec, or 2.13m (the first major peak of the detection envelope). However, the global maximum results from a reflection. This is possible because of the direction dependant attenuation characteristics of the ultrasound transducers. This problem shows that a search for the global

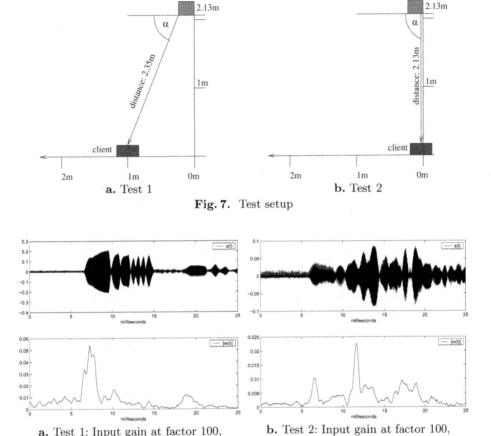

Fig. 7. Test setup

a. Test 1: Input gain at factor 100,   **b.** Test 2: Input gain at factor 100,
horizontal distance 1m   horizontal distance 0m

Fig. 8. Test results

maximum cannot be used in practice. We are working on a heuristic method to find the first peak in the envelope. Leaving out these misdetections, the overall accuracy of the system is within 10cm. This may not look like an improvement over the comparator based approach, however the error bound of 10cm holds for all correctly detected ultrasound pulses. Misaligned transceivers do not lead to growing errors in distance measurements.

5   Improving the Position Update Rate

The most obvious way to improve the position update rate compared to Cricket is, to synchronize the beacons among each other. We chose a simple TDMA scheme combined with the separation of larger areas into cells [12]. Inside a cell, a beacon is assigned to a single time slot. All time slots have the same predefined length. The beacons listen to the radio channel to sychronize to the beginnings of the time slots and to send the ultrasonic pulse within his assigned time slot. This avoids collisions on the radio channel, as well as collisions on the ultrasound channel. Using a bitrate of 2000Bits/sec and the 13 Bit Barker–Code, the ultrasound pulse has a duration of 6.5msec. Within 60msec this pulse can travel about 20m. After this distance, it is not detectable anymore. That means that we are able to do one distance measurement in 66.5msec, or 15 measurements per second. Because we need three mesaurements to calculate a position, we can achieve a position update rate of 5Hz.

To further improve the position update rate, tests with multiple beacons sending simultaneously were done. Each beacon uses a different PN–Sequence. Because there exists only one Barker–Code for any given code length, we used sequences from the group of *Gold–Sequences*. Due to the limited data rate only short sequences in the range of 8 to 16 Bit can be used. Our experiments show that these sequences were too short to separate the incoming signals. To achieve a high enough probability to successfully separate the sequences, the minimum sequence length is 256 Bits [14], [15].

Using such code length results in the need of significantly more computational power and memory for decoding. Sending 256 Bits at a bitrate of 2000Bits/sec takes 128msec compared to 6.5msec for sending a 13 Bit Barker–Code. This method seems to be usable only for ultrasound transducers with a higher bandwidth, and therefore higher possible bitrate. Subsequently sending a 13 Bit Barker–Code and waiting several milliseconds for it to fade away yields an update rate that is comparable to the method of simultaneously sending long codes. The increased effort in decoding simoultaneously sent ultrasound signals does not achieve enough improvement to justify itself.

As there are limits to the achievable position update rate, work must be done to extrapolate the position from previous measurements and fuse this information with other sources of location "hints" like odometry, acceleration sensors, or gyros. We believe that this will provide a reliable and up–to–date source of location information.

6   Conclusion

In this paper we discussed the problems of a high accuracy localization system based on distance measurements which exploit the diffenrences in the travel times of ultrasound and radio waves. We showed some intrinsic problems of ultrasound which mainly result from the low bandwidth of the transducers and introduced pulse compression techniques to obtain a sufficiently accurate signal detection, crucial for the accuracy of the distance measurement. Secondly, we briefly discussed the problem of improving the position update rate by coordinating beacons and by using orthogogal sequences that allow the ultrasound signals to be send completely cuncurrently. The second method turned out to be feasible only with a high computational overhead and also, because the length of the sequences, the benefits are questionable.

References

1. B. PRIYANTHA, A. CHAKRABORTY, H. BALAKRISHNAN: *The Cricket Location-Support system*, Proc. 6th ACM MOBICOM, Boston, MA, August 2000
2. B. PRIYANTHA, A. MIU, H. BALAKRISHNAN, S. TELLER: *The Cricket Compass for Context-Aware Mobile Applications*, Proc. 7th ACM MOBICOM, Rome, Italy, July 2001.
3. Cricket Project: *Cricket v2 User Manual*, MIT Computer Science and Artificial Intelligence Lab, Cambridge, MA 02139, July 2004, `http://cricket.csail.mit.edu/v2man.html`
4. Cricket Project: *Schematics of a Cricket Mote*, MIT Computer Science and Artificial Intelligence Lab, Cambridge, MA 02139, `http://cricket.csail.mit.edu/software2.0/Cricket_Mote_Schematic_6310-0335-02_b.pdf`
5. H. BALAKRISHNAN, R. BALIGA, D. CURTIS, M. GORACZKO, A. MIU, B. PRIYANTHA, A. SMITH, K. STEELE, S. TELLER, K. WANG: *Lessons from Developing and Deploying the Cricket Indoor Location System*, MIT Computer Science and Artificial Intelligence Lab, November 2003
6. P. BAHL AND V. N. PADMANABHAN: *RADAR: An In–Building RF–based User Location and Tracking System*, Proceedings of the IEEE Infocom 2000, Tel-Aviv, Israel, vol. 2, Mar. 2000, pp. 775–784
7. P. BAHL, A. BALACHANDRAN, V. N. PADMANABHAN: *Enhancements to the RADAR User Location and Tracking System*, Microsoft Research Technical Report, February 2000
8. M. ADDLESEE, R. CURWEN, S. HODGES, J. NEWMAN, P. STEGGLES, A. WARD, A. HOPPER: *Implementing a Sentient Computing System*, IEEE Computer Magazine, Vol. 34, No. 8, August 2001, pp. 50-56.
9. A. HARTER, A. HOPPER, P. STEGGLES, A. WARD, P. WEBSTER: *The Anatomy of a Context-Aware Application*, Proceedings of the Fifth Annual ACM/IEEE International Conference on Mobile Computing and Networking, MOBICOM'99, Seattle, Washington, USA, August 1999, pp. 59-68.
10. R. WANT, A. HOPPER, V. FALCAO, J. GIBBONS: *The Active Badge Location System*, Olivetti Research Ltd, 24a Trumpington Street, Cambridge, England
11. LPRS Data Sheet *Easy–Radio ER400TRS*, available at `http://www.lprs.co.uk`
12. M. SEYFFER: *Lokalisation in Gebäuden mit intelligenten Sensoren*, Diplomarbeit, Universität Ulm, February 2004
13. K. UHL: *In–House Lokalisierungssystem zur Positionsbestimmung von Robotern*, Lab Report, Universität Ulm, May 2003
14. H. D. LÜKE: *Korrelationssignale, Korrelationsfolgen und Korrelationsarrays in Nachrichten– und Informationstechnik, Messtechnik und Optik*, Springer, 1992
15. M. I. SKOLNIK: *Radar Handbook*, McGraw–Hill, New York, 2nd. ed., 1990
16. H. PIONTEK: *Introduction to the Tiny–Board*, Universität Ulm, 2003

Position Estimation of Wireless Access Point Using Directional Antennas

Hirokazu Satoh[1], Seigo Ito[1], and Nobuo Kawaguchi[1,2]

[1] Graduate School of Information Science, Nagoya University
[2] Information Technology Center, Nagoya University,
1, Furo-Cho, Chikusa-ku, Nagoya, 464-8601, Japan
{hsato, seigo}@el.itc.nagoya-u.ac.jp, kawaguti@itc.nagoya-u.ac.jp

Abstract. In recent years, wireless LAN technologies have experienced unprecedented growth, and new services and problems have occurred. In this paper, we propose a position estimation technique using directional antennas to assist the detection of wireless access points. Using an asymmetric model for estimation, our technique can radicalize probability distribution quicker than using a symmetric model. Our technique consists of three steps. The first measures the current position of the user, the direction of the antenna and the received signal strength of a target wireless access point. The second step estimates the position of the wireless access point from measured data using a signal strength model based on directivity. And the final step presents estimated results that assist the user. These steps are repeated for real-time assistance. We also conducted an evaluation experiments to clarify the effectiveness of our proposed technique.

1 Introduction

In recent years, wireless LAN technologies are widely being used and several devices are equipped with wireless LAN communication capabilities: for example, laptop PCs, PDAs, mobile phones[18], printers, and digital cameras. With the spread of wireless LAN technologies, new services and problems have come to light. For example, positioning systems and location-aware services using wireless LAN technologies have been proposed[1, 17, 5, 9, 2, 8]. Some problems, however, are the unauthorized deployment of wireless access points and use of wireless access points without proper security configuration [11], which cause intrusions into networks, communication failure of wireless LANs, and interceptions of ID or password for web site authentication by the access points spoofing.

The need for positioning access points arises on several scenes. For example, some positioning systems based on wireless LAN use access point positions at positioning, therefore, the collection and registration of the positions of access points are needed during the construction[6]. Additionally, when unauthorized wireless access points cause communication failure or security problem, they must be detected and removed. The aim of our research is to realize assistance for searching for wireless access points.

T. Strang and C. Linnhoff-Popien (Eds.): LoCA 2005, LNCS 3479, pp. 144–156, 2005.

Although a number of studies have been made on user's positioning using positions of wireless access points and received signal strength in recent years[1, 7, 4], few studies have been made on positioning of wireless access points using user's position and received signal strength. In this paper, we propose a position estimation technique of wireless access points using user's position, the received signal strength from them, and a direction of the directional antenna the user equips. Our technique uses a pre-observed signal strength model as learning data for estimation.

One characteristic of our technique is using directional antennas, that is, using an asymmetric model for estimation. Therefore, our technique can radicalize probability distribution quicker than by using a symmetric model. Because we use directional antennas, antenna direction affects the received signal strength in addition to the distance between the sender and the receiver. Therefore, we can get useful information from the change of antenna direction. Although our technique uses direction information, our technique differs from triangulation. Triangulation needs measurements at multiple positions for estimation and restricts measurements style, that is, needs rotation for measurements at the position. On the other hand, our technique can estimate a position using the data measured at only one position and set measurements style free.

2   Related Work

Wardriving is defined as driving around looking for wireless networks [14]. An user drives around with a laptop PC or a PDA in the user's vehicle for detecting wireless networks. Some users log and collect the position of the networks they found using GPS. The aim of wardriving is detecting wireless networks, that is, detecting the area in that wireless access is available. Although a lot of softwares for wardriving exist, almost of them present only received signal strength data as information for detecting wireless access points [15, 16]. On the other hand, the concern of our research is positioning wireless access points. Therefore, advanced assistance is needed for it. Our technique can estimate the position of a wireless access point, and visualize the results for assistance.

Wireless Security Auditor is an IBM research prototype of an 802.11 wireless LAN security auditor running on PDA[12]. This tool detects wireless access points, and lists the security configuration information of them. Netwrok administrators can use it for managing their networks. However, it is insufficient to solve the network problems by unauthorized wireless access points whose positions are unknown. This tool also presents only received signal strength data as information for detecting wireless access points like most wardriving tools.

3   Proposal Method

Our technique consists of the following three steps. The first step measures the current position of the user, the directional antenna direction, and the received signal strength from a target wireless access point. The second step estimates an access point's position using observed data and a signal strength model based on directivity. The final step presents the estimated results to the user to assist access point detection. These three steps are repeated for real-time assistance.

3.1   Signal Strength Model Based on Directivity

Our technique uses a signal strength model as pre-observed learning data for estimation. Some models proposed so far are based on only the distance between a sender and a receiver[3]. In addition, our technique uses the direction of a receiver's antenna. That is, our technique uses an asymmetric model for estimation.

In our technique, the signal strength model is defined as the function with the relative position of an access point and an antenna as input patrameters. This function outputs probability density function(*pdf*).

The relative position of access points and antennas is composed of distance l between them and angle a between the reference direction of the antenna and the direction to the access point from the antenna. The *pdf* of the received signal strength follows from the signal strength model and the relative position.

$$pdf_{l,a} = SignalModel(l, a) \tag{1}$$

This model is constructed by measuring received signal strength on a known relative position. The *pdf* is calculated from measured data.

3.2   Position Estimation

In this section, we show the details of our estimation technique. We use posterior probability according to Bayesian inference for estimation. This calculation has a good property for processing data measured repeatedly.

First, we define measured data and the candidate position for estimation. Set O is defined by the set of observation o_i measured repeatedly. Each observation consists of three elements: the user's current position p_i, the direction d_i of the antenna the user equips, and the received signal strength s_i of a target wireless access point.

$$O = \{o_1, o_2, \ldots, o_m\} \tag{2}$$

$$o_i = (p_i, d_i, s_i) \quad i = 1, 2, \ldots, m \tag{3}$$

'Candidate position' is defined as the position of a target that estimates whether a wireless access point exists. Set C is defined by a set of candidate position c. We assume that a target wireless access point exists somewhere in C.

$$C = \{c_1, c_2, \ldots, c_n\} \tag{4}$$

We calculate the posterior probability of each candidate position $P(c_j \mid o_1, \ldots, o_m)$ using observation set O and a signal strength model as the existence probability of a target wireless access point at each candidate position. The caluculation follows.

First, we get the $P(s_i \mid p_i, d_i, c_j)$ of each candidate position c_j and each observation o_i using the signal strength model. Consider a certain observation o_i and a certain candidate position c_j. We calculate the relative position of (l, a) using p_i and d_i as elements of observation o_i and c_j, and get $pdf_{l,a}$ from the signal strength model and (l, a). The $pdf_{l,a}$ and s_i as an element of observation o_i give the posterior probability of s_i, that is, $P(s_i \mid p_i, d_i, c_j)$.

Next, we calculate $P(c_j \mid o_1, \ldots, o_m)$ using the $P(s_i \mid p_i, d_i, c_j)$ of each candidate position and observations. Assuming that the prior probability of each candidate position is even and that each observation is independent, $P(c_j \mid o_1, \ldots, o_m)$ is calculated by $P(s_i \mid p_i, d_i, c_j)$ for each candidate position and observation. Posterior probability $P(o_i \mid c_j)$ is represented as follows using posterior probability $P(s_i \mid p_i, d_i, c_j)$.

$$
\begin{aligned}
P(o_i \mid c_j) \\
&= P(p_i, d_i, s_i \mid c_j) \\
&= \frac{P(p_i, d_i, s_i, c_j)}{P(c_j)} \\
&= \frac{P(s_i \mid p_i, d_i, c_j) \cdot P(p_i, d_i, c_j)}{P(c_j)} \\
&= P(s_i \mid p_i, d_i, c_j) \cdot P(p_i, d_i) \\
&\quad (\text{because } c_j \text{ is independent of } p_i, d_i)
\end{aligned}
\tag{5}
$$

Bayesian inference gives the posterior probability of each candidate position as follows.

$$
\begin{aligned}
P(c_j \mid o_1, \ldots, o_m) \\
&= \frac{P(o_1, \ldots, o_m \mid c_j) \cdot P(c_j)}{\sum_{l=1}^{n} P(o_1, \ldots, o_m \mid c_l) \cdot P(c_l)}
\end{aligned}
\tag{6}
$$

Given that each observation o is independent, the posterior probability of observation set $o_1, \ldots, o_m$ is as follows.

$$
\begin{aligned}
P(o_1, \ldots, o_m \mid c_j) &= \frac{P(o_1, \ldots, o_m, c_j)}{P(c_j)} \\
&= \frac{\prod_{k=1}^{m} P(o_k, c_j)}{P(c_j)}
\end{aligned}
\tag{7}
$$

Given that the prior probability of each candidate position $P(c_j)$ is even, the posterior probability of each candidate position is determined as follows by expressions 5, 6 and 7.

$$
\begin{aligned}
P(c_j \mid o_1, \ldots, o_m) \\
&= \frac{\prod_{k=1}^{m} P(o_k \mid c_j)}{\sum_{l=1}^{n} \left\{ \prod_{k=1}^{m} P(o_k \mid c_l) \right\}} \\
&= \frac{\prod_{k=1}^{m} P(s_k \mid p_k, d_k, c_j)}{\sum_{l=1}^{n} \left\{ \prod_{k=1}^{m} P(s_k \mid p_k, d_k, c_l) \right\}}
\end{aligned}
\tag{8}
$$

Finally, we use the posterior probability of each candidate position as the existence probability of a wireless access point at each candidate position.

This calculation can be done incrementally by buffering $\prod_{k=1}^{m} P(s_k \mid p_k, d_k, c_j)$ of each candidate position. Therefore the computation time and the memory size for the calculation are constant despite the number of observation. This property is better suited for processing data measured repeatedly.

4   Wireless Search Assistant

A lot of devices are needed for using our technique. It is hard for an user to search for wireless access points on carrying the devices by the hand. Therefore, we developed an assistant system based on our technique, named Wireless Search Assistant. All required devices are packaged within this system.

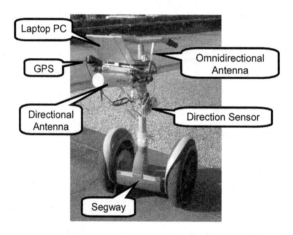

Fig. 1. Wireless Search Assistant Overview

Figure 1 shows an overview of this system, which is composed of a directional antenna, a GPS terminal, four fiber sensors, a direction sensor, a laptop PC, a Head Mount Display(HMD), a Segway[13], and our estimation software. This system measures antenna direction and received signal strength by using direction sensors and directional antennas. In an outdoor environment, this assistant's position is measured by GPS. In an indoor environment, it is measured by dead reckoning using fiber sensors. Because these devices are equipped on a Segway, it has outstanding mobility both indoors and outdoors. Therefore this system is very useful for detecting a wireless access point.

Users searches by looking at estimated results displayed on laptop PCs and HMDs. Our estimation software's GUI consists of the map and signal windows and the access point list(Figure 2). The map window presents the user's position, the antenna direction, and the estimated results that overlap the map. Users

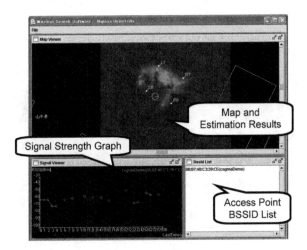

Fig. 2. Estimation Software Screenshot

select a target wireless access point from an access point list and a signal strength model for estimation. This system repeatedly measures data, estimates position, and presents the estimated results on a map, which are represented by colors.

5   Evaluation Experiment

In this section, we evaluate estimation accuracy and search time of our technique. We conducted two cases in each experiment: using a directional antenna and an omnidirectional antenna. 'Directional case' is defined by using a directional antenna, and 'omnidirectional case' is defined by using an omnidirectional antenna. We compared these cases and discussed the results. The Wireless Search Assistant described above was used for this experiment.

5.1   Experiment Environment, Hardware and Setting

Figure 3 shows an overview of the experiment environment conducted in the open air; the user's position was measured by GPS. 16 boxes ware set on a 4 × 4 coordinate grid at 7 meters intervals. Contents of the boxes couldn't be confirmed from outside.

The devices used in the experiment are listed below and Table 1 shows device specification.

- Directional antenna (Buffalo Technology, WLE-HG-DYG)
- Omnidirectional antenna (Buffalo Technology, WLE-NDR)
- GPS terminal (Garmin International Inc., eTrex Vista-J)
- Direction sensor (MicroStrain Inc., 3DM)
- Wireless access point (Buffalo Technology, WHR2-G54)

Fig. 3. Experiment Environment Overview

Table 1. Device Specification

Directional antenna (WLE-HG-DYG)	Polarization Method	Vertical polarity wave
	Directivity, Vertical	$32 \pm 5°$ Half value angle
	Directivity, Horizontal	$32 \pm 5°$ Half value angle
	Antenna Gain	absolute gain 14 dbi
	Frequency Range	2401-2484 MHz (1-11ch)
Omnidirectional antenna (WLE-NDR)	Polarization Method	Vertical polarity wave
	Antenna Gain	absolute gain 2.5 dbi
	Frequency Range	2401-2484 MHz (1-11ch)
GPS terminal (eTrex Vista-J)	Accuracy	<15 meters, 95% typical
Direction sensor (3DM)	Range	Yaw: $\pm$ 180 degrees
	Accuracy	Yaw: $\pm$ 1.0 degrees typical

In each experiment we used the same wireless access point in the construction of a signal strength model. The communication standard used in this experiment was only IEEE 802.11b. We used a digital map (Spatial Data Framework) published by the Geographical Survey Institute[19] for presentation. Estimation setting is the candidate position set C of the lattice position points on the coordinate grid(100 × 100 meters) at 1 meter intervals that covers sufficient experiment space. Measurement intervals are one second.

5.2   Construction of Signal Strength Model

Before the experiment, we constructed signal strength models of directional and omnidirectional antennas. We assumed that a wireless access point is omnidirectional that can be measured without obstacles.

During construction, we assumed that the probability density function of the received signal strength was a normal distribution function and calculated expectation and variance from the measured data using a maximum-likelihood method. In this experiment, we supposed that the interpolation of the probability density function is the interpolation of the parameters by inverse distance weighting(IDW). In particular, given that the interpolation point is p, the sampling points are $q_1 \ldots, q_n$, their respective values are $v_1, \ldots, v_n$, and the distances from p to the sampling points are $l_1, \ldots, l_n$, the value of p, v_p is interpolated using the follow expression. In this experiment, we supposed that inverse distance weighting power w is 1.

$$v_p = \frac{\sum_{i=1}^{n} \frac{v_i}{l_i^w}}{\sum_{i=1}^{n} \frac{1}{l_i^w}} \tag{9}$$

In the directional case, we measured the received signal strength when distance l was 1, 2, 4, 6, 8, 12, 16, 32, 64, and 128 meters and the angle a was 0, 22.5, 45, 67.5, 90, 112.5, 135, 157.5, and 180 degrees. In the omnidirectional case, we measured the received signal strength when the distance l was 1, 2, 4, 6, 8, 12, 16, 32, 64, and 128 meters, assuming that the received signal strength of the omnidirectional antenna is ideal, that is, that antenna direction is independent of received signal strength. Unlike the experiments, the measurements of each case were performed 300 times in 30 seconds at measurement intervals of 0.1 seconds. Figure 4 shows the directional pattern of the used directional antenna at a distance of 16 meters by the measurement.

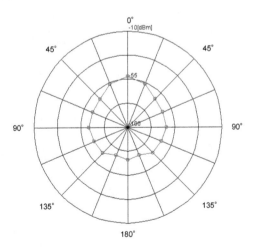

Fig. 4. Directional Pattern of Directional Antenna (16 meters)

5.3   Estimation Accuracy Experiments

We experimented for estimation accuracy with directional and omnidirectional antennas. Here, we supposed that estimation accuracy measurement is the distance between the actual position of a target wireless access point and a position with the highest probability of all candidate positions, wihich is called an 'estimated position'. 'Error distance' is defined as this distance. That is to say, the shorter the error distance is, the higher the estimation accuracy is.

We investigated time transition of error distance. We hid a wireless access point in one of 16 boxes and gave three subjects one minute to search for it. In this experiment, two modes of presentation were conducted. The first mode (called 'half mode') presented only signal changes without estimated results, presenting signal window and access point list (bottom left and bottom right of Fig. 2). The second mode (called 'full mode') presented estimated results in addition to half mode using map window (top of Fig. 2).

Figure 5 and 6 show all results in full and half modes. The vertical axis is error distance and the horizontal axis is time.

At each mode, in each case, error distance shortens as measured data increases. Compared to the omnidirectional case and looking overall at each mode, error distance shortens quicker in the directional case. In the omnidirectional case, half mode error distance shortens quicker than the full mode. The measured data in the experiment show the users' movements between full and half modes. In full mode, users tend to come close to the estimated position presented on a map. On the other hand, in half mode, users tend to move more widely in experiment space than in full mode. It seems reasonable to suppose that this

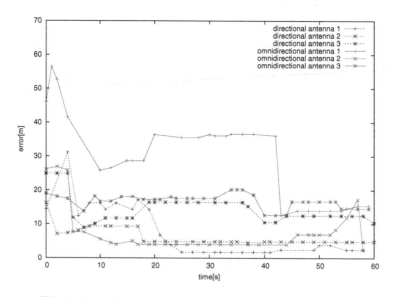

Fig. 5. Time Transition of Estimation Accuracy in Full Mode

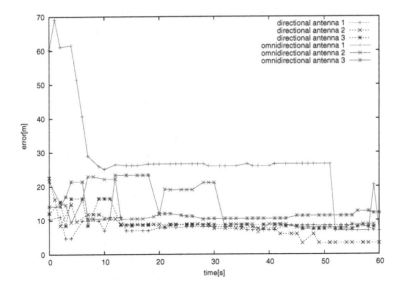

Fig. 6. Time Transition of Estimation Accuracy in Half Mode

difference causes that error distance at half mode shortens eariler than at full mode; users' movements have a effect on estimates, in particular in omnidirectional case.

Differences in the error distance results in the same conditions likey arise because there are no fixed starting search positions.

Perhaps error distance differences between the directional and omnidirectional cases arise because the signal model of the omnidirectional antenna is insufficient, in addition to the effects of our technique. Although we assume that an omnidirectional antenna is truly omnidirectional, perhaps the antenna is only slightly directional amid obstacles.

5.4   Search Time Experiments

Next, we experimented for the search time with directional and omnidirectional antenna. 'Search time' is defined as the time required from starting a search to selecting a box in which a target wireless access point seems to exist.

We investigated search time and whether the selection was correct. We hid a wireless access point in one of the 16 boxes for which three subjects searched. The presentation mode of the Wireless Search Assistant was full.

Tables 2 and 3 show results in directional and omidirectional cases. Compared to omnidirectional cases, search time is about half as long as in directional cases. It seems reasonable to suppose that search time is shortened by effective direction information. The selections were all correct in the directional case, although they are all incorrect in the omnidirectional case.

All trajectories of subject 1 in the directional and omnidirectional cases are shown in Figures 7 and 8. The box is described as square with numbering and

Table 2. Directional Case Results

subject's ID	search time(seconds)	correct or incorrect
1	30	correct
2	33	correct
3	36	correct

Table 3. Omnidirectional Case Results

subject's ID	search time(seconds)	correct or incorrect
1	52	incorrect
2	66	incorrect
3	96	incorrect

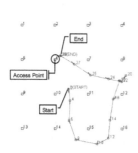

Fig. 7. Trajectory(Subject 1, Directional)

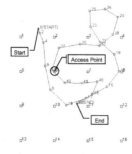

Fig. 8. Trajectory(Subject 1, Omnidirectional)

the subject's position is as circle. In the directional case, the antenna direction is also described. The wireless access point existed in the 6th box. Subject 1 selected correctly in the directional case, although he incorrectly selected the 2th box in omnidirectional case. We suggest that subject 1 moved near the wireless access point in the directional case because the estimated results converged with fine accuracy. On the other hand, subject 1 moved in the space widely in the omnidirectional case because the estimated results did not converge with a high enough percentage of fine accuracy.

It seems reasonable to suppose that search time is shortened by effective direction information.

6 Conclusions and Future Work

In this paper, we proposed a position estimation technique of a wireless access point for assistance in detection. Our technique is based on directional antennas.

Using an asymmetric model for estimation, our technique can radicalize probability distribution quicker than using a symmetric model. Our technique consists of the following three steps. The first step measures the current position of an user, the antenna direction, and the received signal strength of a target wireless access point. The second step estimates the position of the wireless access point from measured data using a pre-observed signal strength model based on directivity. And the final step presents estimated results to assist the user. These steps are repeated for real-time assistance. We conducted experiments with our technique and clarified its effectiveness.

Future work includes an estimation of wireless access point position on the other floors. We intend to use a 3-dimentional signal model for solving the problem. We also plan to consider estimations using an indoor map for more assistance. In an indoor environment, however, a lot of problems for estimation are remained, such as, reflection, diffraction and multipath.

References

1. Paramvir Bahl and Venkata N. Padmanabhan, RADAR:An In-Building RF-based User Location and Tracking System, in Proc. of IEEE Infocom, Tel Aviv, Isreal, 2000.
2. Teruaki Kitasuka, Tsuneo Nakanishi, and Akira Fukuda, Wireless LAN based Indoor Positioning System WiPS and Its Simulation, 2003 IEEE Pacific Rim Conference on Communications, Computers and Signal Processing (PACRIM'03), pp. 272-275, August 2003.
3. Aleksandar Neskovic, Natasa Neskovic and George Paunovic, Modern Approaches in Modeling of Mobile Radio Systems Propagation Environment, IEEE Communications Surveys, Third Quarter 2000.
4. Bill Schilit, Anthony LaMarca, Gaetano Borriello, William Griswold, David McDonald, Edward Lazowska, Anand Balachandran, Jason Hong and Vaughn Iverson, Challenge: Ubiquitous Location-Aware Computing and the Place Lab Initiative, In Proc. of WMASH 2003, San Diego, CA. September 2003.
5. Jason Hong, Gaetano Borriello, James Landay, David McDonald, Bill Schilit and Doug Tygar, Privacy and Security in the Location-enhanced World Wide Web, In Proc. UbiComp 2003, Seattle, WA. October 2003.
6. Bill Schilit, Anthony LaMarca, David McDonald, Jason Tabert, Eithon Cadag, Gaetano Borriello, and William G. Griswold, Bootstrapping the Location-enhanced World Wide Web, In Proc. UbiComp 2003, Seattle, WA. October 2003.
7. Paramvir Bahl, Anand Balachandran, Allen K.L. Miu, W.Russell, Geoffrey M. Voelker and Yi-Min Wang, PAWNs: Satisfying the Need for Ubiquitous Connectivity and Location Services, IEEE Personal Communications Magazine (PCS), Vol. 9, No. 1, February 2002.
8. Jeffrey Hightower and Gaetano Borriello, Location Systems for Ubiquitous Computing, Computer, Vol. 34, No. 8, pp.57-66, IEEE Computer Society Press, Aug. 2001.
9. Cheverst K., Davies N., Mitchell K. and Friday A., Experiences of Developing and Deploying a Context-Aware Tourist Guide: The GUIDE Project, Proc. of MOBICOM 2000, pp.20-31, Boston, ACM Press, August 2000.

10. Rashmi Bajaj, Samantha Lalinda Ranaweera and Dhara P. Agrawal, GPS:Location-Tracking Technology, IEEE Computer, Vol 35, No.4, pp.92-94, 2002.
11. Guideline for wireless LAN security, Japan Electronics and Information Technology Industries Association, http://it.jeita.or.jp/perinfo/committee/pc/wirelessLAN2/index.html (Japanese)
12. Wireless Security Auditor, Global Security Analysis Lab, IBM Research, http://www.research.ibm.com/gsal/wsa/
13. Segway, http://www.segway.com/
14. wardriving.com, http://www.wardriving.com/
15. NetStumbler, http://www.netstumbler.com/
16. Kismet, http://www.kismetwireless.net/
17. WiFi positioning system, Hitachi AirLocation(TM), http://www.hitachi.com/
18. N900iL, http://www.nttdocomo.co.jp/
19. Geographical Survey Institute, http://www.gsi.go.jp/

Exploiting Multiple Radii to Learn Significant Locations

Norio Toyama[1], Takashi Ota[1], Fumihiro Kato[1],
Youichi Toyota[1], Takashi Hattori[2], and Tatsuya Hagino[2]

[1] Graduate School of Media and Governance, Keio University,
[2] Faculty of Environmental Information, Keio University,
Endoh 5322 Fujisawa, Kanagawa, Japan
{next, takot, fumihiro, wisteria, hattori, hagino}@sfc.keio.ac.jp

Abstract. Location contexts are important for many context-aware applications. A significant location is a specialized form of location context for expressing a user's daily activity. We propose a method to cluster positions measured by cellular phones into significant locations with multiple radii. Cellular phones we used are equipped with a positioning system, where data can be taken in low frequency with wide-varying estimated errors. In order to learn significant locations, our system exploits multiple radii for coping with these characteristics and for adapting to a variety of users' spatial behavioral patterns. We also discuss appropriate parameters for our clustering method.

1 Introduction

1.1 Background

Our study is motivated by the increasing sophistication of mobile information devices, such as smart phones or PDAs, and the spread of ubiquitous information environments. Such sophistication of devices can be classified into two points: improvements in network connectivity, and embedding of sensor devices (such as GPS receivers, barcode readers, and infrared interfaces). Over the past few years, study of ubiquitous environments also has brought great advancements; now one can manipulate information appliances and embedded sensors through a home network.

The progress of the combination of advanced mobile information devices and ubiquitous environments enables us to use such a device as an interface to interact with its surroundings. Our final goal is a system which helps us to develop and utilize context-aware applications in such an environment.

1.2 Goal

A context-aware application consists of a set of rules expressing relationships between contexts and operations. For example, applications can consist of rules like "When a user is at home in the morning, get a weather report" or "When

T. Strang and C. Linnhoff-Popien (Eds.): LoCA 2005, LNCS 3479, pp. 157–168, 2005.

a user is going to a station, look up a train schedule". These rules consist of a context as prerequisite condition and a consequent operation. Operations of the rules may be making phone calls, sending e-mails and browsing web sites. Contexts may be where you are, where you go, when, schedule, weather, etc. A location context is one of the most important context to compose these rules.

Our final objective is to create a system which enables end users to develop and execute such context-aware applications on smart phones or PDAs. However, it is difficult to do this with the limited user interface of such devices. In addition, what is obtained as a result of the positioning by GPS systems is just one position in space: a latitude and a longitude. We need more abstracted locations which contain a certain amount of range flexibility like "at home", "around the station", and so on. To make a condition for rules like "at home", you need to define the area range which can be considered home, which is difficult to do with these devices.

1.3   Concept of Significant Locations

In this paper, we discuss a system to learn an abstracted expression of a location from a user's position log. This system should be able to raise the degree of abstraction from just a position expressed by latitude and longitude, to a level meaningful one for a user and an application.

For such kind of expressions, we use the concept of significant locations which appears in [1, 2]. We define the term significant location based on the concept. A significant location consists of a center point and a radius from the center. When a user stays within a location for a certain time, the location is marked as a significant location.

A contribution of our work is an improvement in techniques to utilize multiple radii from positioning logs acquired from cellular phones with positioning functions. Using multiple radii provides two benefits. One is to adapt the system to varying ranges of users' spatial behavioral patterns. For example, we may need expressions for both a whole area of a university campus and each building on the campus. Multiple radii enable the system to handle these multi-step ranges of areas. Another benefit is to cope with characteristics of a positioning log from our target phones. These phones with a GPS based positioning function generate a positioning log which contain widely varying errors. Errors of measurements range from a few meters to a few kilometers.

1.4   Previous Work

Our method of significant location learning is an improvement over the method of Ashbrook et al. [1, 2], in order to make it suited to the characteristics of the measurement log acquired by the measuring function of a cellular phone with positioning functions.

Fox et al. [3] surveyed application methods of location positioning using a variety of Bayesian filters with a scenario of utilizing two or more sensors, such as infrared rays and ultrasonic waves, especially in the local-area.

Liao and Patterson [4, 5] performed quite an accurate location estimate based on the use of GPS logs and GIS data. They applied a dynamic Bayesian network to an issue of location estimation. Their model combines a less-abstracted layer and a more-abstracted layer. The less-abstracted layer is composed of a GPS logged position and speed, and the more-abstracted layer is composed of concepts of a destination and a transportation type. It is difficult to apply their method as is to our system, but this idea could be integrated for our system.

2   System

In order to show that our system can be put into a casual portable device which an ordinary user can use without a special instruction, we implemented a prototype system on a target cellular phone. This phone features a GPS-based positioning function, which has common properties with other standalone GPS devices like Garmin eTrex products, but also has different characteristics. Therefore, we should solve the following issues arising from these characteristics.

2.1   Prototype System

Our prototype system constitutes a client-server model. Here we summarize functions of a client and a server.

Client. W11K, a KDDI cellular phone, is our target client. This mobile device is characterized by the J2ME [7] environment and a positioning function by the Qualcomm's gpsOne [6] technology. We implemented a logging application using J2ME MIDlet which has the functions described below.

- Periodic Positioning
  measures current location at specific intervals, and records it.
- Operation Logging
  records history of incoming/outgoing calls, e-mail and web browsing.
- Sending a Log to the Server
  sends a log of positions and operations to the server.

Server. The server stores and manages logs sent from the client to analyze and learn from those. Perl scripts and a PostgreSQL DBMS are used to build this server. A CGI script gets a log from the client and stores it on the DBMS. Another script analyzes stored logs and generates significant locations. We also created viewers for our data.

2.2   Character of Positioning by Our System

As compared with generic GPS devices, our system with gpsOne technology has two major different characteristics: position anywhere with varying accuracy and lower frequency of positioning.

Position Anywhere with Varying Accuracy. Even indoors where one can not receive GPS signals, one can measure a position supported from base stations of a cellular phone system, as long as cellular signals are reachable.

For standard GPS devices, we could expect the accuracy of measurement within a fixed extent depending on their properties. Meanwhile, accuracy of measured positions of gpsOne vary widely in exchange for the ability to carry out positioning anywhere. With enough GPS satellites, it can detect a location within a few meters of accuracy. However, an error may exceed 1km without enough GPS satellites since a rough location is guessed from nearby cellular base stations.

A presumed measurement error is available from the API as an elliptic shape with three values of *majoraxiserr*, *minoraxiserr*, and *axisangle*. We use *majoraxiserr* meter as a presumed error in our system.

Lower Frequency of Positioning. If we need to satisfy both continuous recording and practical battery life, our system can measure positions only at intervals of several *minutes* to about 10 *minutes*, while standalone GPS systems can usually keep recording over about one day with an interval of several *seconds*.

In our case, table 1 shows the influence of measurement frequencies on battery life. In order to record a log continuously without battery charge, you could only record less frequently as the table shows.

Table 1. Frequency of location measurements and time to battery running out

Interval	Battery Life
10 minutes	21h 20m
5 minutes	14h 30m
2 minutes	9h 40m

2.3   Examples

Figure 1 illustrates significant locations in Fujisawa city generated from the log of the user of test case 1. Circles indicate significant locations. This figure includes three main areas which are around his university on the top-left side, around his home and the nearest station for him on the top-right side, and a downtown of Fujisawa city on the bottom-right side.

3   Methodology

We describe a method for learning significant locations from a position log in this section. We summarize an original method by Ashbrook et al. at first and then explain our modifications to their method. We use multiple radii for two purposes: to describe various scales of spatial behaviors and to adapt varying accuracy of positioning.

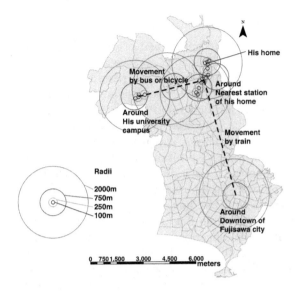

Fig. 1. Movement of a user within Fujisawa City

3.1 Original Algorithm

Ashbrook [1, 2] et al. proposed the basic idea of using a variety of clustering algorithms to generate significant locations. Here we summarize the original idea.

The original algorithm of Ashbrook has three steps. The first step is a kind of "filtering" process to extract staying "places" where a user stopped moving, from a log of positions. The next step is "clustering" where neighboring places are clustered into a single "location". The last step is to learn "sublocations". After these steps, they applied a 2nd order Markov model to predict movements between locations. We do not discuss details of prediction issues in this paper.

To find staying places, they paid attention to the nature of GPS. When a user is in a building, a GPS receiver will lose signals from satellites, so GPS signal loss indicates staying in a building. They concluded from these observations that "places" were points where GPS receiver lost signals for $10 minutes$.

The process of "clustering" is required because even if a user stayed in the same location twice, positions will not be recorded as exactly the same. Places which neighbor each other within an area with a specified radius should be considered to be the same location. They proposed to use a variant of the k-means clustering algorithm in this step.

In the third step, points consisting of the location are split into "sublocations" by using a smaller radius. This step enables one to distinguish campus-level locations and building-level locations. They proposed a basic idea of this step, which they did not discuss in detail. Our idea of using multiple radii is an extension of this concept of sublocations.

3.2   Our Method

In the beginning, we tried to apply their original method for our purpose, but we found some problems. Therefore, we needed to adapt the method to the characteristics of our system which we described in section 2.2.

First of all, we have to reconsider using signal loss as an indication of staying. Our equipment has an advantage, an ability to get position data almost all of the time. However, to make the best use of this advantage, we needed to find alternative method to finding staying places. This method should adapt to the properties of our system's position logs: low-frequency and widely-varying accuracy. These properties make it difficult to distinguish staying or moving from two consecutive positions in an area where positioning accuracy is low. In this kind of area, such as in a building, a distance of consecutive logged positions could be a few kilometers even if a user stayed in exactly the same place.

Our main point of modifications is reordering the filtering step and the clustering step to cope with this issue. In our method, the first step is "clustering". This step clusters all positions in a whole log and generates "candidates" for significant locations. Next, the "filtering" step counts logged positions in candidate locations and removes candidates which include only few logged positions. The basic idea of the method is this; if measured locations are dense in an area with a specified radius, it means a user stays in the area for enough time that the area should be marked as a significant location.

Our algorithm has two major parameters. One is the threshold density of a logged positions in a significant location candidate. This threshold is used in the filtering step to remove noisy significant location candidates. Another is the radius of a significant location for the clustering step. The question we have to ask is how to find appropriate parameters to decide a threshold density and a radius. We will discuss this question in the 4 section.

3.3   Using Multiple Radii

Our system utilizes multiple radii to cluster positions into significant locations. We use multiple radii for two purposes. One is to describe a difference of scales between spatial behaviors, e.g. between walking and using vehicles. Another purpose is to cope with characteristics of a positioning log from our system, widely-varying accuracy.

Scales of Spatial Behaviors. Figure 2 shows an example of an assumed typical daily activity scenario of a user. In larger size locations, a user may move between locations via cars or public transportations in a scale of kilometers. A user might transfer at some locations. After leaving a vehicle, a user may walk to their final destination. This step should be expressed as a medium-scale movement between different locations within walking distances. For example, (s)he may move from the nearest station to home. Finally, (s)he will stay within smaller locations like a campus building or his/her home. In this scenario, we need at least three scales of location radius: large, medium and small.

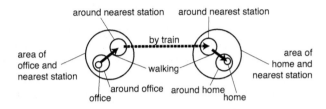

Fig. 2. The typical scenario of daily activity

To adapt varying Accuracy of Positioning. As we described section 2.2, our system generates a position log which has widely-varying accuracy. To adapt to this characteristic, we introduce accuracy-based cut-off processing. The cut-off processing works as follows; when the system clusters measured positions to locations with a radius r, it discards measured positions which have larger presumed errors than the radius r. We regard *majoraxiserr* as the presumed error margin, so all points which have *majoraxiserr* $> r$ will be discarded. Because values of *majoraxiserr* vary from a few meters to a few kilometers, we need to process multi-step radii to make this processing work well.

If we did not introduce this processing, we would have ersatz smaller-radius location generated from positions with larger errors biased to a particular direction. Sometimes this kind of ersatz location is generated even if the user has never been to the area. It seems that a relative location of a phone and base antennas may affect a trend of these biases when GPS signals are weak.

4   Discussion

We have to find appropriate values for two parameters. One is a threshold density of measured positions to form significant locations. The other is a set of radii of significant locations, which is intended to represent scales of human spatial behaviors, and must fit to restrictions posed by characteristics of the positioning system at the same time. In this section, we discuss these parameters based on position histories of test users.

4.1   Test Cases

We have conducted test cases with two users, who have always carried cellular phones equipped with our data collecting system. In the case 1, 3130 positions were measured, which corresponds to 21 days. In the case 2, 4020 positions were measured, which corresponds to 27 days. We discuss these test cases in the remaining part of this section.

To observe basic properties of our method, we applied our algorithm with some sample radii and threshold densities. Sample radii are 50, 100, 250, 500, 750, 1000 and 2000m. Sample threshold densities are from 3 to 9. The table 2 shows the number of generated significant locations for each combination of the sample radii and threshold density. For simplicity, it shows only three threshold densities including the best value, where the best value of threshold density will be discussed in the next section. The numbers shown in the table help us to examine the appropriate values of the parameters, but they are not sufficient. We had to individually check each generated significant location to see if it was truly significant for the user.

Table 2. Test cases : the numbers of significant locations for sample radii and threshold values around best threshold values, and ratios to the numbers of positions

test case	positions	threshold	ratio t/p	50m	100m	250m	500m	750m	1km	2km
		4	0.00128	18	29	23	20	18	17	21
1	3130	5	0.00160	13	25	20	17	17	14	17
		6	0.00191	12	23	19	16	14	14	17
		6	0.00146	21	29	27	26	30	25	26
2	4020	7	0.00170	18	24	23	22	27	23	24
		8	0.00194	14	19	22	21	26	22	21

4.2   Threshold of Measured Positions Density

As described in section 3.2, to produce significant locations with our algorithm, we have to adjust a threshold number of measured positions in a significant location candidate. If we set the threshold too low, results may include too many false locations where the user passed several times by chance. On the other hand, if we set the threshold too high, many useful candidate locations may be discarded.

To find out the best value of the threshold, we made a subjective evaluation for each generated location. As a result of these evaluations of the test cases described in section 4.1, we found that the best value for case 1 is 5, the value for case 2 is 7. Threshold values under these best values are too low to cut off false location candidates. Values over these are too high so that some useful candidates are discarded.

The cause of the difference between the cases seems to be the fact that case 2 has logs which are almost 30 percent longer than case 1. To take the log length into account, let us consider the ratio of the threshold value to the number of positions in the log. The column of *ratio t/p* in table 2 shows these values. We can see the ratios are nearly equal for both cases.

From these observations, we can conclude an appropriate threshold value could be expressed as a ratio, which is around 0.00160 and 0.00170. The best threshold may well depend on the district, the user's behavior, and so on. We must re-investigate this issue after collecting more cases in the future.

4.3   Study of Spatial Behavior Patterns

The next issue is to find an appropriate set of radii size. We assume human movement patterns can be classified into three modes: staying, walking and vehicle, as discussed in section 3.3. It may be possible to regard 10 minutes as a gauge of human movement. This value originally comes from the battery consumption restriction as described in section 2.2, but it is not strange to make an assumption that we use some vehicles if we need to walk more than 10 minutes.

Table 3 shows supposed speed and distances in 10 minutes for each mode. Walking, people will move by the rate of $4km/hour$ on average. This is average, so the lowest threshold of walking mode is around $3km/hour$. Under this value, people are assumed to be staying. An upper threshold of walking mode is around $6km/hour$. Over this value, they are assumed to be moving in a vehicle.

Table 3. Moving modes, speed and distance in 10 minutes moving

Moving mode	Speed	Distances/10 minutes
Staying	$s < 3km/h$	$d < 500m$
Walking	$3km/h < s < 6km/h$	$500m < d < 1km$
Vehicle	$6km/h < s$	$1km < d$

We need appropriate representative radii for each of three modes. The discussion of the GeoOnion RDF/XML vocabulary [8] is informative for establishing a scale of a human spatial behavior. This vocabulary provides the number of properties which relate spatial things together based on their distance in meters. They discussed what is the most appropriate scale, and their current proposal is to use powers of three meters. Their proposal includes wide ranges from 1m to 387km. For our purpose, 5, 6 and 7 powers of three: 243m, 729m and 2,187m are suitable. By rounding off the values to multiples of 50 meters for convenience, we get the radii sizes 250m, 750m and 2,200m, which we decided to use as representatives for staying, walking and vehicle mode respectively.

4.4   Effect of Cut-Off Processing

The next issue is how we can use multiple radii to deal with varying accuracy of the measured positions. Let us consider factors which affect the number of significant locations. We have two major factors. The first factor is "merging". If a radius size gets larger, several locations with a smaller radius may be merged into a single location. The other factor is the cut-off processing we discussed in section 3.3. These two factors are complementary. The merging effect: increasing the radius decreases the the number of significant locations. The cut-off processing effect: increasing the radius increases the number of significant locations.

To understand the effect of cut-off processing, see figure 3 which shows percentages of measured positions of which the presumed error is smaller than the

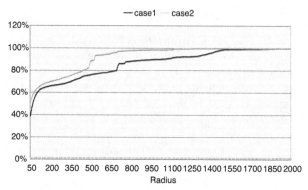

Fig. 3. Percentage of positions which *majoraxiserr* is smaller than the radius

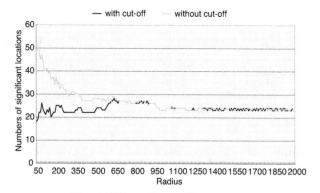

Fig. 4. Effect of cut-off processing

specific radius for two test cases. The horizontal axis shows sizes of radii in meters. The vertical axis indicates percentages of positions of which the presumed error is smaller than the radius which the horizontal axis value specifies. When we run the algorithm at a certain radius, only these portions of positions will be used to generate significant locations; other positions will be discarded.

Figure 4 shows the numbers of generated significant locations for the case 2. To indicate the effect of cut-off processing, the figure shows two lines with and without cut-off processing. To draw this graph, we executed clustering with radii of 10 meters interval and the best threshold value 7, described in section 4.2.

The line of no cut-off processing indicates only the trend caused by merging effects. This line shows the trend that the number of significant locations decreases as the size of radius increases. On the other hand, with cut-off processing, the number of significant locations shows a more complex and interesting tendency. As you can see, there are obvious differences between lines with and without cut-off processing. To understand this effect, we may compare these differences with percentages shown in figure 3. The source of these differences comes mostly from possible ersatz locations which we described in section 3.3.

Let us now consider good representative sizes of radii from the viewpoint of cut-off processing. In general, a smaller radius gives more detailed information. With cut-off processing, however, this is not the case.

Figure 4 illustrates that a radius of about 100m provides the largest number of significant locations in the small radius range with cut-off processing. In addition, as shown in figure 3, the cut-off processing eliminates 50% or more of positions when we use a 50m radius. It means that significant locations of 50m radius have lost large amount of information. By investigating each generated location of the test cases we found that almost all of significant locations of 50m radius have corresponding 100m radius locations. As a consequence, we can conclude 100m is an appropriate for the smallest radius of significant locations.

On the larger side, the numbers of significant locations are nearly same for radii over 1,200m. According to larger radii of figure 3, if we would like to get the largest significant location which uses 99% or more of the user's log data, we should use a radius of 1,840m or larger.

4.5   Set of Representative Multiple Radii

Upon consideration of these aspects, we conclude it is reasonable to use four step radii: 100m, 250m, 750m, and 2,200m. In addition to the representative of three moving modes as described in section 4.3, we introduce representatives from the viewpoint discussed in section 4.4. Namely, we add 100m as the smallest radius because it gives most informative locations. The radius 2,200m could also be understood to be a large enough radius to include whole positioning logs because it is larger than 1,840m.

The last question is whether four step radii are enough or not. To examine the effect when we introduce more steps, we considered intermediate radii of 500m and 1,000m. Almost all of significant locations of 500m and 1,000m radii have one-to-one relationship to significant locations of 750m radius. As a consequence of this consideration, we concluded that 750m radius can be a representative both of 500m and 1,000m and we do not have to introduce intermediate radii.

The example in figure 1 is generated with these values. (However, we used 2,000m instead of 2,200m in this example.) This result shows that these values work well for expressing hierarchical and adjoining relationships of locations.

4.6   Future Work

The parameters of the algorithm should be considered again after more examples are collected. In particular, we should find a better method to filter out significant location candidates. We plan to reconsider a relationship between a radius size, a staying time and a density of measured positions in a location.

Based on this research, we are now implementing a prediction system of likely locations and operations. We plan to use a probabilistic model to describe transitions of locations and relations between different radii size locations. We are thinking to create a probabilistic model where a significant location at a given time can be determined from the significant location at the previous time and the significant location at the same time with larger radius. The model may be used not only to predict future locations, but also to improve accuracy of significant locations of smaller radii because we can choose the most probable location among several candidates.

5   Conclusion

We discussed a method to utilize multiple radii to generate significant locations. At the outset, we expressed our purpose and summarized features of our system which used a cellular phone with gpsOne positioning function. Then, we showed our method in comparison with original method proposed by Ashbrook et al. and explained its modification to fit our requirements. We discussed appropriate parameters for our algorithm by examining two test cases. We found that threshold densities of positions in a significant location could be expressed as ratio of a whole position log for these test cases. We discussed a good set of radii from two viewpoints, one to cope with varying accuracy and another based on study of spatial behaviors of users. Finally, we concluded that it is reasonable to use 4 step radius, 100m, 250m, 750m and 2,200m radii from these discussions.

References

1. Ashbrook, D., Staner, T.: Learning Significant Locations and Predicting User Movement with GPS, Proceedings of the IEEE International Symposium on Wearable Computers (2002) 101-108
2. Ashbrook, D., Starner, T.: Using GPS to Learn Significant Locations and Predict Movement Across Multiple Users, Personal and Ubiquitous Computing, Vol.7(5). (2003) 275-286
3. Fox, D., Hightower, J., Liao, L., Schulz, D., Borriello, G.: Bayesian Filters for Location Estimation, IEEE Pervasive Computing Magazine, Vol.2(3). (2003) 24-33
4. Liao, L., Fox, D. , Kautz, H.: Learning and inferring transportation routines, Proceedings of the National Conference on Artificial Intelligence (2004) 348-353
5. Patterson, D., Liao, L., Fox, D., Kautz, H.: Inferring High-Level Behavior from Low-Level Sensors, Proceedings of the International Conference on Ubiquitous Computing (2003) 73-89
6. gpsOne, http://www.cdmatech.com/
7. Java 2 Platform, Micro Edition (J2ME), http://java.sun.com/j2me/, Sun Microsystems.
8. Brickley, D., et.al.: GeoOnion ESW Wiki, http://esw.w3.org/topic/GeoOnion

Modeling Cardinal Directional Relations Between Fuzzy Regions Based on Alpha-Morphology

Haibin Sun and Wenhui Li

College of Computer Science and Technology, Jilin University,
Changchun 130012, China
Offer_sun@hotmail.com

Abstract. In this paper, we investigate the deficiency of Goyal and Egenhofer's method for modeling cardinal directional relations between simple regions and provide the computational model based on the concept of mathematical morphology, which can be a complement and refinement of Goyal and Egenhofer's model for crisp regions. To the best of our knowledge, the cardinal directional relations between fuzzy regions have not been modeled. Based on fuzzy set theory, we extend Goyal and Egenhofer's model to handle fuzziness and provide a computational model based on alpha-morphology, which combines fuzzy set theory and mathematical morphology, to refine the fuzzy cardinal directional relations. Then the computational problems are investigated. The definitions for the cardinal directions are not important and we aim to present the methodology and power of using fuzzy morphology to model directional relations. We also give an example of spatial configuration in 2-dimentional discrete space. The experiment results confirm the cognitive plausibility of our computational models.

1 Introduction

Many researchers have extended, and have investigated the computational properties for, Goyal and Egenhofer's model [1] for cardinal direction relation between simple spatial regions. This model considers the effect of the region's shape on their directional relations, but the reference region is still approximated by the minimum bounding rectangle, which leads to some anomalous instances. Our work is based on the dilation operation in mathematical morphology, and a region is dilated by a structuring element (ray) with an angle. We consider the intersection of the dilated reference region with the target region to define the cardinal direction relations between them. We find that this method is natural and can avoid those anomalies.

Goyal and Egenhofer [1] introduced a direction-relation model for extended spatial objects that considers the influence of the objects' shapes. It uses the projection-based direction partitions and an extrinsic reference system, and considers the exact representation of the target object with respect to the reference

T. Strang and C. Linnhoff-Popien (Eds.): LoCA 2005, LNCS 3479, pp. 169–179, 2005.

frame. The reference frame with a polygon as reference object has nine direction tiles: north(N_A), northeast(NE_A), east(E_A), southeast(SE_A), south(S_A), southwest(SW_A),west(W_A), northwest(NW_A), and same(O_A). The cardinal direction from the reference object to a target is described by recording those tiles into which at least one part of the target object falls (Fig. 1).

At a finer level of granularity, the model of Goyal and Egenhofer [1] also offers the option to record how much of a region falls into each tile. Such relations are called cardinal direction relations with percentages and can be represented with cardinal direction matrices with percentages. This model can more precisely describe the cardinal direction relations between regions than the model approximating regions by their Minimum Bounding Rectangles (MBRs). But the model still approximates the reference region with its MBR, which leads to some anomalous instances. Figure 2 is taken as an example for illustration. According to the above model the cardinal direction relation between the target region B and the reference region A is O, i.e., the location of B is the same as A. Obviously we can see that B is north of, east of and northeast of A, i.e. B is partially surrounded by A. Our work is based on the dilation operation in

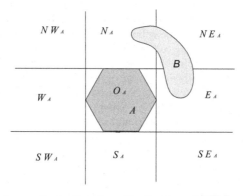

Fig. 1. Capturing the cardinal direction relation between two polygons, A and B, through the projection-based partitions around A as the reference object

mathematical morphology, by which a region is dilated by a structuring element (ray) with an angle. We consider the intersection of the dilated reference region with the target region to define the cardinal direction between them. We find that this method is cognitively plausible.

The importance of modeling for vague regions has been realized by more and more researchers. Generally the vagueness is captured by a broad boundary. The vagueness can be classified as uncertainty and fuzziness. In this paper, we will focus on fuzziness. Cicerone and Felice [14] has investigated the cardinal relations between regions with a broad boundary qualitatively. We present the computational model for cardinal direction between fuzzy regions after we introduce the concept of fuzzy set and fuzzy morphology and present the previous works on modeling directions between fuzzy regions. A fuzzy region can be regarded as

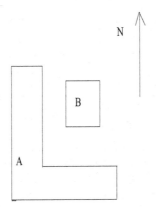

Fig. 2. Illustration for anomalous cardinal direction relation defined by Goyal and Egenhofer's model

a set of α-cut level regions (crisp regions), on which the computational method for cardinal direction relation between crisp regions can be applied. To illustrate our method, we give several examples for computing cardinal directions between crisp regions and fuzzy regions, respectively. We compare the results with human perception.

In this paper, the crisp regions are regular, connected and non-empty closed point sets in the Euclidean space $\Re^2$. Accordingly the fuzzy regions are regular, connected and non-empty closed fuzzy point sets in the Euclidean space $\Re^2$. Schneider [11] has given the definition of fuzzy region based on the framework of fuzzy set theory and fuzzy topology. In the next section, we present the mathematical morphological model for refining cardinal directional relations between crisp regions. Section 3 presents the extended Goyal and Egenhofer's model to handle fuzziness and combines the fuzzy set theory with mathematical morphology to produce the computational model for refining cardinal directional relations between fuzzy regions. An example is given to examine the properties of the models in section 4. Some conclusions are given in the last section.

2   Mathematical Morphological Model

Mathematical morphology is a well-known body of methods and theories, which has been proven valuable in many image analysis applications. Recently it has been used to represent spatial relationship knowledge [2, 3]. The major part of morphological operations can be defined as a combination of two basic operations, dilation and erosion, and non-morphological operations like difference, sum, maximum or minimum of two sets. The operation of interest in this paper is mainly dilation.

The definitions for dilation and erosion of a set X by a structuring element B in a space S (n-dimensional continuous or discrete space), denoted respectively by $D_B(X)$ (or $X \oplus B$) and $E_B(X)$ (or $X \Theta B$), are as follows:

$$D_B(X) = \{x \in S | \check{B}_x \cap X \neq \emptyset\} \ , \tag{1}$$

$$E_B(X) = \{x \in S | B_x \subseteq X\} \ , \tag{2}$$

where B_x denotes the translation of B at point x, and $\check{B}_x$ denotes the reflection of B_x about its origin. To define the cardinal directional relations, of particular interest to us is the class of structuring elements that we refer to as "ray". In the continuous case, they are line segments with one end at the origin as shown in Fig.3. Let Θ denote the angle between a ray and the horizontal line. We will refer to these rays as $ray(r, \theta)$. In the discrete case, we must use an appropriate digitization of a line segment. If a direction is defined as an angle interval, the structuring element is a sector.

Before we use these rays as structuring elements to define the cardinal directional relationships, we should introduce the concept of Hausdorff Distance (HD) metric first. For two non-empty, closed sets X and Y in space S, let S_r denote a closed (super) sphere (a sphere in 3-dimensional space or a circle in 2-dimentional space) centered at origin and whose radius is r. The Hausdorff Distance between X and Y $HD(X, Y)$ is defined as follows:

$$HD(X, Y) = inf\{r | X \subseteq D_{S_r}(Y) \cap Y \subseteq D_{S_r}(X)\} \tag{3}$$

By considering the degree of intersection of a region A dilated by $ray(r, \theta)$ $(r \geq HD(A, B))$ with another region B, we can derive the degree of the relationship B is in the direction Θ relative to A. The degree of intersection can be defined as

$$Area\left(D_{ray(r, \theta)}(A) \cap B\right) / Area(B) \ . \tag{4}$$

From Fig. 3, we can see that the region B is completely included in the dilated region A by a ray with $\theta = \pi/2$, i.e. the region B is completely north of A. If we want to know the degree to which region B is east of A, we can dilate A by a ray with $\theta = 0$ and consider their intersection. If we want to know the degree to which region B is rightly northeast of region A (assuming we consider the right northeast corresponding to $\theta = \pi/4$), we can dilate A by a ray with $\theta = \pi/4$ and consider their intersection. Other direction relations of interest can be defined similarly.

Goyal and Egenhofer's model describes a complete partition of the whole space and defines the cardinal direction relationships more precisely than previous models (e.g. model based on MBRs). But it is rough when compared to the morphological method. These relations can be described as a hierarchy. Considering the regular cardinal directional relations, i.e. N, E, S and W, the morphological model can lead to the same results as Goyal and Egenhofer's model. But when the diagonal cardinal directional relations (i.e. NW, NE, SE, SW) are examined, Goyal and Egenhofer's model presents the rough partition and cannot

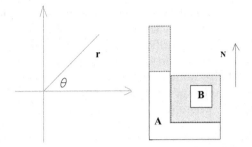

Fig. 3. Example of dilation with a ray of $\theta = \pi/2$ (the part filled with gray color is the dilated region)

represent detailed information, which can be computed using the morphological model. For example, we can not differentiate between the cardinal directional relations of c relative to a and of b relative to a (see Fig. 4) by Goyal and Egenhofer's model, which are all *NE*, but we can know that the cardinal directional relation of B relative to A is right *NE* but it is a little bit *NE* for C relative to A when the morphological model is used.

Fig. 4. Example for cardinal directional relations from which Goyal and Egenhofer's model can not differentiate but the morphological model can

3 Modeling Cardinal Directional Relationships Between Fuzzy Regions

So far, spatial data modeling implicitly assumes that the extent and hence the boundary of spatial objects is precisely determined and universally recognized. This leads exclusively to determinate spatial models. Increasingly, researchers are beginning to realize that there are many spatial objects in reality which do not have sharp boundaries or whose boundaries cannot be precisely determined. Erig and Schneider [4] has identified two kinds of vagueness or indeterminacy concerning spatial objects: uncertainty and fuzziness. In this paper, the fuzzy region is based on a finite-valued (multi-valued) logic, i.e. it is associated to an n-valued membership function for representing a wide range of belonging of a point to a fuzzy region, where $n > 3$.

3.1   Fuzzy Set Theory

Fuzzy set theory [5] is an extension and generalization of Boolean set theory. Let X denote the set of objects, called the universe of discourse (it is 2-dimensional space in this paper), and $\widetilde{A}$ denote a fuzzy subset.

Instead of the characteristic function in classical set theory, which maps an element to $\{0,1\}$, the membership function maps every element to the interval $[0,1]$, which means the degree to which an element belongs to a fuzzy subset, i.e. the membership function is defined as follows:

$$\mu_{\widetilde{A}}: X \rightarrow [0,1] \ .$$

The set $\widetilde{A} = \{(x, \mu_{\widetilde{A}}(x)) \,|\, x \in X\}$ is called a fuzzy set in X. From structured point of view, a fuzzy region can be described in terms of nested α-level sets. The [strict]α-cut level region of a fuzzy region $\widetilde{A}$ is defined by

$$A_\alpha[A_\alpha^*] = \{x \in X | \mu_{\widetilde{A}}(x) \geq [>]\alpha \cap 0 \leq \alpha \leq [<]1\} \ .$$

Clearly, A_α is a crisp region whose boundary is defined by all points with membership value α. The strict α-cut level region for $\alpha=0$ is called the support of $\widetilde{A}$, i.e., $supp(\widetilde{A}) = A_0^*$. The α-cut level regions of a fuzzy region are nested, i.e., for membership values $1 = \alpha_1 > \alpha_2 > \cdots > \alpha_n > \alpha_{n+1} = 0$, one has $A_{\alpha_1} \subseteq A_{\alpha_2} \subseteq \cdots A_{\alpha_n} \subseteq A_{\alpha_{n+1}}$.

α-cuts give a very convenient way for linking fuzzy concepts and crisp concepts. By using α-cuts, all standard operations of fuzzy sets can be derived from their crisp counterparts. Alpha-Morphology can be derived by combining mathematical morphology and fuzzy set theory. Based on the proposed morphological model for cardinal directional relationships between crisp regions and Alpha-Morphology, we can handle the refinement of cardinal directional relationships between fuzzy regions.

3.2   Modeling and Refining Cardinal Directional Relationships Between Fuzzy Regions

In Alpha -Morphology, for two fuzzy sets U and V, the fuzzy dilation of U by fuzzy structuring element V is defined as [6]:

$$(U \oplus V)_\alpha = U_\alpha \oplus V_\alpha \ . \tag{5}$$

This definition is from the field of image processing. The resulting fuzzy set can be obtained using an aggregation schema. Bloch and Maitre [7] presented a formula to compute the result image in a comprehensive way as follows:

$$(U \oplus V)(x) = \sup_{y \in X} \min[U(x - y), V(y)] \ . \tag{6}$$

Koppen et al. [6] proved the two formulae were equal for image processing. Gader [8] defined fuzzy spatial relations between two crisp images, one of which was dilated by a fuzzy structuring element (fuzzy ray), using formula (6) in fuzzy morphology, and showed the experiments were more optimistic than previous methods. Based on these results, we extend the computational model in [8] to handle our case, i.e. cardinal directional relations between fuzzy regions.

For simplicity, we consider the structuring elements to be crisp ones, i.e. $ray(r, \theta)(\theta=0, \pi/4, \pi/2, 3\pi/4, \pi, -3\pi/4, -\pi/2$ and $-\pi/4$, corresponding to East, NorthEast, North, NorthWest, West, SouthWest, South and SouthEast, respectively). For a fuzzy region A and a crisp structuring element B, the formula (5) can be modified as:

$$(A \oplus B)_\alpha = A_\alpha \oplus B \ . \tag{7}$$

To allow for the computation of the area of a fuzzy region, we adopt the definition in [9], where the area of a fuzzy region F is defined as the scalar cardinality of F, i.e.,

$$Area(F) = \sum_{x \in X} \mu_F(x) \ . \tag{8}$$

To aggregate the α-cut level regions and use the aggregated measurements to determine binary cardinal directional relations between two fuzzy regions, we adopt the concept of basic probability assignment in [10]. A basic probability assignment $m(A_{\alpha_i})$ can be attached to each α-cut level region A_{α_i}. $m(A_{\alpha_i})$ can be interpreted as the probability that A_{α_i} is the "true" representative of A. The value of $m(A_{\alpha_i})$ is defined as follows:

$$m(A_{\alpha_i}) = \alpha_i - \alpha_{i+1} \ , \tag{9}$$

which satisfies $\sum m(A_{\alpha_i}) = \alpha_1 - \alpha_{n+1} = 1 - 0 = 1$.

Then, the degree to which fuzzy region $\widetilde{B}$ is located in the direction θ relative to fuzzy region $\widetilde{A}$ can be defined as follows:

$$\mu_\theta \left(\widetilde{A}, \widetilde{B} \right) = \left(\sum_{i=1}^n m(A_{\alpha_i}) \left(\frac{Area((A_{\alpha_i} \oplus ray(r, \theta)) \cap B)}{Area(B)} \right)^p \right)^{1/p} , \tag{10}$$

where i enumerates all the levels $\alpha \in [0, 1]$ that represent distinct α-cuts of a given fuzzy set $\widetilde{A}$, p is used to suit the required degree of optimism or pessimism, and $\alpha_i > 0$.

When compared to the formula (16) in [8], our formula (10) is almost similar to it, but ours is based on operations on regions instead of points. Our method is applicable to regions both in continuous space and discrete space, and the computational cost is cheap while the fuzziness of both the reference region and target region is considered. Moreover if A and B are crisp regions and $p=1$, the formula is the same as the formula (4), so our model provides a unified framework for refining cardinal directional relations between regions. When the structuring element $ray(r, \theta)$ is fuzzy like in [8], we can apply formula (6) to the item $A_{\alpha_i} \oplus ray(r, \theta)$ and the intersection operation in formula (10) becomes a fuzzy one, which has been discussed in fuzzy set theory.

To enable Goyal and Egenhofer's model to handle fuzziness, we define the following formula similar to [13] to compute the degree to which $\widetilde{B}$ is in the direction C relative to $\widetilde{A}$:

$$\mu_C\left(\widetilde{A}, \widetilde{B}\right) = \sum_{i=1}^{n} m(A_{\alpha_i})\mu_C(A_{\alpha_i}, B) , \qquad (11)$$

where $\mu_C(A_{\alpha_i}, B)$ denotes the percentage of relation C between B and A_{α_i} computed by Goyal and Egenhofer's model, and the area of B is computed using formula (8), and i enumerates all the levels $\alpha \in [0, 1]$ that represent distinct α-cuts of a given fuzzy set $\widetilde{A}$, and $\alpha_i > 0$.

3.3   Computational Problems

In this paper, we regard the reference point set as a region, because the point set is conceptually unitary. This kind of point set can be represented by its convex hull. Formally, the convex hull is the smallest convex set containing the points; Informally, it is a rubber band wrapped around the "outside" points. For a point set X, we use $CH(X)$ to denote its convex hull. The algorithm for computing $CH(X)$ has been well studied in computational geometry, and many fast algorithms have been put forward (for example in [12]). The convex hull can reserve the main shape of a region and can simplify the computation. So the formula (10) can be reformulated as

$$\mu_\theta\left(\widetilde{A}, \widetilde{B}\right) = \left(\sum_{i=1}^{n} m(A_{\alpha_i})\left(\frac{Area\left((CH(A_{\alpha_i}) \oplus ray(r, \theta)) \cap B\right)}{Area(B)}\right)^p\right)^{1/p} . \quad (12)$$

To this end, we just need to compute the dilation of the convex hull of the reference region, which leads to a convex region. When the convex region is dilated by a ray, the resulting region is still a convex region whose boundary is the convex hull of the new vertexes resulting from the translating of the original vertexes along the ray plus the original vertexes. So we just need to compute the new vertexes and combine them with the original vertexes to form the resulting region. Obviously the new vertexes are computed from the original vertexes and some may become the inner point of the new region while others form the new vertexes of the new region. An original vertex that leads to inner points when translating along the ray can be decided by checking if the ray going through it intersects with the original region at any other point. So the computation can be further simplified by only considering part of the original vertexes.

We then consider the intersection between two regions, i.e., the intersection between the dilated reference region and the target region (point set). When the target region is based on vector model, the intersection of these two regions can be seen as the intersection of two polygons, which has been investigated widely in computational geometry. When the target region is based on raster model (e.g., in our experiment), we only need to consider the points that fall into the dilated reference region, which is also a well-studied computational geometry problem.

4   Simulation Experiment

To examine the properties of the presented computational model of cardinal directional relations between fuzzy regions, we give an example of spatial configuration in 2-dimensional discrete space (see Fig. 5), which can be a special case of 2-dimensional Euclidean space. In this kind of space, a non-trivial boundary of a point set S is a directed line l_α with the direction α such that at least two point lie in l_α and all other points of S lie in the right half-plane of l_α. A region can be defined by the convex closure, which is formed by a set of non-trivial boundaries. There are three discretized regions A, B and C , which are fuzzy point sets composed of many points labeled with the degree to which they belong to regions A, B and C , respectively. The points that have no labels definitely belong to their regions.

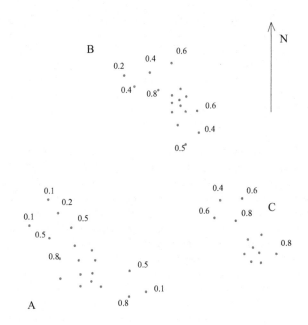

Fig. 5. An example for evaluating our computational model

In this example, we use $p{=}1$. The cardinal direction relations RNE, RSE, RSW and RNW denote right northeast, right southeast, right southwest and right northwest, respectively. We postulate they correspond to $\theta = \pi/4, -\pi/4, -3\pi/4$ and $3\pi/4$. We first use formula (11) to compute the eight cardinal directional relations, and then use formula (12) to refine the results considering RNE, RSE, RSW and RNW. The results of computing the cardinal directional relations between regions A, B and C are listed in table 1. As expected, the results computed using our model are optimistic and conform to human perception and the mathematical model can refine the direction relation to a finer level. For example, the degree to which the fuzzy region C is located southeast relative to the fuzzy region B is 1, which means that region C is definitely southeast of region B regardless of their fuzziness, and the degree to which the fuzzy region C is located rightly southeast of the fuzzy region B is 0.19, which means that the possibility of region C being right southeast of B is small. It can also be seen that our model measures the cardinal directional relationships quantitatively by taking into account of the fuzziness of regions.

Table 1. Computation results using our model for Fig. 5

θ / $\mu_\theta(X,Y)$	N	NE/RNE	E	SE/RSE	S	SW/RSW	W	NW/RNW
$\mu_\theta(A,B)$	0.22	0.98/0.40	0	0	0	0	0	0
$\mu_\theta(A,C)$	0	0.30/0	0.70	0	0	0	0	0
$\mu_\theta(B,A)$	0	0	0	0	0.02	0.98/0.48	0	0
$\mu_\theta(B,C)$	0	0	0	1/0.19	0	0	0	0
$\mu_\theta(C,B)$	0	0	0	0	0	0	0	1/0.18
$\mu_\theta(C,A)$	0	0	0	0	0	0.54/0	0.43	0.03/0

5   Conclusions

Computational models for computing and refining cardinal directional relation between fuzzy regions have been put forward and their usefulness is shown by the results in the experiment. We show that the two models are also compatible with the crisp ones. The morphological model can be a refinement of the conventional model to distinguish more detailed information and avoid some anomalies. The two models are very useful in modeling knowledge in GIS, content-based image retrieval system and computer vision, etc. The application of this technique to one of these systems is the ongoing research.

References

1. R. Goyal and M. Egenhofer: Cardinal Directions between Extended Spatial Objects. IEEE Transactins on Knowledge and Data Engineering,(2000) (in press)

2. Isabelle Bloch: Fuzzy Spatial Relationships from Mathematical Morphology for Model-based Pattern Recognition and Spatial (2003)16-33
3. Isabelle Bloch: Unifying Quantitative, Semi-quantitative and Qualitative Spatial Relation knowledge Representation Using Mathematical Morphology. In: T.Asano et al.(Eds): Geometry, Morphology, and Computational Imaging, 11th International Workshop on Theoretical Foundations of Computer Vision Dagstuhl Castle, Germany, April 7-12, 2002, LNCS, 2616, (2003)153-164
4. Martin Erig and Markus Schneider: Vague Regions. In:5th Int. Symp. On Advances in Spatial Databases (SSD'97), LNCS 1262, (1997)298-320
5. L.A. Zadeh: Fuzzy Sets. Information and Control, (1965)8:338-353
6. M.Koppen, K. Franke and O. Unold: A Tutorial on Fuzzy Morphology. http:// vi-sionic.fhg.de/ipk/publikationen/pdf/fmorph.pdf
7. I. Bloch and H. Maitre: Fuzzy Mathematical Morphologies: A Comparative Study. Pattern Recognition, (1995)28(9):1341-1387
8. P. D. Gader: Fuzzy Spatial Relations Based on Fuzzy Morphology. In FUZZ-IEEE 1997 (IEEE Int. Conf. on Fuzzy Systems), Barcelona, Spain, (1997) 2:1179-1183
9. A. Rosenfeld: Fuzzy geometry: An overview. In: Proceedings of First IEEE Conference in Fuzzy Systems (San Diego), March, (1992)113-118
10. Dubios D and Jaulent M-C: A general approach to parameter evaluation in fuzzy digital pictures. Pattern Recognition Lett, 6, (1987)251-259
11. Markus Schneider: Uncertainty Management for Spatial Data in Databases: Fuzzy Spatial Data Types. In: R.H.Guting, et al. (Eds.): SSD'99, LNCS 1651, (1999)330-351
12. R.L. Graham: An Efficient Algorithm for determining the Convex Hull of a Finite Planar Set. Information Processing Letters, vol. 1 (1972) 73-82
13. F.B.Zhan: Approximate analysis of binary topological relations between geographic regions with indeterminate boundaries. Soft Computing 2, Springer-Verlag (1998) 28-34
14. Serafino Cicerone and Paolino Di Felice. Cardinal Relations Between Regions with a Broad Boundary. 8th ACM Symposium on GIS, (2000) 15-20

Commonsense Spatial Reasoning for Context–Aware Pervasive Systems

Stefania Bandini, Alessandro Mosca, and Matteo Palmonari

Department of Computer Science, Systems and Communication (DISCo),
University of Milan - Bicocca
{bandini, alessandro.mosca, matteo.palmonari}@disco.unimib.it

Abstract. A major issue in Pervasive Computing in order to design and implement context–aware applications is to correlate information provided by distributed devices to furnish a more comprehensive view of the context they habit. Such a correlation activity requires considering a spatial model of this environment, even if the kind of information processed is not only of spatial nature. This paper focuses on the notions of place and conceptual spatial relation to present a commonsense formal model of space supporting reasoning about meaningful correlation. The model consists of a relational structure that can be viewed as the semantic specification for a hybrid logic language, whose formulas represent contextual information and whose satisfiability procedures enhance reasoning, allowing the local perspective typical of many approach to context–awareness.

1   Introduction

Ubiquitous computing can be viewed as a paradigm concerned with a new way of conceiving the interaction among humans (users) and computational devices. Mobile devices, sensors and integrated environments depict a scenario in which users will interact with embedded devices, dynamically connected with each other and almost disappearing in the environment.

Thanks to the improvement and growing availability of information acquisition and delivery technologies (sensors, personal devices, wi-fi, and so on) computational power can be embedded almost in every object populating the environment. Nevertheless, technological evolution is not combined with an equally rapid evolution of the conceptualization necessary to understand and govern the new situation [1]. The term *context–aware* has been introduced to represent new challenges and possibilities, but it is usually interpreted in technological terms, mainly, of physical localization and available resources (e.g. network connectivity).

Context can be defined by a set of different and heterogeneous information concerning the *device properties* (configuration, settings, status, and so on), the *presence of other devices*, their features, their position and function in the environment, and other *abstract and physical information about the environment* itself (predefined or acquired). Perceiving, representing and manipulating contextual information is necessary to perform high-level tasks that devices need to carry out in order to behave as much autonomously as possible according to the basic idea of pervasive computing paradigm.

T. Strang and C. Linnhoff-Popien (Eds.): LoCA 2005, LNCS 3479, pp. 180–188, 2005.

Sensors and devices are located in the environment and computation performed locally by them makes use of information that is related to space and physical environment in different ways: location and other spatial information are, thus, a primary aspect of every model of context, at least as far as a pervasive computing scenario is considered.

Different technological tools, specific devices and techniques, provide the capability to acquire meaningful information about both localization of devices in an environment and relevant features of the environment itself (sensors). Many problems are still open, ranging from basic technical issues (e.g. localization technologies) to protocols and software level issues (e.g. self-configuration of wireless devices), and to high level conceptual considerations (e.g. models of contexts). In fact, a first issue in context–awareness concerns the dynamic "perception" of context (such as localization, communication, collection of data from the environment, and so on); nevertheless, once those information have been acquired, a further challenging problem concern the exploitation of this information.

This exploitation primarily concerns representational issues, according to a formal model of the spatial environment, and the definition of suitable inferential capabilities. In fact, devices localization and context dependent information provided by those devices should be integrated with domain theories specifying knowledge about what can be done with the available information, that is, how this information can be processed according to the system's goal. In particular, from a logical point of view, this processing is a *meaningful correlation* of information provided by devices (that can be a result of local interpretation of raw data, as shown in [2]). A meaningful correlation of heterogeneous data collected from different networked sources consists in exploiting relations among data in order to provide a more comprehensive and informative view on the set of significant properties characterizing the environment.

This correlation task can be achieved by endowing the devices with the suitable inferential power; nevertheless, a preliminary step in order to enable such inferential capabilities is to define a model of context allowing to integrate an explicit representation of the environment with information provided by devices (including their position). According to [3], in order to be enough descriptive, modeling of context information needs to be general, semantically rich and formal.

From this perspective, meaningful correlation can be viewed as a form of commonsense spatial reasoning, where reasoning is grounded on the topology emerging from spatial disposition of the different information sources. Commonsense spatial reasoning presents some specific capabilities, that is, not only to reason about properties of space, but also to exploit spatial information in order to support activities related to various other types of task.

The aim of this paper is to present a logical approach to correlation of information coming from networked devices distributed in the environment: the topological model arising from the devices network can be viewed as a relational structure and, thus, as semantics specification for a hybrid modal language, and reasoning tasks are carried out by means of domain dependent axioms.

In the following section the commonsense spatial concepts of *place* and *conceptual spatial relation* are introduced as the basis of Commonsense Spatial Models, while the formal model is described in Section III. In Section IV it is shown how the defined model can be exploited as kripkean-like semantics for a specific logical language, that is a Multi-Modal Hybrid Language. Concluding remarks end the paper.

2   Basic Concepts: Places and Conceptual Spatial Relations

The literature about space modeling, supporting computational frameworks to be adopted in order to develop reasoning capabilities, is wide and distributed in several areas of Artificial Intelligence such as Automated Vision, Robotics, Knowledge Representation, and so on. Within a rough classification two main classes of approaches can be distinguished: a first one tends to justify commonsense spatial inference with mathematical models such as Euclidean geometry, trigonometry, differential equations systems and so on [4]; in the second one different topological approaches can be considered, ranging from point set and algebraic topology, with the choice of different kinds of primitive entities and relationships (e.g. RCC calculus [5], modal logics [6]), to topological route maps (see [7, 8], and [9]).

Within the second conceptual framework, correlation as commonsense spatial reasoning can be supported by defining a formal model of space that exploits the basic notions of place and conceptual spatial relation. Spatial disposition of information sources distributed in the environment (e.g. close circuit cameras, smart home or complex industrial plant sensor networks) can be mapped into a set of relations among interesting places (i.e. a topology) and high-level reasoning beyond low-level sensors' capabilities can be carried out by reasoning about properties holding at different places.

Suppose to have a sensor platform installed in a building in order to monitor a significant portion of it (and, eventually, to take suitable control actions). Sensors distributed in the environment return values that can be interpreted in order to provide *local* descriptions, possibly generating alerts or alarms, of what is happening in the range of each sensor. Architectural issues are out of the scope of this paper, but in [2] the advantages of distinguishing the detection, local interpretation and correlation levels have been widely discussed, and a four-leveled architecture, which had been fruitfully exploited in the traffic monitoring domain [10], has been presented.

An example of such an environment, e.g. an apartment, is given in Figure 1. Here, different types of sensors are located into separated rooms: in the corridor, for example, there can be a camera, a smoke/fire detector and a broken-glass sensor. Sensors and rooms are related together by means of orientation relations, such as "*to be at north of*"; rooms are linked together by means of proximity relations; and, finally, rooms and sensors are linked together by means of containment relations. In the example proximity between rooms has been defined taking into account "direct access", but the proximity relation can be interpreted differently as well (e.g. as the relation between adjoining rooms). Here sensors and rooms and their reciprocal relations define a commonsense model of space of the monitored area.

A commonsense model of space supporting reasoning about the environment emerges therefore as a topology whose nodes are identified by interesting *places* and

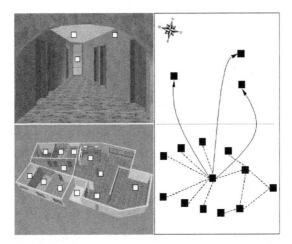

Fig. 1. The emergence of a commonsense spatial model in the context of a monitored apartment. On the left a 3D model of the apartment and a cross-section of its corridor are presented. In the right side, the generation of the corresponding spatial model is represented: the nodes are the interesting places (rooms and sensors), while proximity and containment relations are represented by dashed and unbroken lines respectively. Orientation relations can be guessed but have been omitted for sake of clarity

whose relations are *conceptual spatial relations* (CSR) arising from an abstraction of the spatial disposition of these places. A place is a conceptual entity completely identified by an aggregation of attributes/properties of different kind; examples are the type of place (e.g. a place can be a sensor or a room), its internal status properties (e.g. "is_faulty"), its functional role (e.g. a kitchen or a living room), and so on.

Observe that a CSR is grounded on physical space but not "founded" on it: no necessary relationship among CSRs and any objective physical representation of space needs to be assumed as primitive. Nevertheless, theoretical considerations about the epistemological relevance of this notion of "emergent topology" based on these two basic concepts concerns controversial philosophical issues, which would deserve a deeper analysis that goes beyond the aims of this paper.

Once a topological model has been defined, properties holding at different places can be correlated together to provide a more comprehensive understanding of the environment (e.g. neither a broken glass nor a person detected by the camera are per se a proof of intrusion, but those two facts considered together may lead to infer that a stranger is entered into the house passing through the window and walking in the corridor). Observe that a fundamental characteristic of a commonsense model of space is finiteness, that is, the number of places is always limited; this issue is significant for computability and tractability but is also sound with the fact that, when considering a specific situation, any reasoner necessarily selects a limited portion of the context. As it will be stressed out in the conclusions, this work does not deal with dynamical aspects of the environment yet: the interesting places may change in time; nonetheless, this problem is related to the places selection process and concerns how the model forms and changes, but it does not hinder the model finiteness.

3 CSM, A Model for Commonsense Spatial Reasoning

From a representational perspective the conceptual framework introduced naturally recalls the definition of a relational structure, whose nodes are places and relations are CSRs. A relational structure is a non-empty set on which a number of relations have been defined; they are widespread in mathematics, computer science and linguistics. In particular, according to the epistemological framework specified in the previous section, only finite structures are considered and this is a fundamental characteristic with respect to the computational tractability problem as mentioned in the previous section. A general commonsense spatial model is thus defined as follows:

Definition 1. *A commonsense spatial model* $CSM = \langle P, R_\sigma \rangle$ *is a relational structure, where* $P = \{p_1, ..., p_i\}$ *is a finite set of places, and* $R_\sigma = \{R_1, ..., R_n\}$ *is a finite non-empty set of binary conceptual spatial relations, labeled by means of a set of labels* N.

Finiteness and cardinality of P (the domain must contain at least two places) are minimal requirements to have a well-founded commonsense model of space according to the observation reported at the end of the previous section. An *edge labeled multigraph*(a graph with admitted multiple edges between nodes as in [11]), whose nodes and labeled edges are respectively places and CSRs, is a powerful instance of a CSM.

A place can be anything that satisfies the informal definition of the previous section. As for R_σ, although R_σ can be any arbitrary set of binary CSRs, some classes of relations significant for a wide reasoning domain will be characterized in the following paragraphs. As far as a commonsense model of space is concerned, it is not possible (nor useful) to identify a minimal set of primitives relations (as for RCC). In fact, this approach is not aimed at providing a mathematical model of space, but rather to define the basic elements for the specification of axioms defining relevant properties of specific environments.

Nevertheless, there are some significant classes of relations that provide a model enough powerful but still general. In particular, a place can be "oriented by" the presence of an other (distinct) place, a place can be "contained in" or can be "proximal to" an other place. Although many different relations can fit here, according to different application domains, it seems natural to identify in *Orientation, Containment*, and *Proximity*, the archetypes of any form of commonsense spatial arrangement among entities.

Orientation. First of all, we need some relations to ensure *orientation* in space: assuming reference points is a rudimentary but fundamental way to start orienting into space. Assuming points of reference consists in ordering entities with respect to these particular points. Since many different sources of orientation can be found (stars, magnetic fields, a subjective set of mnemonic sites, and so on), a further step is to choose *good* reference entities and this can be achieved by means of the traditional four *cardinal points*: North, East, South, West. The latter suggests the definition of a set of orientation relations R_N, R_E, R_S, and R_W among places (observe that, from a formal perspective, only two of these relation symbols need to be taken as primitive).

Thus, two relations $R_N \subseteq P \times P$ and $R_E \subseteq P \times P$ are introduced and interpreted in the following way. Let p and q be two places, the relation $R_N(p,q)$ holds iff p is *at north of* q ($R_E(p,q)$ is defined analogously). Orientation relations are both *strict partial orders* on the set of places that is, they are *irreflexive, asymmetric* and *transitive* relations; the order is "partial" because two places might be incomparable. Moreover, both relations have a *superior* and an *inferior* that coincide respectively with North and South, and with East and West. The relations R_S and R_W are defined as the inverse respectively of R_N and R_E. Other non-primitive relations such as *at north-east of* (R_{NE}), *at north-west of* (R_{NO}), and so on, can be defined by means of usual set theoretic operators from the previous ones, e.g. $R_{NE} = R_N \cap R_E$.

It is important to observe that, for what concerns orientation, the notion of order among entities is more fundamental than the contingent choice of particular reference points in order to enable that ordering. The choice of cardinal points seems quite intuitive, but, if different perspectives are needed, reference points can be easily changed or added preserving the basic structure (a lattice with superior and inferior) and the relations' properties (irreflexivity, asymmetry, transitivity). For instance, higher/lower relations can be represented by orientation relations with suitable entities as superior and inferior of the lattice.

Containment. Since places are arbitrary entities, possibly with different shapes, dimensions and nature (e.g. a room and a printer can both be places), a physical inclusion relation $R_{IN} \subseteq P \times P$ is needed in order to relate different types of places: an object may be in a room that may be in a building (where the object, the room and the building are interesting place of the same topology). The relation $R_{IN}(p,q)$ is interpreted as stating that the place q is *contained* in the place p; R_{IN} is a typical mereological relation: it is a partial order, and, more precisely, a *reflexive, antisymmetric* and *transitive* relation. Here, the stronger antisymmetry (i.e. $\forall p, q (R_{IN}(p,q) \land R_{IN}(q,p) \to p = q)$) holds because this can be exploited to infer identity between two places for which is said that one is in another and vice versa.

Proximity. Another basic relation useful to characterize space concerns the possibility of accessing one place from another (in both physical and metaphorical sense). Two places are said to be proximal if it is possible to go from one to the other without passing through another place: a *proximity* relation $R_P \subseteq P \times P$ is then introduced, whose meaning is that the place q is directly reachable from place p. This relation can be modeled as an adjacency relation since is *irreflexive* and *symmetric*. However, different criteria of reachability can be adopted to define an adjacency proximity relation. In a network of radio transmitter/receiver devices proximity is a very different notion from the one adopted in crowding dynamic analysis or in molecular morphogenesis.

Therefore, according to the above observations about orientation, containment, and proximity relations, it is possible to define an *elementary* Conceptual Spatial Model CSM_e as a CSM where, at least $\{North, South, East, West\} \in P$ (the upper and lower bounds of the orientation relations), and $R_\sigma = \{R_N, R_E, R_{IN}, R_P\}$.

4   Reasoning into Space: A Hybrid Logic Approach

Since the commonsense spatial model just introduced is a relational structure, it can be naturally viewed as the semantic specification for a modal logical language. According to a well known modal logic tradition, which relates to Kripke "possible worlds" semantics, classes of relational structures (such as $CSMs$) can be considered as "frames", structures whose relations define the meaning of specific sets of modal operators.

Therefore, modal languages turn out to be very useful as far as reasoning about relational structures is concerned, and have been exploited for temporal and spatial logics, for logic of necessity and possibility and many others (see [12]). Nevertheless, recent studies in Modal Logic lead to further improve its expressiveness and power according to issues coming mainly from research in the Knowledge Representation area. One of the most notable results has been the development of Hybrid Logic. Hybrid languages are modal languages that allow to express (in the language itself) sentences about satisfiability of formulas, that is to assert that a certain formula is satisfiable at a certain world (i.e. at a certain place in our framework). In other words, its syntactic side is a formidable tool to reason about what is going on at a particular place and to reason about place equality (i.e. reasoning tasks that are not provided by basic modal logic).

The definition of a hybrid logic for commonsense spatial reasoning according to the presented CSM requires the assumption of a specific *sort* of atomic formulas (i.e. "nominals") to refer to the interesting selected places. As usual, each place-nominal is true at exactly one place of the CSM and the introduction of the so-called "satisfaction operators" $@_i$ provides the capabilities of reasoning globally on the universe of places. Given a model $W = \langle CSM, V \rangle$, where CSM is the frame and V is an hybrid valuation, the true condition for a formula $@_i\phi$ (where ϕ can be any arbitrary formula), is given as follows:

$$W, w \models @_i\phi \text{ if and only if } W, w' \models \phi,$$

where the place w' is the denotation of i, i.e. $V(i) = w'$. A complete set of symbols for modal operators is then given according to the classification of the basics conceptual spatial relations introduced above. Thus, with respect to the CSM_e, the operators $\Diamond_N$, $\Diamond_E$, $\Diamond_S$, $\Diamond_W$, $\Diamond_{IN}$, and $\Diamond_P$ are introduced; their groundedness in the CSM is guaranteed by the fact that their accessibility relations are defined, respectively, by the CSM's relations R_N, R_E, R_{IN}, and R_P (the semantics of $\Diamond_S$, $\Diamond_W$ is defined over the inverse of the R_N and R_E relations).

According to the aims of the modeled correlation task, a domain dependent set of properties can be chosen and represented in the formal language by means of a suitable set of symbols for propositional letters (e.g. the information "there is a man", coming from a local data processing, can be represented with a proposition "is_man", true or false at some place of the model).

The combination of the multimodal and hybrid expressiveness provides a powerful logical reasoning tool to shift perspective on a specific place by means of a $@_i$ operator, which allows checking properties holding over there; for instance, with respect to Figure 1, when a system devoted to intrusion detection need to query if "a glass is broken" at the place corresponding to the broken-glass sensor, the satisfiability of the formula $@_{window\_sensor}broken\_glass$ must be checked. Moreover, exploiting this operator, it is possible to define local and internal access methods to explore the spatial model according to the other defined operators - e.g. checking the satisfiability of the formula $@_{kitchen} \lozenge_W \lozenge_{IN} smoke$ formally represents the verification, for the system, that "in" some room "at west of" the kitchen some "smoke" has been detected.

Hybrid Modal logic is particularly useful to model reasoning about correlation in pervasive computing environments and, especially, when correlation is exploited for context–awareness, thanks to the double perspective over reasoning that this logic introduces, that is, both local and global.

In modal logic, in fact, reasoning and deduction start always from a given point of the model, i.e. from what is taken as the "current world". In terms of the interpretation of worlds as places, this means that reasoning is performed by a local perspective, and precisely, from the place in the environment taken as the current one. Since, according to the presented model, devices are places, each device can reason about context from its local perspective but exploiting a shared model of the environment. Taken a device, checking the satisfiability of the formula $\lozenge_P (sensor \wedge broken\_glass)$ from this current place means to query if a broken glass has been detected by a sensor adjacent to the current one (an adjacent place on which $sensor$ and $broken\_glass$ are true). On the other hand, hybrid modal logic, still preserving the same local attitude to reasoning of classic modal logic, allows global queries such as $@_{window\_sensor} broken\_glass$. This, in fact means, that whatever is the device on which reasoning is performed, the query regard a specific place/device, that is, the $window\_sensor$.

This double approach to knowledge representation and reasoning typical of hybrid logic (which has been well described in [13]) allows correlation to be modeled as performed both by a central processing unit that reason globally and by single devices locally: this is consistent with different technological approaches to context–awareness, from more centered–based approaches such as blackboard approaches, to approaches stressing more the autonomy of devices, such as multi-agent based approaches.

5   Concluding Remaks

In this paper we presented a commonsense spatial model of space supporting correlation of information coming from distributed sources, which does not assume a strong mathematical ontology, but focuses on the commonsense concepts of place and spatial conceptual relation.

· We have shown that the proposed model can suitably provide a formal semantics for a hybrid modal language, whereas the axiomatization and the definition of a complete calculus is object of current work. It is easy to observe that a CSM is not a closed model, in the sense that, although some basic conceptual spatial relations have been formally characterized, the definition of new arbitrary relations is left open, still preserving the basic model definition (def. 1). A similar modal approach to correlation as commonsense spatial reasoning has been already applied to design and implement the Alarm Correlation Module of SAMOT, a monitoring and control system mainly devoted to traffic anomalies detection (as shown in [14]). In this system the representation of space is mono-dimensional, but correlation is performed along both space and time dimensions.

Actually, there are many domains in which time dimension is crucial and a very interesting problem for further formal and theoretical work is how to consider time and dynamism integrated with CSM. On one hand, in fact, considering the dynamical evolution of a system, correlation may need to relate facts true at different places at different time (properties holding over a place change in time). On the other hand, in domains characterized by the presence of wireless technologies, interesting places, properties holding over them and the relations' extension may change, since new interesting places can be discovered (e.g a mobile object is identified as a place) and known places can move.

References

1. Zambonelli, F., Parunak, H.: Signs of a revolution in computer science and software engineering. In: Proceedings of Engineering Societies in the Agents World III (ESAW2002). Volume 2577., Springer-Verlag (2002) 13–28
2. Bandini, S., Mosca, A., Palmonari, M., Sartori, F.: A conceptual framework for monitoring and control system development. In: Ubiquitous Mobile Information and Collaboration Systems (UMICS'04). Volume 3272., Springer-Verlag (2004) in press
3. Joshi, A., Finin, T., Yelsha, Y.: Me-services: A framework for secure & personailzed discovery, composition and management of services in pervasive environments. LNCS **2512** (2002) 248–259
4. Davis, E.: Representations of commonsense knowledge. Morgan Kaufmann Publishers (1990)
5. Randell, D.A., Cui, Z., Cohn, A.G.: A spatial logic based on regions and connection. In: Proc. 3rd Int. Conf. on Knowledge Representation and Reasoning, San Mateo, CA, Morgan Kaufmann (1992) 165–176
6. Aiello, M., Benthem, J.V.: A modal walk trough space. Journal of Applied Non-Classical Logics **12** (2002) 319–363
7. Kuipers, B.: Modelling spatial knowledge. Cognitive Science **2** (1978) 129–154
8. Leisler, D., Zilbershatz, A.: The traveller: A computational model of spatial network learning. Environment and Behaviour **21** (1989) 435–463
9. Gopal, S., Klatzky, R., Smith, T.: Navigator: A psychologically based model of environmental learning through navigation. Journal of Environmental Psychology **9** (1989) 309–331
10. Bandini, S., Bogni, D., Manzoni, S., Mosca, A.: A ST-modal logic approach to alarm correlation in monitoring and control of italian highways traffic. In: Proceedings of The 18th International Conference on Industrial & Engineering Applications of Artificial Intelligence & Expert Systems. Bari, June 22-25, 2005, LNCS (in press)
11. Harary, F.: Graph Theory. Addison-Wesley, Reading, MA (1972)
12. Blackburn, P., de Rijke, M., Venema, Y.: Modal Logic. Cambridge University Press (2000)
13. Blackburn, P.: Representation, reasoning and realtional structures: a hybrid logic manifesto. Logic Journal of the IGPL **8** (2000) 339–365
14. Bandini, S., Bogni, D., Manzoni, S.: Alarm correlation in traffic monitoring and control systems: a knowledge-based approach. In van Harmelen, F., ed.: Proceedings of the 15th European Conference on Artificial Intelligence, July 21-26 2002, Lyon (F), Amsterdam, IOS Press (2002) 638–642

Contextually Aware Information Delivery in Pervasive Computing Environments

Ian Millard, David De Roure, and Nigel Shadbolt

School of Electronics and Computer Science,
University of Southampton,
Southampton SO17 1BJ, UK
{icm02r, dder, nrs}@ecs.soton.ac.uk

Abstract. This paper outlines work in progress related to the construction of a system which is able to deliver information in a contextually sensitive manner within a pervasive computing environment, through the use of semantic and knowledge technologies. Our approach involves modelling of task and domain as well as location and device. We discuss ideas and steps already taken in the development of prototype components, and outline our future work in this area.

1 Introduction

The pervasive computing vision leads us to believe that future working environments may well feature a wide range of interconnected portable and/or personal devices, in conjunction with static displays populating the surroundings. While this infiltration of technology into everyday life is aimed at improving access to information, communication, and the ease of work, there is a danger that users may become inundated with data or distracted from the task at hand to such an extent that productivity begins to suffer.

The authors propose that by enabling a system to understand and reason about the activities of the occupants of such an environment, then that environment can be significantly more supportive of those working within it. Given the correct knowledge relating to both the general environment and the current situation, a contextually aware system may provide access to resources required for undertaking a given task, and offer interesting or related information, while at the same time removing unwanted or inappropriate data or distractions until such a time that they are more suitable.

In our earlier paper [11], we outlined the notion of a contextually aware environment which aims to present the right information to the right users, at the right time and in the right place. In order to achieve this, the system must clearly have a sufficient understanding of its environment, the people and devices that exist within it, their interests and capabilities, and the tasks and activities that are being undertaken. That is to say, the system must be able to identify where, and under what context each person is engaged with their current task.

T. Strang and C. Linnhoff-Popien (Eds.): LoCA 2005, LNCS 3479, pp. 189–197, 2005.

2   Contextual Modelling

As we have discussed, central to any contextually aware system is the need to represent and collect a wealth of information such as current activities, skills, interests, personal preferences, privacy requirements and the relationships between people. However, the authors believe that attempting to create a generic notion of context would be a hugely complex and difficult task, and would probably be of little use as it is likely that it would not afford the levels of detail required by an implemented system. Instead, we put forward the case that the creation of generic systems or frameworks, which can be specialised for use within a given domain, would be more appropriate in the majority of cases.

2.1   Location Modelling

One area identified as a key component for the representation and understanding of activities within a general working environment is that of location. We must have good knowledge of where people and devices are, though not necessarily in a spatial representation as it is more important to comprehend semantic information regarding locations, such as the type and/or purpose of different areas or spaces, their relative proximity and positioning, facilities offered, or activities usually undertaken in that area. We have created a location ontology [8] which permits these kinds of relationships to be described, and can be used as a basis for combining sensed location data and inferring additional facts about the environment, as discussed in [11].

2.2   Task Modelling

Another important area is representing what people are doing at any particular time. We propose a task-oriented model, as most working days can be conceptualised as a sequence of different tasks, such as a project meeting, document review, teleconference, patient consultation, or student supervision. Each of these tasks may have a variety of important properties, ranging from the date/time/location, to meaningful relationships between people and/or resources that feature in or are required for a particular task.

   In addition, many tasks are repetitive or recurring, such as project meetings, preparing accounts, or performing routine maintenance. Each instance of these types of event is likely to be very similar, exhibiting many of the same properties. In recognition of this, each instance of a task or event may be identified as complying with one or more hierarchical templates, implying that common properties, features and/or resources are required in a similar fashion to implementing an interface in an object oriented programming language. High level templates may provide the basis for a hierarchy of more specialised instances of those tasks, defined specifically for the domain in which they are applicable. They may additionally include data such as the location, persons attending, topics of interest, importance and/or 'interruptability' relating to a particular instance of an event or class of events.

2.3 Domain Modelling

For any given system deployment, it is likely that an additional array of information, concepts and relationships will be essential for representing domain or background knowledge. This data can only properly be represented in detail by domain specific ontologies, utilised within the more general notions of location and task. In our prototype work, which focuses on the academic domain, these properties are ontologically represented by the AKT Reference Ontology [1] and extensions thereof.

2.4 Device Modelling

Finally, to facilitate the display of information in situ to a person's location, a contextually aware system must also be able to comprehend what display resources are available, their capabilities, and location. To achieve this, a device ontology [4] has been created which permits a particular set of features relating to a computer interface to be described. These focus mainly on the input/output and user-interaction capabilities of the device, rather than the typical system hierarchy approach of describing processor, disk, and memory specifications, although the ontology could easily be extended (or others incorporated) if these details were thought necessary at a later date.

Properties are defined to express the visual output resources in some detail, including accepted content types and the dimensions, resolution, type of the display, available screen layouts, and current status. Given these properties, it should be possible for services wishing to present information to locate suitable display resources, taking into account the intended recipient(s) and the format/sensitivity of the data. In addition, provision is made for representing relationships between devices and their users,

Existing work in this area, such as Composite Compatibilities/Preference Profiles [3], has generally looked at techniques for adapting content for a particular display based around the inclusion, exclusion, or modification of various components within a document, given a minimalist representation of available screen resources. While such capabilities may be useful for repurposing content across many different types of device, the content must be generated with such intentions in mind. In addition, representation of display resources is not sufficient to satisfy the requirements discussed above.

3   Acquisition of Contextual Data

In addition to having the capability of modelling the events and activities within a pervasive environment, we must be able to obtain data in near real-time to populate these models for a system to have any chance of success.

Sensing the location of people and devices within a building is non-trivial. Coordinate position technologies range from GPS, offering relatively low accuracy over a very large (outdoor) area, to small scale localised systems offering much higher resolution. Conversely, there are many off-the-shelf systems that can be used to sense the presence of a tag, fob or card in close proximity to a specific receiver unit at a fixed-point location.

The different forms of physical location technology offer presence detection with a variety of different accuracies, reliabilities, and ranges in which those observations can be made. A location system is likely therefore to have to take into account data from a number of different information sources, potentially including both spatial (coordinate) and symbolic point information. Combining these two fundamentally different types of system is non-trivial and is an important research topic. [7]

For applications and services to operate efficiently within a pervasive environment, some form of middleware is required to monitor these potential sources of location information, and present that data through an integrated interface. To achieve this, it is proposed that the location ontology is used in conjunction with a real-time OWL inference engine, permitting location assertions to be instantiated and retracted as and when appropriate sensor data is received. The capabilities offered by the ontology and OWL engine permit data from various sources to be easily combined, and further facts inferred.

Determining contextual information relating to people within an environment is also a non-trivial task. Many indicators offer snippets of information, and knowledge of the environment may permit other inferences to be made.

For example, let us consider the determination of which activity in a schedule the user is currently undertaking. Clearly the time of day is a strong indicator of when we can anticipate an activity to have started. However schedules often slip, and we can achieve a more thorough assertion by additionally considering the location of individuals due to participate in an event, as we know, for example, that a meeting between two people cannot have commenced if they are not co-located (either physically or by some virtual means).

Other indicators may be inferred from monitoring computing devices, such as ascertaining an idle state and/or availability of the user through observing screensavers or instant messaging clients.

4 Prototype System

A prototype system is being developed, with the aim of being capable of delivering messages to users within an academic environment in a contextually-sensitive manner.

As described in [11], we have already constructed some components which will be useful in achieving our goals. We have developed a combined RDF repository and OWL inference engine, 'OwlSrv', which is capable of executing custom inference rules and handling near real-time updates. This plays a central role, providing all of the data storage, query and inferential capabilities required. OwlSrv consists of a Jena-based [10] dynamic repository and inference engine, operating on a number of OWL ontologies together with the set of custom rules, and it presents a similar interface and function as the 3Store [6] system with which we have worked previously.

Our earlier demonstrator application consists of a client program designed to assist academics by presenting their schedule and information relevant to their current task, automatically determined based on custom inference rules running over data held in OwlSrv. It offers a strong indication of when users switch task context.

Our prototype system focuses only on delivering messages to users, as opposed to additionally presenting documents and resources, as issues surrounding real-time event generation are more interesting, and our previous application covers the latter areas reasonably well.

The conceptual architecture of our prototype is shown below in Fig. 1 below.

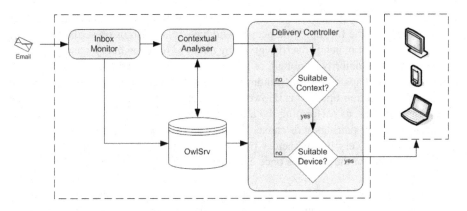

Fig. 1. Conceptual Architecture of the Contextually Aware Message Delivery System

4.1 Message Injection

To enable messages to be easily fired into the system, email has been chosen as the source message format. A daemon process monitors a specified IMAP inbox for the user, asserting RDF representations of each message received in the OwlSrv repository. Upon arrival of a new message, a 'contextual analyser' is invoked to prepare information concerning persons related to the message.

4.2 Analysis of Message

Many pieces of information within the data repository can be used to build up the contextual picture for a given domain. To assist in this task the contextual analyser has been created, which permits domain specific queries to be executed, with each result contributing to a metric or weighting for a number of different concepts relating two entities. For example, this tool can be used to give a value indicating whether two individuals are of a personal or professional acquaintance, or to give a notion of superiority based on line management or academic status. Given appropriate understanding of the domain in which a contextual message delivery system is deployed, suitable queries should be definable to identify the important factors which make up the contextual picture of that particular domain.

In the prototype system, the contextual analyser is realised through a process which, on request, executes a number of RDQL queries read from a configuration file in order to identify specific relationships within domain repositories relating to a sender-recipient pair. The queries are arranged and weighted to calculate a number of metrics relevant to the domain, with each successful query result contributing a decimal increment to one or more of these metrics. The weighting of each query result for a given metric is also specified in the configuration file, hence realising the different levels of relevance a single query may have on different metrics.

The result of any individual RDQL query is a bag of variable/value bindings, to which we have permitted the application of simple set operations. Set union can be used to (somewhat crudely) merge results from multiple repositories, and set subtraction enhances the expressiveness of our metric calculations by chaining together RDQL queries in the form of 'find people who have relationships prescribed by *query x*, but which do not match properties given by *query y*' (or even *query z,* etc).

In contrast the majority, if not all, of these queries could be performed by the OWL reasoner, and perhaps with some minor extensions it could also be able to perform the weighting summation operations. However, this approach is less favourable than the on-demand queries, as performing the analysis arbitrarily in the reasoner for all possible person–person pairings will cause a vast overhead, which is likely to be incomputable or lead to very unsatisfactory performance. Performance reasons also lead us to store the result of the analyses once calculated, for further reuse if required again at a later stage.

Other systems which analyse large data sets in order to extract underlying relationships, such as that of Community of Practise analysis [2], often use a search or 'growing' algorithm which starts from a given node and follows links between nodes propagating weightings as they do so. While these methods can be used to provide good indication of relationships of an individual within in a community, it does not guarantee us the analysis of the relationships between a pair of individuals. This kind of analysis may play a part as a background low-priority task, quietly performing analysis in case it is one day required, although it should be noted that by relaxing the constraint of either the sender or recipient in our query/metric system, a very similar function could be performed.

4.3 Delivery of Message

Having performed the analysis to generate metrics classifying important relationships between the sender and recipient(s) of a message, and other information within the repository, the 'delivery controller' then makes an assessment by applying further domain-specific rules as to whether it is appropriate in the current context to deliver that message. These rules may consider factors such as the type of activity currently being undertaken, properties expressed directly in the message, e.g. 'personal' or 'urgent', or a manual indication from users indicating their availability or willingness to be disturbed, in conjunction with the domain metrics just calculated.

For example, in the academic domain, one factor we may look to is the relative superiority of the sender and recipient, such that when the recipient is engaged in activities deemed to be of a certain importance, only urgent interruptions from their superiors or line managers are permitted. In the case of where the recipient is co-located with other persons, for example in a meeting, we should perhaps consider the relationship between the sender and meeting participants as a whole. Again, a thorough understanding of the environment into which the contextually aware system is to be deployed is essential for the creation of these rule sets.

The location and execution process of these rules is yet to be determined. They are likely to reside as part of the custom rule set loaded into the OWL reasoner, acting when data relating to a message is asserted, and information relating to its sender and recipient(s) is available. The rules needed to model the required delivery behaviour may be numerous and complex, though the data over which they are applicable is likely to be small, and each rule will seldom be fired. However, experimentation may indicate that an external 'on-demand' processing cycle acting in a similar fashion to the contextual analyser may be required on grounds of performance.

If a message is deemed suitable or relevant for delivery given the current contextual representation, the next stage of analysis for the delivery controller is to determine if that message can actually be delivered to the recipient in an appropriate manner. Using the descriptions of devices within our pervasive environment, as described earlier, suitable device(s) may be notified to display the message. This selection must consider which devices are currently being used by, or are in proximity to the recipient, the type and format of the message, its urgency, and whether or not the message is of a personal nature. Similar implementation issues arise here as seen with the determination of whether a message is suitable for delivery.

Messages received but deemed to be inappropriate for delivery in the current context, or those which are unable to be delivered suitably, shall remain queued indefinitely until such a time that their delivery is both appropriate and achievable.

5   Future Work

The delivery controller is the next component to be implemented, and a number of architectural options present themselves. It may be possible to integrate the required functionality tightly into OwlSrv, utilising custom hooks into the inference engine to perform more complex tasks as rules are fired. However, building monolithic applications often leads to poorly adaptable or extendable systems, hence encapsulating the required functionality in an external service or application seems a more logical solution.

Coordination of data and instructions between the display controller, OwlSrv and display devices may be achieved through a number of mechanisms, including publish/subscribe message space models such as Elvin [13] or EQUIP (a platform developed for the Equator project), through the existing HTTP query and update interfaces to OwlSrv, or through Web Service models.

Several of the components of the prototype system use custom data formats, which do not tend towards interoperability. However, ongoing work of particular interest in the field of standardisation includes the SPARQL query language [12] for accessing

RDF repositories, and the Semantic Web Rules Language (SWRL) [14]. The use of both of these standards, once released, could greatly enhance the potential for the reuse of the inherent domain knowledge and behaviours built into a specific contextually aware system.

As a step towards a fully distributed implementation of our system, we adopt a service-oriented view. Descriptions will then be needed not only for the devices but also for publishing and discovering the services. For this we can turn to Semantic Web Services approaches such as the OWL Web Ontology Language for Services (OWL-S) [9], which is an OWL-based Web service ontology designed to facilitate fuller automation of Web service tasks, such as Web service discovery, execution, composition and interoperation.

6   Conclusions

In this paper we have discussed the novel application of semantic technologies to pervasive computing scenarios in order to enable the development of contextually aware environments. Our notion of context pays particular attention to modelling the user's task, and we believe this to be an essential view, to be coupled with a systems perspective on context [5]. Our ideas and work to date on the construction of a prototype application have been presented.

Acknowledgements

This work is supported under the Advanced Knowledge Technologies (AKT) Interdisciplinary Research Collaboration (IRC) in conjunction with the Equator IRC, which are sponsored by the UK Engineering and Physical Sciences Research Council under grant numbers GR/N15764/01 and GR/N15986/01 respectively.

References

1. AKT Reference Ontology – http://www.aktors.org/ontology/
2. Alani, H., Dasmahapatra, S., O'Hara, K. and Shadbolt, N. *Identifying Communities of Practice through Ontology Network Analysis.* IEEE Intelligent Systems, 18 (2). 18-25. (2003)
3. Composite Capabilities/Preference Profiles – http://www.w3.org/Mobile/CCPP/
4. Device Ontology – http://signage.ecs.soton.ac.uk/ontologies/device
5. Fallis, S., Millard, I. and De Roure, D., *Challenges in Context*, W3C "Mobile Web Initiative" Workshop, Barcelona, Spain, November 18-19 (2004).
6. Harris, S. and Gibbins, N., *3store: Efficient Bulk RDF Storage.* in 1st International Workshop on Practical and Scalable Semantic Web Systems, Sanibel Island, Florida, USA (2003).
7. Leonhardt, U. *Supporting Location-Awareness in Open Distributed Systems*, Imperial College of Science, Technology and Medicine, University of London (1998).
8. Location Ontology – http://signage.ecs.soton.ac.uk/ontologies/location

9. Martin, D., Paolucci, M., McIlraith, S., Burstein, M., McDermott, D., McGuinness, D., Parsia, B., Payne, T., Sabou, M., Solanki, M., Srinivasan, N., and Sycara, K., *"Bringing Semantics to Web Services: The OWL-S Approach,"* presented at First International Workshop on Semantic Web Services and Web Process Composition (SWSWPC 2004), San Diego, California, USA. (2004).

10. McBride, B., *Jena: Implementing the RDF Model and Syntax specification.* in 2nd International Semantic Web Workshop (2001).

11. Millard, I., De Roure, D. and Shadbolt, N. *The Use of Ontologies in Contextually Aware Environments.* In Proceedings of First International Workshop on Advanced Context Modelling, Reasoning and Management, pages pp. 42-47, Nottingham, UK. (2004)

12. SPARQL Query Language for RDF – http://www.w3.org/TR/rdf-sparql-query/

13. Sutton, P., Arkins, R. and Segall, B., *Supporting Disconnectedness - Transparent Information Delivery for Mobile and Invisible Computing*, CCGrid 2001 IEEE International Symposium on Cluster Computing and the Grid, 15-18 May, Brisbane, Australia.(2001).

14. SWRL: A Semantic Web Rule Language combining OWL and RuleML
    http://www.daml.org/rules/proposal/

Classifying the Mobility of Users and the Popularity of Access Points

Minkyong Kim and David Kotz

Department of Computer Science,
Dartmouth College
{minkyong, dfk}@cs.dartmouth.edu

Abstract. There is increasing interest in location-aware systems and applications. It is important for any designer of such systems and applications to understand the nature of user and device mobility. Furthermore, an understanding of the effect of user mobility on access points (APs) is also important for designing, deploying, and managing wireless networks. Although various studies of wireless networks have provided insights into different network environments and user groups, it is often hard to apply these findings to other situations, or to derive useful abstract models.

In this paper, we present a general methodology for extracting mobility information from wireless network traces, and for classifying mobile users and APs. We used the Fourier transform to convert time-dependent location information to the frequency domain, then chose the two strongest periods and used them as parameters to a classification system based on Bayesian theory. To classify mobile users, we computed *diameter* (the maximum distance between any two APs visited by a user during a fixed time period) and observed how this quantity changes or repeats over time. We found that user mobility had a strong period of one day, but there was also a large group of users that had either a much smaller or much bigger primary period. Both primary and secondary periods had important roles in determining classes of mobile users. Users with one day as their primary period and a smaller secondary period were most prevalent; we expect that they were mostly students taking regular classes. To classify APs, we counted the number of users visited each AP. The primary period did not play a critical role because it was equal to one day for most of the APs; the secondary period was the determining parameter. APs with one day as their primary period and one week as their secondary period were most prevalent. By plotting the classes of APs on our campus map, we discovered that this periodic behavior of APs seemed to be independent of their geographical locations, but may depend on the relative locations of nearby APs. Ultimately, we hope that our study can help the design of location-aware services by providing a base for user mobility models that reflect the movements of real users.

1 Introduction

Wireless networks have become popular and are getting more attention as a way to provide constant connectivity over a large area in cities and as an inexpensive way to provide connectivity to rural areas. The growing popularity of wireless networks

T. Strang and C. Linnhoff-Popien (Eds.): LoCA 2005, LNCS 3479, pp. 198–210, 2005.

encourages the development of new applications, including those that require quality of service (QoS) guarantees. To provide QoS, it is often useful to predict user mobility. We also need simulators of wireless network environments to test these new applications and these simulators require user mobility models.

As more mature wireless networks become available, several studies of wireless networks have been published, including studies of a campus [4, 5], a corporate environment [2], and a metropolitan area [8]. Although these studies help us to understand characteristics of different network environments and user groups, it is often difficult to apply the findings of these studies to other applications.

In this paper, we introduce a method to characterize real wireless network traces and classify different mobile users based on their mobility. We transform our traces using the Discrete Fourier Transform (DFT) to make them independent of the particular time that traces were gathered. This transform exposes periodicity in traces.

We then use AutoClass [3], an unsupervised classification tool based on Bayesian theory. Classification is important because user mobility differs widely from user to user [2]. Thus, it is difficult to describe diverse user mobility patterns with a single model. Classification breaks down this complex problem into several simpler ones, by dividing users into groups that have common characteristics and thus might be modeled similarly. Moreover, classification is important because a collection of individual cases has little predictive power for new cases.

In the second part of this paper, we focus on the behavior of access points (APs). We apply our method to extract information from real wireless network traces and classify APs. Understanding the behavior of APs is important for many applications, such as traffic engineering for APs and resource provisioning for QoS sensitive applications.

An important benefit of using the Discrete Fourier Transform is that it is easy to compute the inverse DFT to obtain the time series. After clustering instances based on the information extracted from DFT, we can construct a sequence of numbers corresponding to the power spectrum representative of each class. We can then use an inverse DFT to obtain the time series that represents that class. This method is also used by Paxson [6] to synthesize approximate self-similar networks. We leave this modeling process as future work.

2   Methodology

In this section, we describe our traces and the parameters that we have chosen to represent user mobility and behavior of APs. We then describe how we converted our traces from the time domain to the frequency domain using a Fourier Transform and how we classified users and APs using AutoClass.

2.1   Trace Collection

We collected syslog traces of APs from the Dartmouth College campus-wide wireless network. The APs record client events (such as authenticating, deauthenticating, associating, disassociating, and roaming) by sending syslog messages to a central server, where the logs are timestamped with a one-second granularity. Currently, most of the

APs on our campus are Cisco 802.11b APs. Although they are in the process of being replaced by Aruba APs, we focus on Cisco APs because at the time of the study they were still the dominant set of APs and covered most of the campus.

We have been collecting syslog records since 476 Cisco APs were installed in 2001. In this paper, we focus on four weeks of traces collected from October 3 to October 30, 2004. During these four weeks, we saw 7,213 devices (i.e., MAC addresses) visiting 469 APs. In the following discussion, we refer to a MAC address as a user, although a user may own more than one device with a wireless network interface. We expect that most of the devices are laptops, based on the previous study over the traces collected at Dartmouth [4]. We saw roughly 4.5 million syslog events, of which 1.9 million events represent devices associating or reassociating with APs.

2.2   Parameter Selection

To cluster users or APs we must choose an appropriate parameter.

Diameter as Mobility Measure. One limitation of our study is that we do not have the exact geographical location of a user. We only have the information about the location of APs on our campus and the AP which a user is associated with. Thus, we approximate a user's location using the location of the AP with which the user is associated. Because many areas are covered by more than one AP, some clients change association from an AP to another even when they do not physically move. Sometimes a client associates repeatedly with a fixed set of APs, a phenomenon we call the *ping-pong* effect.

The ping-pong effect cannot happen across two APs that are apart farther than a certain distance because APs have limited coverage, but this distance is often hard to pinpoint. The Cisco specification states that the indoor range at 11 Mbps is 39.6 meters and the outdoor range is 244 meters. Obviously, a ping-pong effect is extremely unlikely between two APs that are more than 244 meters apart, but choosing this value as the threshold is too aggressive, filtering out too many user movements. Because different APs are configured differently and located in different environments, it is hard to define a precise distance threshold to decide whether a change between two APs is due to the ping-pong effect or not. Although Henderson [4] defined the limit as 50 meters, in our traces we found that some clients ping-pong between two APs more than 50 meters apart. Thus, we do not use a threshold to filter out ping-pong effects, but choose a parameter that is less sensitive to them.

Our goal is to classify wireless network users based on their mobility patterns. Our traces list events at a particular AP with a particular mobile user. We first gathered the events associated with each user. Although the events are recorded with a one-second granularity, we aggregated them into one value for each hour. We considered several alternatives to represent this value. Because of the ping-pong effect, the total distance traveled (the sum of the distance between APs visited, in sequence) often does not reflect user mobility. A user may appear to travel a long distance if he experiences many ping-pong effects even though he did not move at all. A better measure is the *diameter*, defined as the maximum Euclidean distance (i.e., the straight line distance between two points) between any two APs visited during a fixed time period. Although we still cannot tell whether a diameter is due to real user movements or ping-pong effects when it is short, we can at least be confident that it is caused by real movements when a diameter is longer than a certain distance.

Number of users to Describe APs. For APs, we used the same set of traces, but gathered the events associated with each AP. Then, we counted the number of unique users visiting each AP during each hour. By counting the number of unique users instead of the number of user visits, we remove noise caused by ping-pong effects.

2.3   Filtering Traces

We found it was necessary to filter the traces to select the most meaningful data.

Mobility. In our traces, many users do not move at all, and many others appear in the traces for a short time and disappear. Because we want to find meaningful patterns of user mobility, we need to remove these stationary and transient users. We removed any user who did not move or did not connect to wireless network for a 3-day or longer period. We chose three days based on the assumption that regular mobile users are unlikely to stay at one place for more than three days. They may stay at one place for the weekend; thus using two days as the filtering limit may be too aggressive. We also filtered out the users whose hourly diameter never exceeded 100 meters. Note that we did not filter out the *diameters* shorter than 100 meters; we filtered out the *users*. This filtering reduced the number of users from 7,213 to 246; thus our study focuses on the relatively rare "mobile users."

APs. There are many APs on our campus that are not actively used. To remove these APs, we filtered out the APs that never had more than 50 visitors during a hour. This filtering reduced the number of APs from 469 to 216.

2.4   Discovering Periodic Events

For each user, we create a 672-element vector that represents the user mobility (i.e., diameter) of each hour for four weeks. Our goal is to classify users according to their mobility patterns. Finding similar patterns by comparing these diameter vectors directly is not trivial. For example, the same mobility patterns may appear with more than one user, but they may be shifted in time or scaled. Also, we are not interested in discovering the exact value of diameter at a physical time.

To preserve the diameter but discount for shifts in absolute time, we used the Discrete Fourier Transform (DFT) to transfer our parameters from the time domain to the frequency domain. Since the Fourier Transform is well known, we only briefly describe it here, borrowing a description from *Numerical Recipes in C* [7]. Suppose that we have a function with N sampled values:

$$h_k \equiv h(t_k), \quad t_k \equiv k\Delta, \quad k = 0, 1, 2, ..., N - 1. \tag{1}$$

Δ denotes the sampling period; it is one for our case. The DFT estimates values only at the discrete frequencies:

$$f_n \equiv \frac{n}{N\Delta}, \quad n = -N/2, -(N/2 - 1), ..., N/2 - 1, N/2 \tag{2}$$

where the extreme values of n correspond to the lower and upper limits of the Nyquist critical frequency range. Then, the DFT of N points h_k is defined as following:

$$H_n \equiv \sum_{k=0}^{N-1} h_k e^{2\pi i f_n t_k} = \sum_{k=0}^{N-1} h_k e^{2\pi i k n / N}. \tag{3}$$

Agrawal [1] has shown that a few Fourier coefficients are adequate for classifying Euclidean distances. He chose the first two strong, low frequency signals. Based on this study, we chose the two strongest frequency (or period) signals as our parameters for our classification of user mobility.

2.5   Clustering

To classify user mobility patterns, we use AutoClass [3], a classification system based on Bayesian theory. A key advantage of this system is that it does not need to specify the classes beforehand, allowing *unsupervised* classification. We had, and needed, few preconceptions about how our mobility data should be classified.

AutoClass takes fixed-size, ordered vectors of attribute values as input. Given a set of data X, AutoClass seeks maximum posterior parameter values $\vec{V}$ and the most probable T irrespective of $\vec{V}$, where $\vec{V}$ denotes the set of parameter values instantiating a pdf and T denotes the abstract mathematical form of the pdf. First, for any fixed T specifying the number of classes and their class models, AutoClass searches the space of allowed parameter values for the maximally probably $\vec{V}$. Second, AutoClass performs themodel-level search involving the number of classes J and alternate class models T_j. It first searches over the number of classes with a single pdf T_j common to all classes. It then tries with different T_j from class to class.

3   User Mobility

In this section, we present the result of user mobility patterns converted from the time domain to the frequency domain. We then show the classification of mobile users generated by AutoClass.

3.1   Mobility Patterns

To illustrate our method, we choose one typical user from our traces. The diameters of this user in the time domain and frequency domain are shown in Figure 1 and Figure 2, respectively.

Figure 1 shows the diameter of each hour of one user and the number of unique APs visited by the user during each hour over four weeks. The x-axis shows the dates for Sundays, and the y-axis shows the diameter and the number of APs. This user often had a diameter of 40 meters. By looking into the traces, we found that the user was visiting a fixed set of APs repeatedly: the ping-pong effect. While shorter diameters are due to ping-pong effects, longer ones represent real movements.

Note that the number of unique APs does not necessarily correlate with the diameter: although the number of APs may indicate mobility, it cannot distinguish whether an

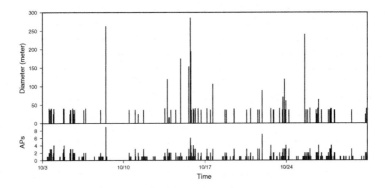

Fig. 1. Hourly diameter and APs visited by one user. This figure shows the user's hourly diameter and the number of unique access points visited by this user during each hour. Labels on the x-axis indicate the dates for Sundays

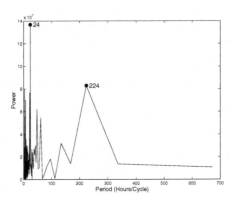

Fig. 2. Diameter in frequency domain. Two dots denote the two most strongest periods. In this example, they are approximately 24 hours and 224 hours

increase in number is due to real movements or due to the ping-pong effects. Even when this user associated with up to four APs, the diameter was still around 40 meters. On the other hand, in the third largest peak where the user moved around 240 meters, he only visited two unique APs. Thus, the number of APs visited by the user is not appropriate to describe mobility.

Figure 2 shows the DFT of this users' vector of diameters. The two most significant periods are 24 and 224. This implies that user mobility patterns are likely to repeat in these periods.

We transformed all of our users' diameter vectors using the DFT and recorded the two strongest periods. Figure 3 shows the cumulative fraction of users with different periods as their first and second strongest periods. For the strongest period, the biggest jump is approximately around 24 hours. The distribution also has smaller jumps at the following hours: 84 (3 days and 12 hours), 168 (one week), 224 (9 days and 8 hours), and 336 (two weeks). Note that by using the DFT, we can observe a jump only at the

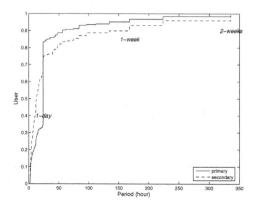

Fig. 3. Significant periods of user mobility. Cumulative distribution of the number of users versus period. From the power spectrum density graphs, we recorded the two most significant periods for each user

Table 1. Classes of user mobility. Mean, standard deviation and coefficient of variation (%) of each parameter are listed. Period is in hours and diameter is in meters

Class	Instances (#)	Instances (%)	Key Parameter	Period 1			Period 2			Diameter		
				Mean	Std	CV	Mean	Std	CV	Mean	Std	CV
0	74	30.1	p2	43.1	67.8	157.3	19.4	7.8	40.2	279.1	94.1	6.0
1	75	30.5	p1	23.7	3.8	16.0	5.8	3.3	56.9	312.6	101.0	5.8
2	42	17.1	p1	23.8	4.6	19.3	41.0	34.7	84.6	184.9	90.2	8.7
3	23	9.2	p1	3.0	0.7	23.3	3.8	1.9	50.0	324.7	113.4	6.3
4	13	5.3	p2	103.9	81.7	78.6	118.2	55.9	47.3	228.7	88.5	6.9
5	15	6.1	p2	23.0	3.4	14.8	264.7	80.4	30.4	318.6	105.7	5.9
6	4	1.7	p2	5.6	0.7	12.5	209.7	28.0	13.4	255.1	118.9	8.4

period that is an integer fractions of the input length (672). We were not surprised to see users with one day, one week, or two weeks as their primary periods. But, it is interesting to observe more users with 3-days-and-12-hours than 4 days. The users with the period of 9-days-and-8-hours instead of 9 or 10 days may be an artifact from using the DFT because neither the period of 9 nor 10 days is an integer fraction of 4 weeks while that of 9-days-and-8-hours is an integer fraction; it is nonetheless interesting to observe users with this period as their primary or secondary periods.

3.2   Classification

We use the two strongest periods as our first two elements of three-element input vectors to AutoClass. In addition to these two periods that we gathered from the DFT, we also measured the maximum hourly diameter (d_{max}) observed over our traces for each user. As described in Section 2.3, we filtered out users whose d_{max} was less than 100 meters; this removed most of the stationary users.

AutoClass classified mobile users into seven classes. Table 1 shows the number of instances that fell into each class and the parameters that most influenced class assign-

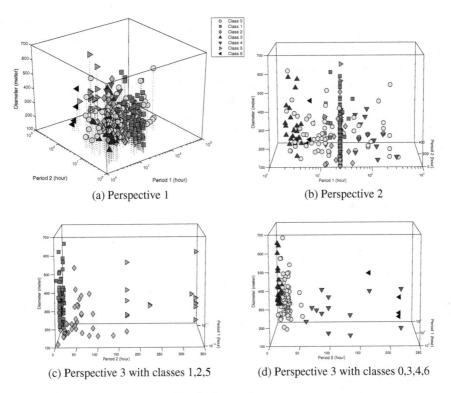

(a) Perspective 1

(b) Perspective 2

(c) Perspective 3 with classes 1,2,5

(d) Perspective 3 with classes 0,3,4,6

Fig. 4. Clustered users

ment. The table also shows the mean and standard deviation of parameters of members within each class. Although parameters with smaller coefficient of variation (CV) often play an important role in class assignment, this is not necessarily true. It is how much the parameter value of an instance is different from those of others that determines whether the parameter plays a critical role in class assignment. Note that our third parameter d_{max} never played the major role in assigning instances to classes.

Figure 4 shows how classes are clustered in three dimensions in different perspectives for a better view. We first notice that there are many users tightly clustered around one day as their primary period. At the same time, there are many others for which one day was not their strong period. The first group of people with a strong one-day period make up classes 1, 2, and 5, while the second group of people make up the rest of classes.

First, we consider the group of users that have a strong one-day period. This group of people are divided into three classes based on the secondary period; classes 1, 2, and 5 correspond to small, mid-range, and big secondary periods as shown in Figure 4(c). Class 1 represents users who have one day as their strongest period and a small secondary period. Students who have regular classes may exhibit this kind of mobility behaviors. The average second period for class 2 is close to two days. The average for class 5 is close to 11 days, but this value is misleading; secondary periods of this class are bimodal around one week and two weeks. Thus, class 5 can be described as a cluster of users with one day and either one or two weeks as their strong periods. Note that mobile users with one day as their strongest period and a small secondary period are most prevalent—Class 1 is the biggest class.

Second, we look into the group of users whose primary period is not one day. These users are divided into four classes. As shown in Figure 4(d), classes 3, 0, 4, and 6 have smallest to biggest secondary periods, respectively. Class 6 consists of users with the very small primary periods and 9-days-and-8-hours as the secondary period. It is interesting to note that most of the users whose primary period is not one day have their secondary period close to one day—Class 0 is the biggest class among these four classes.

4   Access Points

We now use the same method to classify APs based on how busy they are.

4.1   Periodicity

Figure 5 shows the cumulative distribution of the number of APs with primary and secondary periods: 85% of APs had their primary period at one day (24 hours); 25% of APs had their secondary period at 1 week (168 hours). Compared to the mobility traces, more APs have their primary period at one day and the secondary period at one week.

4.2   Classification

As input to AutoClass, we used three parameters: the period at which power is maximum, the period at which the power is second to maximum, and the maximum number of users that an AP serviced during any hour, u_{max}.

Table 2 shows the number of cases that resulted in each class. AutoClass classified the input cases into four classes. The last parameter (u_{max}) did not make any difference

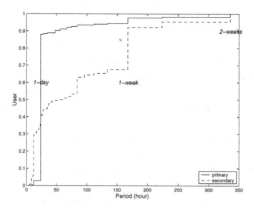

Fig. 5. Significant periods of APs. Cumulative distribution of APs versus period. From the power spectrum density graphs, we recorded the two most significant periods for each AP

in classifying the input cases. Thus, we do not include it in the table. The determining parameter for the first three classes was the secondary period (p2). This is because the primary period (p1) was equal to 24 hours for most of the cases, and therefore did not play a critical role in determining to which class a case belongs.

Table 2. Classes of access points

Class	Instances (#)	Instances (%)	Key Parameter	Period 1			Period 2		
				Mean	Std	CV	Mean	Std	CV
0	99	45.8	p2	23.8	1.7	7.1	158.6	67.9	42.8
1	68	31.5	p2	24.0	0.0	0	11.6	2.3	19.8
2	28	13.0	p2	25.4	10.4	40.9	28.3	6.9	24.4
3	21	9.7	p1	165.1	97.4	59.0	90.0	97.7	108.6

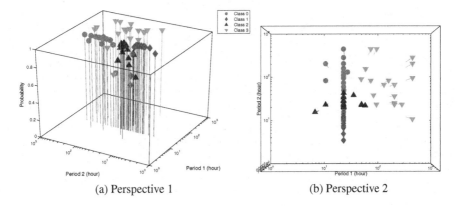

(a) Perspective 1   (b) Perspective 2

Fig. 6. Clustered access points

Figure 6 shows each instance in three dimensions in two different perspectives. Because u_{max} did not play a major role for classification, we do not include it in this graph. Instead, we include the probability of an instance being in a particular class as the third axis. AutoClass computes this probability, for each instance, which indicates the likelihood that an instance is a member of a class. If this probability is one, that instance is a strong member of the class. Not surprisingly, the probability drops for the instances in the regions where different classes meet.

Figure 6 shows that most APs had their primary period at one day. It is also clear that classes 0, 1, and 2 had distinct secondary periods. Note that among these three classes, class 0 had the most instances; this means that APs with one day as their primary period and around one week as their secondary period were the dominant category. Class 3's primary period is much bigger than one day; its secondary period is also big.

Figure 7 shows the geographical location of the APs on our campus. Many of the Cisco APs on our campus have recently been replaced by Aruba APs. Because we focus

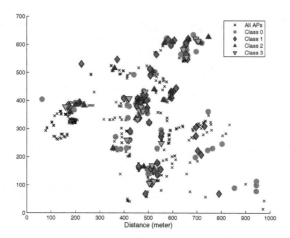

Fig. 7. Map of access points on campus

only on Cisco APs, many APs on the map did not appear in our traces and therefore were not classified. Also, APs who never had more than 50 users per hour are not classified.

There are two things to note in Figure 7. First, APs within a small geographical location, even within the same building, often had different patterns of behavior. Thus, characterizing APs based on their geographical locations or type of building may be erroneous. Second, class 0 and 1 are located all over the campus, but class 2 and 3 are located only where many APs are deployed. We do not have a clear explanation of why this is happening, but it is still interesting to note that deploying too many APs within a limited space sometimes prevents APs from having a strong period of one day.

5   Lessons Learned

In the Fourier Transform, it is important to truncate data so that the input data is a multiple of the period of the signal. This is the reason that we used 4-week traces instead of one-month; we truncated data to be multiple of one week (i.e., 168). For access points, we tried both 4-week traces and one-month traces. With 4-week traces, an AP had one day as the strongest period and one week as the second. When we used one-month traces, we got the same value of one day for the first maximum, but got *one week and 12 hours* for the second maximum instead of exactly one week.

After clustering data, it was important to visualize the result. Visualization helped understanding how classes are divided and how each parameter contributes in distinguishing instances. But, it was not trivial to find the 'right' way to present clustered data. We expect it will even be harder with longer traces and more input parameters for classification.

6   Conclusion and Future Work

In this paper, we present a method to extract information from real wireless network traces and transform the time series to the frequency domain using the Fourier Transform. We then extracted the two most significant periods and clustered instances using a Bayesian classification tool. Our study is unique in using Fourier Transform and Bayesian theory to provide insights into user mobility and behavior of access points.

This paper presents ongoing work, and we plan to pursue several extensions. First, we would like to try our method with longer traces. We expect the trend will be similar to our study presented here although there may be varieties depending on the long-term academic schedules, such as when a term starts and ends. Second, we want to expand our study of APs to include the newly deployed Aruba APs, but we must first update our map data. Third, we plan to build generalized models for user mobility and activities of APs. We believe that our method will help us build models by identifying the most significant characteristics, by clustering users into groups that need different models or different parameters, and by abstracting traces. Finally, after successfully modeling user mobility based on our real traces, we would like to build a simulator for wireless network environments using our mobility model.

Acknowledgments

This project was supported by Cisco Systems, NSF Award EIA-9802068, and Dartmouth's Center for Mobile Computing. We are grateful for the assistance of the staff in Dartmouth's Peter Kiewit Computing Services in collecting the data used for this study. We would like to thank Songkuk Kim for the insightful suggestions throughout the process of developing our method. We also thank Tristan Henderson for commenting on draft versions of this paper.

References

1. Rakesh Agrawal, Christos Faloutsos, and Arun N. Swami. Efficient similarity search in sequence databases. In D. Lomet, editor, *Proceedings of the 4th International Conference of Foundations of Data Organization and Algorithms (FODO)*, pages 69–84, Chicago, Illinois, 1993. Springer Verlag.
2. Magdalena Balazinska and Paul Castro. Characterizing mobility and network usage in a corporate wireless local-area network. In *Proceedings of MobiSys 2003*, pages 303–316, San Francisco, CA, May 2003.
3. Peter Cheeseman and John Stutz. Bayesian classification (AutoClass): Theory and results. In Usama M. Fayyad, Gregory Piatetsky-Shapiro, Padhraic Smyth, and Ramasamy Uthurusamy, editors, *Advances in Knowledge Discovery and Data Mining*, Philadelphia, PA, USA, 1996. AAAI Press/MIT Press.
4. Tristan Henderson, David Kotz, and Ilya Abyzov. The changing usage of a mature campuswide wireless network. In *MobiCom '04: Proceedings of the 10th Annual International Conference on Mobile Computing and Networking*, pages 187–201, Philadelphia, PA, USA, 2004. ACM Press.

5. Ravi Jain, Anuparma Shivaprasad, Dan Lelescu, and Xiaoning He. Towards a model of user mobility and registration patterns. *MC²R*, 8(4):59–62, October 2004. MobiHoc 2004 poster abstract.

6. Vern Paxson. Fast approximation of self similar network traffic. *Technical Report LBL-36750*, 1995.

7. William H. Press, Saul A. Teukolsky, William T. Vetterling, and Brian P. Flannery. *Numerical Recipes in C: The art of scientific computing*. Cambridge University Press, Cambridge, 1992.

8. Diane Tang and Mary Baker. Analysis of a metropolitan-area wireless network. *Wireless Networks*, 8(2-3):107–120, 2002.

Prediction of Indoor Movements
Using Bayesian Networks

Jan Petzold, Andreas Pietzowski, Faruk Bagci, Wolfgang Trumler,
and Theo Ungerer

Institute of Computer Science,
University of Augsburg,
Eichleitnerstr. 30, 86159 Augsburg, Germany
{petzold, bagci, trumler, ungerer}@informatik.uni-augsburg.de

Abstract. This paper investigates the efficiency of in-door next location prediction by comparing several prediction methods. The scenario concerns people in an office building visiting offices in a regular fashion over some period of time. We model the scenario by a dynamic Bayesian network and evaluate accuracy of next room prediction and of duration of stay, training and retraining performance, as well as memory and performance requirements of a Bayesian network predictor. The results are compared with further context predictor approaches - a state predictor and a multi-layer perceptron predictor using exactly the same evaluation set-up and benchmarks. The publicly available Augsburg Indoor Location Tracking Benchmarks are applied as predictor loads. Our results show that the Bayesian network predictor reaches a next location prediction accuracy of up to 90% and a duration prediction accuracy of up to 87% with variations depending on the person and specific predictor set-up. The Bayesian network predictor performs in the same accuracy range as the neural network and the state predictor.

1   Introduction

We investigate to which extend the movement of people working in an office building can be predicted based on room sequences of previous movements. Our hypothesis is that people follow some habits, but interrupt their habits irregularly, and sometimes change their habits. Moreover, moving to another office fundamentally changes habits too.

Our aim is to investigate how far machine learning techniques can dynamically predict room sequences, time of room entry, and duration of stays independent of additional knowledge. Of course the information could be combined with contextual knowledge as e.g. the office time table or personal schedule of a person, however, in this paper we focus on dynamic techniques without contextual knowledge.

Further interesting questions concern the efficiency of training of a predictor, before the first useful predictions can be performed, and of retraining, i.e. how long it takes until the predictor adapts to a habitual change and provides again useful predictions. Predictions are called useful if a prediction is accurate with a certain

T. Strang and C. Linnhoff-Popien (Eds.): LoCA 2005, LNCS 3479, pp. 211–222, 2005.

confidence level (see [14] for confidence estimation of state predictors). Moreover, memory and performance requirements of a predictor are of interest in particular for mobile appliances with limited performance ability and power supply.

The predictions could be used for a number of applications in a smart office environment. We demonstrate two application scenarios:

- In the Smart Doorplate Project [17] a visitor is notified about the probable next location of an absent office owner within a smart office building. The prediction is needed to decide if the visitor should follow the searched person to his current location, go to the predicted next location, or just wait till the office owner comes back.
- A phone call forwarding to the current office location of a person is an often proposed smart office application, but where to forward a phone call in case that a person just left his office and did not yet reach his destination? The phone call could be forwarded to the predicted room and answered as soon as the person reaches his destination.

Our experiments as part of Smart Doorplate Project yielded a collection of movement data of four persons over several months that are publicly available as Augsburg Indoor Location Tracking Benchmarks [12, 13]. We use this benchmark data to evaluate several prediction techniques and compare the efficiency of these techniques with exactly the same evaluation set-up and data. Such a comparison of context prediction techniques has to our knowledge never been done. Moreover, we can estimate how good next location prediction works - at least for the Augsburg Indoor Location Tracking Benchmark data.

Several prediction techniques are proposed in literature — namely Bayesian networks [6], Markov models [2] or Hidden Markov models [16], various neural network approaches [5], and the State predictor methods [15]. The challenge is to transfer these algorithms to work with context information.

For this paper we choose the Bayesian network approach, because Bayesian networks are well suited to model time, and compare the results with the best results from the state predictor method described in [15] and the multi-layer perceptron predictor defined in [18]. The benchmark data allowed next location prediction and duration of stay prediction based on previous room sequences, previous duration of stays, and time and date of room entry. The prediction accuracies of the Bayesian predictor are compared with state and multi-layer perceptron predictor data based on room sequences only.

The next section states related work on context prediction except for our own techniques outlined in section 5.5. Section 3 introduces the application scenarios and the applied benchmarks, and section 4 shows the chosen dynamic Bayesian network model of the application scenario. Section 5 gives the evaluation results. The paper ends with the conclusions.

2   Related Work

The Adaptive House project [10] of the University of Colorado developed a smart house that observes the lifestyle and desires of the inhabitants and learned to anticipate and accommodate their needs. Occupants are tracked by motion detectors and a neural network approach is used to predict the next room the person will enter and the activities he will be engaged. Hidden Markov models and Bayesian inferences are applied by Katsiri [8] to predict people's movement. Patterson et al. [11] presented a method of learning a Bayesian model of a traveller moving through an urban environment based on the current mode of transportation. The learned model was used to predict the outdoor location of the person into the future.

Markov Chains are used by Kaowthumrong et al. [7] for active device selection. Ashbrook and Starner [1] used location context for the creation of a predictive model of user's future movements based on Markov models. They propose to deploy the model in a variety of applications in both single-user and multi-user scenarios. Their prediction of future location is currently time independent, only the next location is predicted. Bhattacharya and Das [3] investigate the mobility problem in a cellular environment. They deploy a Markov model to predict future cells of a user.

An architecture for context prediction was proposed by Mayrhofer [9] combining context recognition and prediction. Active LeZi [4] was proposed as good candidate for context prediction.

All approaches perform location prediction with specific techniques and scenarios. None covers a smart office scenario and none compares several prediction techniques. Moreover, none of the evaluation data is publicly available. Therefore the applied techniques are hard to compare.

3   Application Scenarios and Benchmarks

The Smart Doorplate application [17] acts as testbed for the implementation and evaluation of the proposed Bayesian predictor. A Smart Doorplate shows information about the office owner like a traditional static doorplate. The Smart Doorplate, however, additionally shows dynamic information like the presence or absence of the office owners. If an office owner is absent from his office the doorplate directs a visitor to the current location of the absent office owner. Furthermore it predicts the next location of the absent office owner and the entering time of this location. This additional information can help the visitor to decide whether he follows the office owner or waits for him.

The predicted location information can also be used for switching over the phone to the next location of a clerk. That means when the clerk leaves his office, the system predicts the next location of the clerk and switches over the phone call to this location.

To evaluate prediction techniques in the two described scenarios we needed movement sequences of various clerks in an office building. Therefore we recorded the movements of four test persons within our institute building and packaged the data in the *Augsburg Indoor Location Tracking Benchmarks* [12].

We collected the data in two steps, first we performed measurements during the summer term and second during the fall term 2003. In the summer we

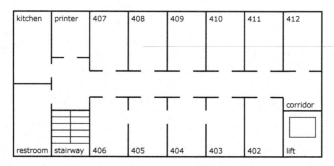

Fig. 1. Floor plan of the institute building

recorded the movements of four test persons through our institute over two weeks. The floor plan of the institute building is shown in figure 1. The summer data range from 101 to 448 location changes. Because this data was too short we started a further measurement with the same four test persons in the fall. Here we accumulated date over five weeks. The fall data range from 432 to 982 location changes. These benchmarks will be used for evaluating the Bayesian predictor in the described scenarios.

4   Bayesian Network Modeling and Implementation

A Bayesian predictor uses the conditional likelihood of actions represented by variables applying the Bayesian formula on a Bayesian network model. A Bayesian network is a directed acyclic graph of nodes representing random variables (X_i) and arcs representing dependencies between the variables. In case there is an arc from X_1 to X_2 then node X_1 is a parent of node X_2. Each variable takes values from a finite set and specific probabilities for those values. To calculate the joint probability distribution the following chain rule is used:

$$P(X_1, ..., X_n) = \prod_{i=1}^{n} P(X_i|Parents(X_i))$$

In order to predict a future context of a person, the usage of a dynamic Bayesian network was chosen. This network consists of different time slices which all contain an identical Bayesian network. The nodes between time slices are connected with arrows to represent dependencies among these time slices.

In our case we predict future locations of a person and additionally the duration of stay and the time when the person is probably changing to a new location. Our application scenario is modelled by the dynamic Bayesian network shown in figure 2. This network exemplarily shows three time slices at time $t-1$, t and $t+1$ but actually there is no limit of time slices in the past or in the future. Since a Bayesian network is assigned to each person in the system, the person doesn't appear as a variable in the network.

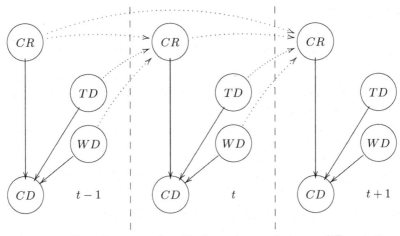

Fig. 2. Dynamic Bayesian network with dependencies between different time slices (dotted arrows)

In each time slice the current duration (CD) basically depends on the current room (CR) of the person. The current room essentially depends on the sequence of the last n rooms visited by the person. Thus the CR's from previous time slices are connected to the CR in the current time slice. The time of day (TD) and the weekday (WD) are also important for the prediction of a person's specific behavior. For this reason CD is closely linked to the current TD and the current WD. The current room also depends on those two influences but from the previous time slice.

5   Evaluation

5.1   Location Prediction

Our first set of evaluations concerns the prediction accuracy and the quantity of performed predictions for next room prediction including and excluding predictions from own office. To predict the next location when somebody leaves his own office is particularly hard, but important for the scenario of phone propagation. Otherwise, for the Smart Doorplate scenario of a visitor standing in front of an office with an absent office owner, it is only interesting if (and when) an office owner comes back or proceeds to another location - and not where a present office owner will go when he leaves his office.

There are two additional factors that influence the evaluation results. First, at the very start the prediction table is totally empty and a useful prediction cannot be done. When a move from a certain office to another has never been done, we cannot predict it. Thus we exclude from the prediction results given below all cold start predictions where we find an empty entry in the prediction table. As consequence the total number of predictions – called quantity – decreases, but the prediction accuracy increases since the prediction accuracy is defined as ratio of the number of correct predictions to the total number of predictions.

Second, the prediction will still be unconfident, when only very little data about previous moves of a person is known to the predictor. In many application cases it is better to perform no prediction instead of a wrong prediction. A predictor trained with data of several weeks will be better than an untrained predictor. We will start our evaluations with untrained predictors using the fall benchmark data only and show how much a predictor trained with the summer benchmarks data improves for the predictions of the fall data.

Evaluation set-up 1: Next location prediction without training, predictions from own office included. Table 1 shows the results of next location prediction of the four test persons excluding empty predictor entries and without training. The prediction accuracies (and quantities) with a history of 1 to 5 rooms are shown separately.

Table 1. Prediction accuracy of location prediction in percent (quantity of predictions in parentheses in percent) with predictions from own office included

	1 room	2 rooms	3 rooms	4 rooms	5 rooms
Person A	55.39 (95.35)	55.61 (87.91)	48.65 (71.63)	51.79 (54.42)	35.38 (33.49)
Person B	56.93 (97.15)	53.65 (90.24)	50.42 (76.22)	53.94 (58.33)	48.10 (38.62)
Person C	43.69 (97.59)	44.73 (87.72)	38.71 (67.98)	48.28 (47.15)	35.48 (28.07)
Person D	50.14 (97.24)	50.61 (87.56)	50.79 (70.51)	51.74 (53.23)	45.28 (36.64)

The results show for most persons an improvement of accuracy if the room sequence is increased from one to two rooms. For person D the accuracy increases up to 4 rooms, whereas the correct predictions of persons A and C decreases after 2 rooms, and for person B already decreases after 1 room. The explanation is simply that person B repeats less often a long room sequence – perhaps a specific habit of person B.

For predictions based on longer room sequences the quantity of predictions decreases because the cold start predictions, which are excluded from the table, concern also the predictions that could be done with less rooms. In particular the number of predictions for person C decreases extremely for 5 room sequences. A low quantity means that most of the predictions are empty because the system is in the learning process. A high quantity shows that the system already knows many patterns and delivers a prediction result. Because of this a larger data base could improve the quantity.

Evaluation set-up 2: Next location prediction without training, predictions from own office excluded. The second evaluation set-up ignores predictions if a person is in his own office. The results in table 2 show that the prediction accuracy improves significantly in this case.

The predictor of person A reaches an accuracy of about 90%, but the accuracy decreases for a longer previous room sequence. In contrast, the accuracy increases with a longer room sequence for person D, but the accuracy decreases if the

Table 2. Prediction accuracy of location prediction in percent (quantity of predictions in parentheses in percent) without predictions from own office

	1 room	2 rooms	3 rooms	4 rooms	5 rooms
Person A	90.19 (91.15)	89.47 (83.19)	90.47 (59.29)	87.27 (48.67)	88.24 (19.47)
Person B	77.65 (94.83)	80.17 (85.98)	78.18 (64.94)	81.25 (50.18)	78.57 (24.35)
Person C	66.67 (95.88)	67.45 (82.02)	61.19 (59.18)	72.83 (42.32)	70.97 (18.73)
Person D	75.00 (95.43)	75.84 (80.91)	76.07 (61.83)	76.82 (44.81)	76.74 (27.80)

room sequence is longer than four rooms. These results show that the sequence of previous rooms influences the accuracy of predictions, however, the results exhibit no common rule. However, some persons act in certain patterns and a better prediction is made if the patterns are included. Table 2 shows also that the quantity decreases with a longer room sequence. In these cases the predictors deliver good results but very rare.

Evaluation set-up 3: Trained versus untrained next location prediction for two room sequences, predictions from own office excluded. Up to now we considered the predictors without any previous knowledge about the persons. We want to analyze the behavior of the predictors if the predictors will be trained with the summer data of the benchmarks. After the training the measurements were performed with the fall data to compare the trained with the untrained predictors, because the results of the untrained predictors were reached by using only the fall data. The results in table 3 show that the training improves the prediction accuracy. Also the quantity (in parentheses) is higher with training. With these results we can see that a larger data base has a positive effect because the system retains knowledge about certain behavior patterns.

Table 3. Prediction accuracy of location prediction in percent (quantity of predictions in parentheses in percent) with and without training based on two previous rooms, and without predictions from own office

	Without training	With training
Person A	89.47 (83.19)	89.65 (90.27)
Person B	80.17 (85.98)	83.72 (92.99)
Person C	67.45 (82.02)	72.14 (88.76)
Person D	75.84 (80.91)	76.49 (87.55)

5.2   Duration Prediction

Our second prediction target is the duration of a person's stay in the current room (his own office or a foreign room). Again we consider prediction accuracy and quantity, additional influence factors as well as trained versus untrained.

To improve the prediction accuracy we tested the influence of the time of the day and the weekday. The week consists of seven days, so we used the discrete values "Monday" to "Sunday". For sectioning the day we must find good discrete values. When we are sectioning the day in too many intervals, for some intervals there will not be a sufficient amount of data for a good prediction. So we classified the day in four discrete time intervals: morning (7:00 a.m. to 11:00 a.m.), noon (11:00 a.m. to 2:00 p.m.), afternoon (2:00 p.m. to 6:00 p.m.), and night (6:00 p.m. to 7:00 a.m.). Likewise, the duration is partitioned in nine intervals of 0-5, 5-10, 10-15, 15-20, 20-30, 30-45, 45-60, 60-120, and more than 120 minutes. We also tested the influence of the time of the day and weekday on next location prediction, however, without reaching an improvement of the prediction accuracy. Therefore we omit results of these measurements.

Evaluation set-up 4: Duration prediction based on current room and all combinations of time of day and weekday including own office. In our modeled network (see figure 2) the duration is independent of the previous room sequence of a person. Therefore we investigated the influence of the time of day and the weekday based on the current room only (room sequence of one).

Table 4. Prediction accuracy of duration prediction in percent (quantity of predictions in parentheses in percent) including own office

	None	Time of day	Weekday	Both
Person A	53.67 (95.35)	54.74 (88.37)	46.33 (82.79)	44.37 (66.05)
Person B	60.21 (97.15)	59.83 (93.09)	55.84 (89.63)	54.55 (78.25)
Person C	73.93 (97.59)	73.52 (92.76)	72.41 (89.91)	70.09 (76.97)
Person D	53.55 (97.24)	52.42 (90.55)	48.07 (88.71)	45.89 (72.81)

Table 4 shows the results of the duration prediction where the unknown predictions were effectively ignored. The results don't show any improvement if the time of day and the weekday is considered. In almost all cases the prediction accuracy and the quantity decrease by considering the time parameters. The reason for this behavior can be the small data base with small time structure. The quantity decreases since the number of prediction decreases with a higher number of influence parameters like in the case of next location prediction.

Evaluation set-up 5: Duration prediction based on current room and all combinations of time of day and weekday excluding own office. This evaluation set-up ignores predictions of the duration if the person is in his own office. Also in this scenario we investigated the influence of the time of day and the weekday. Table 5 shows that the prediction accuracy is significantly improved opposite the result including the own office (see table 4). Obviously, it is particularly hard to predict the duration of a person's stay in his own office.

In duration prediction the influence of the time of day improves the prediction accuracy for person A, B, and C. The consideration of the weekday doesn't improve the accuracy for all persons. The combination of the time of day and the weekday delivers again better results for persons A, B, and C as predictions without any time parameter. The reason for the impairment of the consideration

Table 5. Prediction accuracy of duration prediction in percent (quantity of predictions in parentheses in percent) excluding own office

	None	Time of day	Weekday	Both
Person A	77.67 (91.15)	84.62 (80.53)	71.25 (70.80)	80.00 (48.67)
Person B	86.38 (94.83)	87.87 (88.19)	83.48 (82.66)	88.20 (65.68)
Person C	83.59 (95.88)	83.97 (88.76)	83.56 (84.27)	83.62 (66.29)
Person D	68.70 (95.44)	68.63 (84.65)	61.42 (81.74)	63.64 (59.34)

of the weekday could be the small data base which contains no weekly structure. If we include more parameters the quantity decreases as expected. So you must find a good balance between the accuracy and the quantity, e.g. it is not meaningful for person B to increase the accuracy from 86% to 88% when the quantity decreases from 94% to 65%.

Evaluation set-up 6: Duration prediction with and without training based on current room and all combinations of time of day and weekday excluding own office. To investigate the training behavior of duration prediction we use the same set up like in the case of next location prediction. We used the summer data for training. Then we compared the results which were reached with the fall data with the previous results which were reached with the same data sets. Except for person B we can see in table 6 an improvement of the prediction accuracy. The quantity is better with training as without training for all persons.

Table 6. Prediction accuracy of duration prediction in percent (quantity of predictions in parentheses in percent) excluding own office with and without training, using parameter time of day

	Without training	With Training
Person A	84.62 (80.53)	85.58 (92.04)
Person B	87.87 (88.19)	86.54 (95.94)
Person C	83.97 (88.76)	86.77 (96.25)
Person D	68.63 (84.65)	69.78 (93.36)

5.3 Retraining

A problem of prediction techniques which are based on previous behavior patterns is the learning of behavior changes. Most of the techniques need a long retraining process. Therefore we simulated a behavior change similar to the move of a person to a new office by using 60 data sets of person A followed by 140 data sets of person C. We compared the next room prediction of this set-up with the room prediction results of person C on its own. A well-trained Bayesian predictor needs a long retraining to adapt to the habit change. Therefore we investigated the influence of the number of previous location changes called internal storage which will be used to calculate the conditional likelihood. By restriction the internal storage to 100 or 200 data sets only the retraining can be accelerated. By using the internal storage 100 retraining was done after 90 room changes. In the case of internal storage 200 the retraining ended after 130 room changes. There is no universal rule for determining the optimal size of the internal storage. It depends on the application in which the prediction system is used.

5.4   Storage and Computing Costs

Every person has its own predictor. A predictor must store a sequence of the last r room changes where r is the size of the internal storage. For every room change the room, the time of day, the weekday, and the duration must be stored. In our evaluation set-up there are 15 different rooms which can be stored with 4 bits. For the times of day we need 2 bits, and 3 bits for the weekday. For the duration we used nine discrete values which need 4 bits. Thus the storage costs C of the sequence of the room changes of a person are the following:

$$
\begin{aligned}
C &= r \cdot (room + time\_of\_day + weekday + duration) \\
&= r \cdot (4\,bit + 2\,bit + 3\,bit + 4\,bit) \\
&= r \cdot 13\,bit
\end{aligned}
$$

We realized the Bayesian predictor in Java and we tested the predictor on two different systems, a PC with a clock speed of 2.4 GHz and a memory of 1 GB, and a PDA with a clock speed of 400 MHz and memory of 64 MB. The query speed depends on the size of the internal storage for the sequences of room changes. Therefore we used a large internal storage of 2000. The evaluated predictor used a room sequence of five rooms and no time parameters. In the simulation the predictor handled five times the fall data of person B. On both systems we executed this test three times. The results are shown in table 7.

Table 7. Computing time on PC and PDA

	PC	PDA
processor	Intel Pentium 4	Intel PXA250
clock speed	2.4 GHz	400 MHz
memory	1 GB	64 MB
average computing time	5.44 s	1065.41 s
number of predictions	1355	1355
average computing time per prediction	4.01 ms	786.28 ms

5.5   Comparison with Other Techniques

In previous works we investigated other techniques to predict the next location of a clerk. Specially we developed a new prediction technique which is called state predictor method [15]. This method was motivated by branch prediction techniques of current microprocessors and is similar to the well-known Markov predictor. Furthermore we implemented a neural network to predict the next room. The neural network was a multi-layer perceptron with back-propagation learning [18].

Table 8. Prediction accuracy in next location prediction of Bayesian network, Neural network, and State predictor (in percent)

	Bayesian network	Neuronal network	State predictor
Person A	85.58	87.39	88.39
Person B	86.54	75.66	80.35
Person C	86.77	68.68	75.17
Person D	69.78	74.06	76.42

To compare the three different techniques we evaluate them in the same scenario and with the same set-up. We used the trained predictor of set-up 3 (2 room sequence, own office ignored) as basis for all compared techniques. Table 8 shows the prediction accuracy of the different prediction techniques. The Bayesian network delivers the best results for persons B and C and the state predictor performs best for persons A and D.

6   Conclusion

This paper investigated the efficiency of in-door next location prediction and compared several prediction methods. We modelled the scenario by a dynamic Bayesian network and evaluated accuracy of next room prediction and of duration of stay prediction, training and retraining performance, as well as memory and performance requirements. The results were compared with the state predictor and multi-layer perceptron predictor methods. Our results showed that the Bayesian network predictor reaches a next location prediction accuracy of up to 90% and a duration prediction accuracy of up to 87% with variations depending on the person and specific predictor set-up. The Bayesian network predictor performs in the same accuracy range as the neural network and the state predictor.

References

1. D. Ashbrook and T. Starner. Using GPS to learn significant locations and predict movement across multiple users. *Personal and Ubiquitous Computing*, 7(5):275–286, 2003.

2. E. Behrends. *Introduction to Marcov Chains*. Vieweg, 1999.
3. A. Bhattacharya and S. K. Das. LeZi-Update: An Information-Theoretic Framework for Personal Mobility Tracking in PCS Networks. *Wireless Networks*, 8:121–135, 2002.
4. K. Gopalratnam and D. J. Cook. Active LeZi: An Incremental Parsing Algorithm for Sequential Prediction. In *Sixteenth International Florida Artificial Intelligence Research Society Conference*, pages 38–42, St. Augustine, Florida, USA, May 2003.
5. K. Gurney. *An Introduction to Neural Networks*. Routledge, 2002.
6. F. V. Jensen. *An Introduction to Bayesian Networks*. UCL Press, 1996.
7. K. Kaowthumrong, J. Lebsack, and R. Han. Automated Selection of the Active Device in Interactive Multi-Device Smart Spaces. In *Workshop at UbiComp'02: Supporting Spontaneous Interaction in Ubiquitous Computing Settings*, Göteborg, Sweden, 2002.
8. E. Katsiri. Principles of Context Inference. In *Adjunct Proceedings UbiComp'02*, pages 33–34, Göteborg, Sweden, 2002.
9. R. Mayrhofer. An Architecture for Context Prediction. In *Advances in Pervasive Computing*, number 3-85403-176-9. Austrian Computer Society (OCG), April 2004.
10. M. C. Mozer. The Neural Network House: An Environment that Adapts to its Inhabitants. In *AAAI Spring Symposium on Intelligent Environments*, pages 110–114, Menlo Park, CA, USA, 1998.
11. D. J. Patterson, L. Liao, D. Fox, and H. Kautz. Inferring High-Level Behavior from Low-Level Sensors. In *5th International Conference on Ubiquitous Computing*, pages 73–89, Seattle, WA, USA, 2003.
12. J. Petzold. Augsburg Indoor Location Tracking Benchmarks. Technical Report 2004-9, Institute of Computer Science, University of Augsburg, Germany, February 2004. http://www.informatik.uni-augsburg.de/skripts/techreports/.
13. J. Petzold. Augsburg Indoor Location Tracking Benchmarks. Context Database, Institute of Pervasive Computing, University of Linz, Austria. http://www.soft.uni-linz.ac.at/Research/Context_Database/index.php, January 2005.
14. J. Petzold, F. Bagci, W. Trumler, and T. Ungerer. Confidence Estimation of the State Predictor Method. In *2nd European Symposium on Ambient Intelligence*, pages 375–386, Eindhoven, The Netherlands, November 2004.
15. J. Petzold, F. Bagci, W. Trumler, T. Ungerer, and L. Vintan. Global State Context Prediction Techniques Applied to a Smart Office Building. In *The Communication Networks and Distributed Systems Modeling and Simulation Conference*, San Diego, CA, USA, January 2004.
16. L. R. Rabiner. A Tutorial on Hidden Markov Models and Selected Applications in Speech Recognition. *IEEE*, 77(2), February 1989.
17. W. Trumler, F. Bagci, J. Petzold, and T. Ungerer. Smart Doorplate. In *First International Conference on Appliance Design (1AD)*, Bristol, GB, May 2003. Reprinted in Pers Ubiquit Comput (2003) 7: 221-226.
18. L. Vintan, A. Gellert, J. Petzold, and T. Ungerer. Person Movement Prediction Using Neural Networks. In *First Workshop on Modeling and Retrieval of Context*, Ulm, Germany, September 2004.

Geo Referenced Dynamic Bayesian Networks for User Positioning on Mobile Systems

Boris Brandherm and Tim Schwartz

Department of Computer Science, Saarland University,
Stuhlsatzenhausweg, Bau 36.1, D-66123 Saarbrücken, Germany
{brandherm, schwartz}@cs.uni-sb.de

Abstract. The knowledge of the position of a user is valuable for a broad range of applications in the field of pervasive computing. Different techniques have been developed to cope with the problem of uncertainty, noisy sensors, and sensor fusion.

In this paper we present a method, which is efficient in time- and space-complexity, and that provides a high scalability for in- and outdoor-positioning. The so-called geo referenced dynamic Bayesian networks enable the calculation of a user's position on his own small hand-held device (e.g., Pocket PC) without a connection to an external server. Thus, privacy issues are considered and completely in the hand of the user.

1   Introduction

In order to compute the position of a user, some kind of sensory data has to be evaluated. Car navigation or out-door localization systems typically use GPS (Global Positioning System) to determine the coordinates of a user. Unfortunately, GPS does not work properly in in-door environments because of the fundamental physics of GPS satellite signals, so other technologies have to be used. Examples are infrared beacons, radio frequency emitters, laser range scanners or video cameras. These senders (and their respective sensors) differ in various ways, like sending-characteristics, sending-range, accuracy, price, form factor, and power consumption. Of course there is the usual trade-off between the obviously technical important factors (accuracy, power consumption) versus the limiting factors (price, form factor, sending characteristics), so that the decision, which sender to use, depends heavily on the environment and the planned application as well as on financial conditions. It may even be the case that an already existing installation of positioning senders and/or sensors should be improved by installing additional senders with better accuracy. A location system should therefore not only be able to process different kinds of sensory data but should also be able to fuse the data together in order to calculate the best possible position information.

Another problem is the noisy nature of sensors; therefore, a probabilistic approach for calculating user positions is highly preferable. Commonly Bayesian

T. Strang and C. Linnhoff-Popien (Eds.): LoCA 2005, LNCS 3479, pp. 223–234, 2005.

filters [1] are used to estimate the true location of the user out of the received sensor signals. Several techniques are known to implement Bayes filters, which differ from each other in complexity and certain restrictions (see [2] for a detailed description of different Bayes filter techniques). Particle filters are such an implementation and they are widely used in localization systems. Although overcoming most restrictions of other filters, they still have to be trained or calibrated for their environment.

We were trying to find a new way how to fuse sensory data and how to cope with the problem of inaccurate data. This new approach should be low in calculation and memory costs, so it can be implemented and executed on a mobile system (e.g. Hewlett-Packard iPAQ). We came up with the idea of *geo referenced dynamic Bayesian networks* and we want to share and discuss this idea with the research community.

The paper is organized as follows: Section 2 describes some technical issues and design decisions of our localization system. Section 3 explains the idea of geo referenced Bayesian networks, gives a detailed description of the algorithm and illustrates it with an example. Our working implementation of the system is introduced as a proof of concept in Sect. 4. Section 5 resumes our research and gives an outlook to our future work.

2 Technical Issues

Although our localization system is not the main issue of this paper we will first explain a few technical details of it because we think that this may be helpful for understanding the concept of geo referenced dynamic Bayesian networks.

2.1 Exocentric Localization

In localization systems like the Active Badge, the Active Bat or Ubisense (see [3, 4, 5]) the users wear some kind of sender (infrared, ultrasound, ultrawideband radio) and the respective sensors are installed in the environment. We call these systems exocentric localization systems, because the sender actively reveals the user's position to its surroundings (like calling out "I'm here, I'm here"). A user of this system does not know what the collected localization data is used for or who has access to this data. Of course, functions can be implemented that give the user a choice on who will be allowed to access his position data, but this means that the user has to trust the system. If he does not, his only alternative is not to wear the sender (or to switch it off). By doing this, the user will not only prohibit others to gain information about his current position, but he will also be unable to use the system for his own needs (e.g. for navigational purposes).

2.2 Egocentric Localization

An egocentric localization calculates the user's position on a mobile, personal device of the user. This can be accomplished by placing the *senders* in the environment (now, the senders call out "You're near me, you're near me") and by equipping the user's mobile device with the respective sensors. With this technique the position calculation and the location information is literally in the hands of the user and he can decide if he wants to share this information through a WiFi-connection with other people or applications. If he does not trust the systems privacy control, he can switch off his WiFi-Connection and will still be able to determine his own position for navigational purposes.

We favor this approach and thus we designed our example system in that way, using infrared beacons and active RFID tags as senders (that are installed in the environment and *not* worn by the users).

2.3   Infrared Beacons

We use infrared beacons that are manufactured by eyeled GmbH[1]. These beacons are powered by batteries and send out a 16-bit wide identification code that can individually be adjusted for each beacon. The emitted infrared beam has a range of about 6 meters and has, due to the physical attributes of light, a conical sending characteristic. The price for such a beacon is about 80 Euro. The required infrared sensor is often already integrated in a PDA for data exchanging purposes.

2.4   Active RFID Tags

Radio Frequency IDentification (RFID) tags are available as passive and active parts. In both forms, the tags store some identification code that can be read out with a special RFID reader that sends out a radio signal. The passive tags get their power out of the reader's radio signal and therefore have a very low range (usually up to 15 cm). Active RFID tags have their own power supply through a battery. We use active RFID tag from Identec Solutions AG[2], which have a range of up to 10 meters. Due to the physical attributes of radio waves, the sending characteristic is radial. One active tag costs about 20 Euro. The reading devices for active RFID tags come in various form factors. In conjunction with the PDA, we use a PCMCIA reader card, which costs about 1500 Euro. (The high costs mainly arise from the fact that these readers are manufactured in very low quantities.)

In [6] the authors describe a localization system that uses active RFID tags. Their system fits more into the exocentric localization approach (see 2.1), because they install the readers in the environment and equip the user with a tag. The authors also describe an experiment, where they find out that receiving active RFID tags highly depends on different factors, like static obstructions and human movement. Therefore, they also install reference tags in the environment to improve their system's accuracy.

A system that fits in the category of egocentric localization is described in [7]. Their approach is similar to ours, because they install the tags in the environment

---

[1] http://www.eyeled.de

[2] http://www.identecsolutions.com

(on the floor, we place our tags at the ceiling) and let the user wear one or more readers. They derive the user's position at time t by calculating the sum of the known positions of the received RFID tags and the previous known location of the user, followed by a division through the number n of received tags plus one (for the previous location):

$$\text{UserPos}(t) = \frac{1}{n+1} \left(\text{UserPos}(t-1) + \sum_{i=1}^{n} \text{Coord}\left(\text{receivedTag}[i]\right) \right) \quad (1)$$

(The equation is simplified and adapted to our notation, for the original version see [7].) Our experience with active RFID tags shows that the reader often detects tags that are far away from the user. Therefore, we try to cancel out these false readings with the use of dynamic Bayesian networks.

3   Geo Referenced Dynamic Bayesian Networks

In the following, we describe the idea of geo referenced dynamic Bayesian networks and how they can be used for sensor fusion and to cancel out false readings.

3.1   Idea

Bayesian Networks (BNs) are a computational framework for the representation and the inference of uncertain knowledge. A BN can be represented graphically as a directed acyclic graph (see Fig. 1). The nodes of the graph represent probability variables. The edges joining the nodes represent the dependencies among them. For each node, a conditional probability table (CPT) quantifies these dependencies.

Dynamic Bayesian networks (DBNs) are an extension of Bayesian networks. With a DBN, it is possible to model dynamic processes: Each time the DBN receives new evidence a new time slice is added to the existing DBN. In principle, DBNs can be evaluated with the same inference procedures as normal BNs; but their dynamic nature places heavy demands on computation time and memory. Therefore, it is necessary to apply roll-up procedures that cut off old time slices without eliminating their influence on the newer time slices.

Our idea is to let such a DBN represent the characteristics of the used senders (in our case IR beacons and RFID tags). Note that we do not use the DBN to represent all the senders that are actually installed in the environment. We use one small DBN that prototypically describes the reliability of the sender types (assuming that all senders of a certain type have the same reliability). This prototypical DBN gets instantiated several times during the runtime of the system and each instantiation gets assigned to a geo coordinate *GeoPos*.

Figure 1 shows the network that we use in our example application. The top right node (labeled UserPos=GeoPos?) represents the probability that the user is standing at the assigned geo coordinate *GeoPos*. The node to the left of it (UserPos=GeoPos?_1) is the probability that was calculated in the previous time

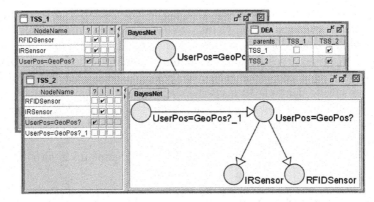

Fig. 1. Screenshot of JavaDBN with directed graph (DEA) and time slice schemata (TSS_1 and TSS_2)

slice. The two bottom nodes (IRSensor and RFIDSensor) represent the probability that an IR beacon and/or a RFID tag installed at *GeoPos* can be detected under the condition that the user is standing at *GeoPos*.

Receiving an infrared signal gives very high evidence that the user is standing near the respective beacon (the infrared sensory data is nearly noise free). Receiving a RFID tag gives smaller evidence that the user is standing near the tag (the reader data is very noisy). These characteristics are coded in the CPTs of the IRSensor- and RFIDSensor-nodes. The actual geoDBN that we use in our example application is described in more detail in Section 3.3.

These networks with their assigned coordinates are the geo referenced dynamic Bayesian networks (geoDBNs). Each geoDBN represents the believe that the user is standing at the associated coordinate.

The active RFID tags have a small internal memory that can be freely used to read and write data. We use this memory to store the coordinate of the tag. IR beacons are always combined with one RFID tag that provides the coordinates of the nearby beacon and the tag itself. When a tag or beacon is sensed by the PDA, geoDBNs are instantiated and associated with the induced coordinates. These induced coordinates depend on the stored coordinate and on the sender type.

The calculation of the user position is somewhat similar to (1) but we weight the coordinates with the calculated probabilities of the existing geoDBNs:

$$\text{UserPos}(t) = \sum_{i=1}^{n} \alpha \, w(\text{GeoDBN}[i]) \, \text{GeoPos}(\text{GeoDBN}[i]). \qquad (2)$$

Here n is the number of existing geoDBNs at time t ($n \geq \#ReceivedSenders_t$), GeoPos(GeoDBN[$i$]) is the coordinate and w(GeoDBN[i]) the weight of the ith geoDBN. α is a normalization factor that ensures that the sum of all weights multiplied with α is one.

To reduce calculation cost and memory usage the number of instantiated geoDBNs must be as low as possible. To achieve this goal, geoDBNs with a weight lower than $threshold_{use}$ are marked as unused (these geoDBNs provide only little evidence that the user is in the vicinity of their geo coordinate). This threshold should match the a priori probability for the geoDBN at its first instantiation. To cope with resource restrictions a maximum number of possible geoDBNs can be specified. If this number is exceeded those geoDBNs that provide the least estimation will be deleted. To keep the overhead for memory management low (or to prevent garbage collection if the system is implemented in languages like Java or C#) GeoDBNs that are marked as unused can be "recycled" by resetting them to initial values and new coordinates.

The following section describes the algorithm and explains it with an example.

3.2   Estimation of the User Position

A new estimation of $\mathcal{U}$'s current position is calculated after each new measurement and is based on the current sensory data as well as on the previous data. A weighted combination of the old and new data is achieved through inference in the respective geoDBN. The schematic approach looks like this:

1. Perform a new measurement and store the received coordinates
2. Extend every existing geoDBN with a new time slice and cut off the old time slice.
3. Insert the new evidences of the sensors:
   (a) If there is not already a geoDBN at a received coordinate, create a new geoDBN and insert the evidence.
   (b) If there is a geoDBN at a received coordinate, insert the evidence in the current time slice.
4. Go through all geoDBNs and calculate the estimation that the user is at the associated coordinate.
5. Sort the geoDBNs in descending order of their belief.
6. Mark geoDBNs as unused that provide an estimation that is lower than $threshold_{use}$.
7. Calculate the user position by considering only those geoDBNs that provide an estimation above $threshold_{consider}$.

An example run illustrates the approach and the algorithm (the shown values are for demonstration purposes only; they do not represent real data):

Let the following table describe the current situation. The array GeoDBN[.] contains the existing geoDBNs $geoDBN_a$ to $geoDBN_d$ with their respective coordinates sorted in descending order of their belief.

	i	GeoDBN[i]	GeoPos(.)	Belief(.)	$w(.)$
	1	$geoDBN_a$	$(10,5,0)$	$(0.7,0.3)$	0.7
	2	$geoDBN_b$	$(12,5,0)$	$(0.7,0.3)$	0.7
$n=3$	3	$geoDBN_c$	$(8,5,0)$	$(0.5,0.5)$	0.5
	4	$geoDBN_d$	$(14,5,0)$	$(0.1,0.9)$	0.1

Using (2) the computed user position is $(10.21, 5.00, 0.00)$. Note that only the first three geoDBNs contribute to the calculation since the belief of the fourth geoDBN lies below $threshold_{consider}$ (Step 7 in the algorithm).

The next measurement (Step 1) receives the senders $RFID_b$, $RFID_d$, IR_d, $RFID_e$, and $RFID_f$. A quick look at the preceding table reveals that geoDBNs already exist for the first three senders whereas the last two are new. The new data is inserted in the geoDBNs (for the first three senders, Step 3b) or new geoDBNs are created respectively (the last three senders, Step 3a). After an update of all geoDBNs (Step 4) and the final sorting (Step 5) we get the following situation:

i	GeoDBN$[i]$	GeoPos(.)	Belief(.)	$w(.)$
1	geoDBN$_b$	$(12,5,0)$	$(0.85, 0.15)$	0.85
2	geoDBN$_d$	$(14,5,0)$	$(0.75, 0.25)$	0.75
3	geoDBN$_a$	$(10,5,0)$	$(0.5, 0.5)$	0.5
$n=4$	geoDBN$_c$	$(8,5,0)$	$(0.25, 0.75)$	0.25
5	geoDBN$_e$	$(20,5,0)$	$(0.06, 0.94)$	0.06
6	geoDBN$_f$	$(16,5,0)$	$(0.06, 0.94)$	0.06

The fifth and sixth geoDBN are below $threshold_{consider}$, so only the first four geoDBNs are used for the calculation (Step 7). Equation (2) evaluates to $(11.78, 5.00, 0.00)$. Another measurement detects senders $RFID_b$, $RFID_d$, and $RFID_f$. Updating and sorting gives the following:

i	GeoDBN$[i]$	GeoPos(.)	Belief(.)	$w(.)$
1	geoDBN$_d$	$(14,5,0)$	$(0.89, 0.11)$	0.89
2	geoDBN$_b$	$(12,5,0)$	$(0.85, 0.15)$	0.85
3	geoDBN$_a$	$(10,5,0)$	$(0.25, 0.75)$	0.25
$n=4$	geoDBN$_f$	$(16,5,0)$	$(0.25, 0.75)$	0.25
5	geoDBN$_c$	$(8,5,0)$	$(0.05, 0.95)$	0.05
6	geoDBN$_e$	$(20,5,0)$	$(0.04, 0.96)$	0.04

The beliefs of geoDBN$_c$ and geoDBN$_e$ are now below $threshold_{use}$, so both are marked unused (Step 6, unused values are gray in the table). Only geoDBN$_d$, geoDBN$_b$, geoDBN$_a$, and geoDBN$_f$ are considered in the calculation, so the user position is now $(13.02, 5.00, 0.00)$.

The following section is a detailed description of the geoDBN that we use in our current implementation of the system.

3.3   The Geo Referenced Dynamic Bayesian Network of the Example Application

To model the time slice schemes of the DBN we use our own developed Java application called *JavaDBN* (see Fig. 1). A directed graph (DEA) determines in which orders the time slice schemes can be instantiated, furthermore the user can specify the query and evidence nodes of each time slice schema (see the tabular in each time slice schema window in Fig. 1). Then JavaDBN generates source code (so far Java and C++) that represents the DBN and that contains routines for the inferences. This code permits a constant-space evaluation of DBNs and contains a non-approximative roll-up method, which cuts off older time slices and that incorporates their impact without loss of information on the remaining time slices of the DBN (see [8, 9] for the theoretical background). To our knowledge, this is the first system that allows DBNs to run on mobile systems.

The source code has the following additional features: All for the computation necessary variables are created on initialization of the DBN. No variables are created or disposed at run-time to minimize overhead for memory management (C++) or to avoid the activation of a garbage-collection (Java). The inference in the DBN is optimized regarding the selected query and evidence nodes. Special functions to set evidence are provided. Since variables are freely accessible from outside the class, evidence of a node can also be set directly.

The geo referenced DBN consists of a time slice schema for instantiation of the initial time slice (TSS_1) and a time slice schema for instantiation of succeeding time slices (TSS_2). TSS_1 and TSS_2 are visualized in Fig. 1. TSS_1 differs from TSS_2 in that it does not contain a node UserPos=GeoPos?_1, which models the influence of the precedent time slice on the current one. The node UserPos=GeoPos? in TSS_2 represents the motion model and contains the two states $a_1 = $ "yes" and $a_2 = $ "no". The a-posteriori probability distribution over both these states (also called the belief) represents the network's estimation of the knowledge of whether $\mathcal{U}$ is located at the associated geo coordinate. The state "yes" stands for "knowing if $\mathcal{U}$ is located at the geo coordinate" and the state "no" for "not knowing if $\mathcal{U}$ is located at the geo coordinate". The conditional probability table (CPT) of the node is as follows: $\begin{bmatrix} & a_1 & a_2 \\ a_1 & 0.7 & 0.001 \\ a_2 & 0.3 & 0.999 \end{bmatrix}$. The conditional probabilities of the node are adapted to the mean walking speed of a pedestrian, which causes an accordingly fast decrease of the network's belief if the respective sender is not received for a few subsequent measurements (i.e. the user is not in the vicinity of the sender). The belief increases in an according way if the user enters (or re-enters) the range of the sender.

The node UserPos=GeoPos? in TSS_1 also contains the two states $a_1 = $ "yes" and $a_2 = $ "no", but it lacks the preceding node, so its CPT reduces to the a-priori probabilities $\begin{bmatrix} a_1 & 0.05 \\ a_2 & 0.95 \end{bmatrix}$.

The nodes IRSensor and RFIDSensor correspond to the sensors in the real world. These nodes will be instantiated with the results of the measurement. The conditional probabilities of the nodes model the reliability of the sensors (perceptual model) according to the real world situation. The node IRSensor has the states $b_1 = $ "yes", and $b_2 = $ "no". The following CPT is associated with this node: $\begin{bmatrix} & a_1 & a_2 \\ b_1 & 0.9 & 0.05 \\ b_2 & 0.1 & 0.95 \end{bmatrix}$. The CPT has the following interpretation: Assumed that the user is in the vicinity of the IR beacon (up to about 2 m distance) then the sensor of the PDA will detect the sender in 90% of all cases and it won't detect in 10% of all cases. In reverse, if the user has a greater distance, the sensor will still detect the IR signal in 5% of all cases but won't detect it in 95% of all cases. Note that the IR beacon sends its signal every 500 ms whereas an RFID tag will send only after receiving a ping-signal from the PDA of the user. Node RFIDSensor, like Node IRSensor, contains the states $c_1 = $ "yes", and $c_2 = $ "no". The lower precision of the RFID-Sensor compared to the IR-Tag is reflected in the associated

CPT: $\begin{bmatrix} & a_1 & a_2 \\ c_1 & 0.6 & 0.3 \\ c_2 & 0.4 & 0.7 \end{bmatrix}$. The imprecision of the RFID signal is, among other things,

due to its high sending range. This accuracy can be increased by decreasing the transmitting power of the RFID reader, which sends the above-mentioned ping-signal.

A new time slice is instantiated whenever data arrives and then the estimation is calculated, followed by the roll-up.

4   Example Application

To test our approach we implemented a localization system and equipped parts of our lab with IR beacons and RFID tags. The software is running on an HP iPAQ h5550 with 128 MB ram and Windows CE 4.2 as operating system. The iPAQ is settled in an expansion pack that provides an additional battery and a PCMCIA slot that hosts the active RFID tag reader card. IR signals are received through the internal IR port of the iPAQ. The system itself consists of two applications, which both run on the iPAQ simultaneously: The first application is the PositionSensor, which takes a measurement of both sensors every 500 ms and that reads the stored coordinates of every received RFID tag. The collected data is then processed via the described calculations and the estimated coordinate of the user is send to the second application, the BMW Personal Navigator (see [10]), which visualizes the position to the user.

Figure 2 shows how the PositionSensor is integrated into a framework where applications only query the layer of the Interaction Manager. This Interaction Manager is responsible for forwarding raw sensor data to its corresponding classifiers or DBNs and synchronizing DBN rollups with processing and reading sensors. It also registers new classifiers and sensors and handles their access. This separates the development of applications from extending sensors, classifiers, and DBN libraries — which can still be used independently from all other modules or the framework itself.

Figure 3 shows a test walk of a user $\mathcal{U}$ from Room 121 (lower left in Fig. 3) to Room 119.3 (upper right in Fig. 3) in our lab. The stickman indicates the system's estimation of $\mathcal{U}$'s position. The corresponding circle around each stickman shows the area of the possible real positions. RFID tags are marked as green squares, IR beacons as red squares with cones that indicate the sending direction. Both rooms are equipped with RFID tags in each of the four corners and additional IR beacons at the entrances. Room 119.3 also contains a bookshelf that bears

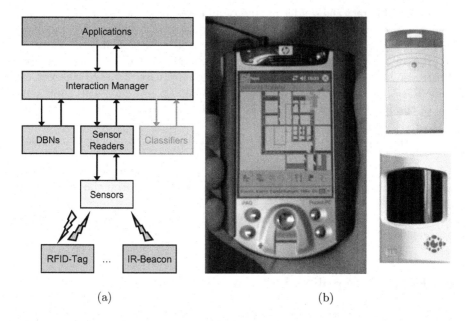

Fig. 2. (a) System Architecture and (b) the iPAQ with PCMCIA RFID-Reader Card, built-in WLAN and IR-Reader (left) and RFID-Tag and IR-Beacon (right)

an IR beacon. The hallway has RFID tags at about every 2 meters and two IR beacons (one in each direction) at about every 4 meters. $\mathcal{U}$ moved with slow to normal walking speed. The user position is very accurate on room level, i.e. if $\mathcal{U}$ is in a particular room, the system will estimate the position somewhere in that room. The position in the room itself is rather coarse but the accuracy can be greatly increased by placing IR beacons at points of special interest (like the bookshelf in Room 119.3). The position estimation in the hallway varies about 1 meter from the actual position.

5   Conclusion and Future Work

We presented a new approach to estimate a user position by probabilistic sensor fusion. Our method is low in space and time complexity and can therefore run completely on a PDA. The calculation of the position does not need a model of the environment and the time-consuming task of calibration is superfluous. Any arbitrary environment can be enhanced by simply putting up tags and beacons and writing the respective coordinates in the memory of these devices. Regions with coarser or finer granularity can be established by using senders with the appropriate precision (e.g. IR beacons for higher precision and RFID tags for lower precision).

We think that the concept of geo referenced dynamic Bayesian networks can be generalized to referenced dynamic Bayesian networks, because DBNs can also

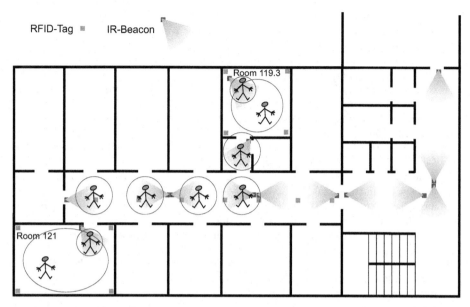

Fig. 3. A test walk through our lab. The stickman indicates the estimated position and the circle the area of the real positions

be referenced to actions, plans or symbolic locations. Future work will be the integration of other sensor types (e.g. video cameras, microphone arrays) and a system to determine the optimal placing of the senders and sensors. Another project at our lab researches the usability of biophysiological sensors. Part of that work (e.g. accelerometers) can be used to adapt the motion model of the geoDBNs to the current state of the user (e.g. sitting, walking fast or walking slow).

Acknowledgments

We would like to thank Anthony Jameson, Antonio Krüger, Jörg Baus, and Christoph Stahl for their many helpful suggestions.

This research is being supported by the German Science Foundation (Deutsche Forschungsgemeinschaft (DFG)) in its Collaborative Research Center on Resource-Adaptive Cognitive Processes, SFB 378, Project EM 4, BAIR, and its Transfer Unit on Cognitive Technologies for Real-Life Applications, TFB 53, Project TB 2, RENA.

References

1. Russell, S.J., Norvig, P. In: Artificial Intelligence, A Modern Approach. second edn. Pearson Education (2003) 492–580

2. Fox, D., Hightower, J., Liao, L., Schulz, D., Borriello, G.: Bayesian filtering for location estimation. IEEE Pervasive Computing **2** (2002) 24–33
3. Want, R., Hopper, A., Falcao, V., Gibbons, J.: The Active Badge Location System. ACM Transactions on Information Systems **10** (1992) 91–102
4. Harter, A., Hopper, A., Steggles, P., Wart, A., Webster, P.: The anatomy of a context-aware application. In: 5th Annual ACM/IEEE International Conference on Mobile Computing and Networking (Mobicom '99). (1999)
5. Ubisense Unlimited: Ubisense. http://www.ubisense.net (2004)
6. Ni, L.M., Liu, Y., Lau, Y.C., Patil, A.P.: Landmarc: Indoor location sensing using active RFID. In: IEEE International Conference in Pervasive Computing and Communications 2003 (Percom 2003). (2003)
7. Amemiya, T., Yamashita, J., Hirota, K., Hirose, M.: Virtual Leading Blocks for the Deaf-Blind: A Real-Time Way-Finder by Verbal-Nonverbal Hybrid Interface and High-Density RFID Tag Space. In: Proceedings of the 2004 Virtual Reality (VR'04), IEEE Virtual Reality (2004) 165–172
8. Darwiche, A.: A differential approach to inference in Bayesian networks. Journal of the Association for Computing Machinery **50** (2003) 280–305
9. Brandherm, B., Jameson, A.: An extension of the differential approach for Bayesian network inference to dynamic Bayesian networks. International Journal of Intelligent Systems **19** (2004) 727–748
10. Krüger, A., Butz, A., Müller, C., Stahl, C., Wasinger, R., Steinberg, K.E., Dirschl, A.: The Connected User Interface: Realizing a Personal Situated Navigation Service. In: IUI 04 - 2004 International Conference on Intelligent User Interfaces, ACM Press (2004)

Issues and Requirements for Bayesian Approaches in Context Aware Systems

Michael Angermann, Patrick Robertson, and Thomas Strang

Institute of Communications and Navigation,
German Aerospace Center,
D-82234 Wessling/Oberpfaffenhofen,
Germany,
firstname.lastname@dlr.de

Abstract. Research in advanced context-aware systems has clearly shown a need to capture the inherent uncertainty in the physical world, especially in human behavior. Modelling approaches that employ the concept of probability, especially in combination with Bayesian methods, are promising candidates to solve the pending problems. This paper analyzes the requirements for such models in order to enable user-friendly, adaptive and especially scalable operation of context-aware systems. It is conjectured that a successful system may not only use Bayesian techniques to infer probabilities from known probability tables but learn, i.e. estimate the probabilities in these tables by observing user behavior.

1 Introduction

Context and context-awareness has become a major field of research in recent years. One of the reasons is that context-awareness is believed to be a promising solution for a couple of problems arising in pervasive computing scenarios [9, 13, 14, 15]. A well designed context model in conjunction with a framework that facilitates context management tasks such as collection, refinement, persistency, distribution, synchronization, provisioning and reasoning based on the model is a prerequisite for the fruitful use of context in any context-aware system [12].

Deriving contextual information about the system's entities, such as the user, the devices, the network or the environment is an essential, yet challenging task. This holds in particular if the contextual information is the outcome of a context refinement process (e.g. sensor fusion) and does not only reflect changes on primary aspects such as location, time, activity or identity.

A major issue in most context-aware systems is how to deal with uncertain, incomplete or ambiguous data. Incomplete contextual information can for instance be the result of temporarily or permanently absent sensors (e.g. due to energy deficiencies), whereas ambiguous context information can for instance be the result of two or more sensors observing the same entity but getting different values, such as two thermometers in the same room showing different temperatures (e.g. 19.1°C and 22.7°C). Both sensors add information to the knowledge about

T. Strang and C. Linnhoff-Popien (Eds.): LoCA 2005, LNCS 3479, pp. 235–243, 2005.

the context. However, since obviously their indicated values are subject to some form of error, some uncertainty about the "true" temperature remains. To care about uncertainty and ambiguity must not be neglected in any serious context-aware system architecture. Many context modelling and retrieval architectures tend to over-simplify this uncertainty by assuming a perfect knowledge in combination with perfect inference. Only few approaches cover uncertainty, e.g. by using heuristic approaches such as attaching quality/validity meta-information to contextual information. Among them are the Extended ORM model [5], the ASC model [13] or the Cues [10].

Over the last 15 years, Bayesian networks have evolved as a major tool in a wide area of scientific disciplines requiring sound statistical analysis, automated reasoning or exploitation of knowledge hidden in noisy data [8]. These range from fields in medical research, genetics, insurance analysis, and fault handling to automation and intelligent user interaction systems. Bayesian networks (BN) combine techniques from graphical models with those from Bayesian analysis to provide a formal framework within which complex systems can be represented and analyzed. A BN encompasses a set of random variables that represent the domain of interest and the BN encodes many of the important relationships between these variables, such as causality and statistical dependence and independence. Specifically, their structure says something about the qualitative nature of these relationships whereas their network parameters encode the probabilistic relationships among the variables of interest. Past areas of work have focused on inference in BN [6] and estimation of network structure and parameters given real observed data (see e.g. [11, 2]). Current work is addressing numerous new areas of applications of BN, as well as special network structures such as dynamic Bayesian networks [7].

In this paper, we want to elaborate on the requirements for a probabilistic approach to cover uncertainty in context. Probabilistic approaches for context-modelling naturally lend themselves to a fruitful combination with Bayesian methods, in particular with Bayesian networks. In conjunction with other statistical techniques they have several advantages for data analysis. Following [3] this includes

- A Bayesian network readily handles situations where some data entries are missing.
- Given observed data, a Bayesian network can be used to learn causal relationships, and hence can be used to gain understanding about a problem domain and to predict the consequences of intervention.
- Because a Bayesian network has both a causal and probabilistic semantic, it is an ideal representation for combining prior knowledge and data.
- Bayesian statistical methods in conjunction with Bayesian networks avoid the overfitting of data and do not need to separate data into training and testing sets. They are also able to incorporate smoothly the addition of new data as it becomes available.

It is obvious that these characteristics of Bayesian networks should qualify them perfectly to model uncertainty in a context-aware system. This is the rationale for working out the requirements and a generic architecture for including Bayesian networks into context-aware systems. We have identified four main issues for this Bayesian approach:

1. Infer contextual knowledge given a network and associated probabilities
2. Estimate network probabilities given observations
3. Deduce network structure given observations
4. Combine (few) data that is valid for a single individual with (much more) data valid for a larger group of users

This paper is organized as follows. In section 2 we discuss several issues to be considered when applying Bayesian information processing for context-aware systems. In particular we show how they result in system requirements. In section 3 we investigate the combination of knowledge derived from observing individual behavior with group behavior. This combination is considered to be essential for providing instant high perceived system performance for users coming freshly into the system.

2   Issues in Applying Bayesian Information Processing to Context-Aware Systems

The following list highlights in more detail the areas in context processing that should involve Bayesian techniques and the issues that arise when doing so.

1. **Human 'domain-expert' modelling of Bayesian Networks' structure and parameters.** In the Bayesian research community, it is well established that domain experts' knowledge can be valuable, if not essential contributions to the Bayesian process, mainly in specifying the structure of the underlying network. Therefore, a Bayesian context-aware system must provide an interface to such domain experts. The experts' contributions can be in the form of

   - identifying and configuring context variables suitable to the domain, this includes choice of suitable quantization levels and mapping to context ontologies, and choice of the total number of variables to be incorporated into the later network,
   - encoding causal relationships in a prior network [3],
   - defining priors on network structure,
   - using an initial network and defining equivalent sample size,
   - confirmation of learned structures, conditional probability distributions and a-priori probabilities,
   - obtaining, and making explicit, new insights into the domain's features by analyzing learned structures and probability distributions.

   In a limited sense, it is conceivable that the actual end-user could be treated as an additional domain expert; at least he or she is the person who often has the best understanding of the personal domain. In traditional context aware systems the user is incorporated by allowing preferences or rules to be established that govern the system behavior. In a context system that uses Bayesian techniques, 'such preferences' could be enriched by letting the user state certain facts about his or her domain and translating these to network and parameter priors. One may even go as far as suggesting a hierarchy of domain experts in which the end-user may play a certain role, in addition to, for instance, company representatives that govern the configuration of systems used in a corporate environment such as services for sales staff. Several, secondary requirements pertaining to integrity, quality control, privacy of user data and security arise from this potential involvement of others into the representation of a user's context.

2. **Automated learning of BN structure using complete or incomplete data** [2, 1]. By incomplete we mean the use of observed data sets where some variables' values are unknown, which will presumably be the case for many context systems where sensor data or other observations can be missing. This learning may apply to subsets of variables (i.e. with the goal of constructing a sub network) or the full set, depending on the domain complexity. It includes the identification of new hidden nodes to simplify the network structure and more accurately represent the domain at hand [16]. The result of automatic learning may be iteratively used to reinforce or expand the modelling done by hand in the first step.

3. **Automated learning of BN parameters** (a priori probabilities, conditional probabilities), using complete and incomplete data sets. Again, there may be an iteration to the first and second stages above. In a context systems application, such learning should ideally be a continuous process as new data is gathered.

4. **Inference given observations and updates to observations.** Depending on the network size and structure, suitable algorithms must be chosen, for instance belief propagation for tree structures [8], loopy belief propagation (see [17]) or join tree algorithms. This will may also apply optional consistency checks (non Bayesian) that the domain experts may define on individual variables or groups of variables (e.g. temporal constraints on location changes or checking whether certain combinations of variables' values simply cannot be the case).

5. **Utility theoretic decision making.** It is clear that in order to reduce the burden on developers or providers of the actual end-user context aware services, and also to improve efficiency, such services will not only exploit context variables' estimated values or their likelihoods, but would like to delegate to the Bayesian processing layer certain decisions based on cost functions and the estimated likelihoods of context variables. An example is a service that wants to influence the choice of a communications network used to transport its application data, based on certain context conditions.

6. **Value-Of-Information evaluation** to gain information on
   - which sensors' readings would be of value
   - which additional (possibly non-Bayesian and more or less costly) context inference would be of value

– which remote context elements should be accessed in a federated system, possibly over a costly network

Such methods must minimize global costs and take into account the aggregate benefit (or cost) of acquiring the new context data (or of not obtaining it) and the costs that are incurred to acquire the context. Suitable interfaces to services are again needed that indicate the value to the service of knowing values of context variables - possibly to a certain accuracy, and also interfaces to a context broker that allow it to indicate costs of determining certain context values - including sensor values and items of only remotely available context. These costs will, of course, depend on the state (including observational status) of other variables in the BN.

7. **Combine (a smaller amount of) data that is valid for a single individual with (much more) data collected from observing a larger group of users.** This difficult task will be addressed in the next section.

8. **Enforce privacy restrictions on users' personal data.** The exploitation of context information is always a balance between users' privacy and benefits gleaned from exploiting personal data. In a system where such information is being gathered not only for the benefit of an isolated user in his isolated BN, but also for many other users, and furthermore may require the inspection of human domain experts, the privacy issue become more pressing. The architecture, design and implementation of all components will need to ensure that this problem is being addressed satisfactorily. These considerations will also have implications on point 6, where, for example, pieces of context pertaining to location may only be available in raw form on a local device and the user would not like them to be passed continuously to a central server.

9. **Scalability and performance issues.** This problem is related to the last three in the sense that it is expected that each active user will own a particular instantiation of a personal BN and that context information may not always be available centrally. With potentially many millions of active users, and the period of an individual's 'activity' in pervasive computing environments being extended to almost 24 hours a day, hosting many such networks will pose memory and computational challenges. It may become necessary to host various parts of a BN on different computing nodes, and take into consideration the topographic availability of context. For instance, a person may carry location sensors and continuous context transfer to a central server might be too ex-pensive or undesired, e.g. for privacy concerns; in this case, local pre-processing on a local sub network could reduce network costs and/or preserve privacy.

An object oriented design approach with proper encapsulation, providing separation of above functions from their actual implementation, may possibly allow future developments in this rapidly evolving field to be incorporated more easily.

For context-aware systems to become key components of commercially viable services, several interfaces and functionalities are necessary to enable a complete

Bayesian context processing "chain", which would be the result of implementing the all major components related to steps 1-7 above.

3   Deriving Individual User Behavior from Group Behavior

Among the most attractive features of the Bayesian approach is its capability to use data, gained from observing a large user community (*collective*), as information about an individual, whose behavior may even not have been observed at all. This is facilitated by deriving suitable a priori distributions ("priors") for the parameters of the individual's Bayesian network. In the domain of context aware applications this seems particularly important since the potential scope of observations of each particular individual is much smaller than the observation of many users and in fact, a system begins with no observations for a new user. Nevertheless, as observations accrue, they should be incorporated fairly with the prior knowledge obtained from the collective data.

An important issue is the identification of suitable clusters to distinguish certain features of users. These clusters effectively define sets of users, e.g. "all male users", "all users of age below 18" years, "all users spending more then 100$ per month on mobile phone charges". Suitable a priori distributions for the individual are then computed by selecting the sets he or she fits into and adding up the probability distributions of all of each set's members and multiplying the sums obtained for all fitting sets.The influence of these priors gradually declines with increasing availability of observations upon the individual. Fig. 1 illustrates the necessary steps from including domain experts' knowledge to the derivation of information about individual user behavior.

1. Define priors on structure or prior collective BN or an initial collective BN
2. Define cluster variables
3. Update collective BN using collective data
4. Update clusters if this is beneficial
5. Update collective BNs parameters using collective data
6. Provide a valid collective BN as a result of steps 1-5
7. Clone this collective BN to become the individual's BN
8. Determine cluster variables' values for the individual (e.g. from individual's preferences or subscription data)
9. Using the clusters, identify in the collective BN which variables these cluster choices have effects upon
10. Using 7 and 9, compute priors of the parameters of conditional probability functions (CPF) and a-priori probabilities, respectively; for instance assuming an unrestricted multinomial distribution model [3]
11. Using the individual's personal data history continually update the parameters of her BNs CPFs and a-priori probabilities
12. Construct the individual's BN using estimates of the individual's BN parameters based on step 11
13. Perform inference and/or decisions based on the individual's BN

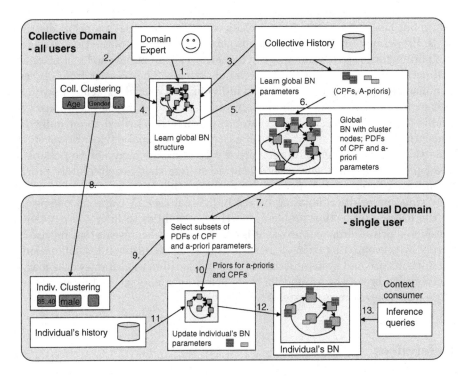

Fig. 1. Necessary steps from including domain experts' knowledge to the derivation of information about individual user behavior

We can see that the actual transfer of knowledge obtained from observations of collective user behavior to individual user behavior is performed in steps 7 and 8, when the collective BN is cloned to become the individual's BN and the values of the individual's cluster variables are determined.

4   Conclusions

In this short paper we have addressed several issues we consider relevant for the development of future context-aware systems that become capable of a) properly operating on the ambiguous/uncertain context of real world applications and b) scaling towards potentially very large numbers of users (e.g. several millions, such as in today's mobile phone networks).

We suggest that domain experts need user interfaces to set up and continuously maintain the structure and eventually certain a priori probabilities of the Bayesian networks. It is proposed that implementors of services need clearly specified and standardized interfaces not only to query for context values, but also for obtaining such (meta-)information as the "value-of-information", probabilities of context values or decision-theoretic computation of utilities. Interfaces to low-level sensors appear necessary which, again, do not only convey the actual values but also meta-information such as error distributions or a priori distributions. Proper algorithms for continuous automatic learning of the BNs' structure and parameters as well as inference of higher-level context have to be in place and well encapsulated inside of the context-subsystems. The transfer from observation made on the collective of users to individuals is crucial to create a favorable user experience from the beginning. In order to protect privacy it appears desirable to keep the observations that can be directly ascribed to an individual under the individual's direct control, e.g. on the personal device. However, despite this and and other efforts for privacy protection it may not be possible to develop an "inherently safe" context-aware system that can absolutely prevent the misuse of context information for activities that may be considered as privacy violating or discriminating by the individual user. A Bayesian perspective on the privacy protection problem and the possibilities to infer context that is considered private from context that is considered public, reveals that misuse can hardly be prevented by technological measures, alone. Instead, a combination of careful software and system design efforts and sensible legal frameworks may be necessary to break the ground for wide acceptance and successful operation of large-scale context-aware systems and services.

References

1. Chickering, D., Heckerman, D.: Efficient Approximations for the Marginal Likelihood of Bayesian Networks with Hidden Variables, Machine Learning, (1997), Vol. 29, pp. 181-212.
2. Cooper, G., Herskovits, E.: A Bayesian Method for the Induction of Probabilistic Networks from Data, Machine Learning, (1992), Vol. 9, pp. 309-347.
3. Heckerman, D.: A Tutorial on Learning with Bayesian Networks. Technical Report MSR-TR-95-06, Microsoft Research, Redmond, Washington, 1995. Revised June 96
4. Heckerman, D., Mamdani, A., Wellman, M.: Real World Applications of Bayesian Networks, Communications of the ACM, (1995), Vol. 38.
5. Henricksen, K., and Indulska, J. Modelling and Using Imperfect Context Information. In WorkshopProceedings of the 2nd IEEE Conference on Pervasive Computing and Communications (PerCom2004), (Orlando, FL, USA, March 2004), pp. 33-37.
6. Lauritzen, S., Spiegelhalter, D. J.: Local Computations with Probabilities on Graphical Structures and Their Application to Expert Systems. Journal of the Royal Statistical Society Series B 50:157-224 (1988).
7. Murphy, K.: Dynamic Bayesian Networks: Representation, Inference and Learning. Ph.D. Dissertation, UC Berkeley, Computer Science Division, 2002.
8. Pearl, J.: Probabilistic Reasoning in Intelligent Systems: Networks of Plausible Inference. Morgan Kaufmann, San Mateo, CA (1988).
9. Satyanarayanan, M.: Pervasive Computing: Vision and Challenges. IEEE Personal Communications Magazine. Vol. 8. No. 4. (2001) 10-17
10. Schmidt, A., Beigl, M., and Gellersen, H.-W. There is more to Context than Location. Computers and Graphics 23, 6 (1999), 893-901.

11. Spiegelhalter, D., and Lauritzen, S.: Sequential Updating of Conditional Probabilities on Directed Graphical Structures, Networks, Vol. 20, (1990), pp. 579-605.
12. Strang, T., Linnhoff-Popien C.: A Context Modeling Survey. Workshop on Advanced Context Modelling, Reasoning and Management as part of UbiComp 2004 - The Sixth International Conference on Ubiquitous Computing, Nottingham/England (2004)
13. Strang, T.: Service-Interoperability in Ubiquitous Computing Environments. PhD thesis, LMU Munich, October 2003
14. Want, R., Hopper, A., Falcao, V., Gibbons, J.: The Active Badge Location System. ACM Transactions on Information Systems. Vol. 10. No 2. (1992) 91-102
15. Xynogalas, S., Chantzara, M., Sygkouna, I., Vrontis, S., Roussaki, I., Anagnostou, M.: Context Management for the Provision of Adaptive Services to Roaming Users, IEEE Wireless Communications. Vol. 11. No 2. (2004) 40-47
16. Boyen, X. , Friedman, N.,Koller, D.: Learning the structure of complex dynamic systems,UAI, 1999.
17. Murphy, K.P., Weiss, Y., Jordan, M: Loopy belief propagation for approximate inference: an empirical study, in Proceedings of Uncertainty in AI, 1999.

Context-Aware Collaborative Filtering System: Predicting the User's Preference in the Ubiquitous Computing Environment

Annie Chen

IBM Zurich Research Laboratory,
Säumerstrasse 4, CH-8803 Rüschlikon, Switzerland
ach@zurich.ibm.com

Abstract. In this paper we present a context-aware collaborative filtering system that predicts a user's preference in different context situations based on past experiences. We extend collaborative filtering techniques so that what other like-minded users have done in similar context can be used to predict a user's preference towards an activity in the current context. Such a system can help predict the user's behavior in different situations without the user actively defining it. For example, it could recommend activities customized for Bob for the given weather, location, and traveling companion(s), based on what other people like Bob have done in similar context.

1 Introduction

In the ubiquitous computing world, computing devices are part of the bigger environment, known as the *pervasive context*. These devices could be aware of various contexts in the environment, such as the location, the surroundings, people in the vicinity, or even the weather forecast. Connectivity, or the implied access to the information highway, is no longer bounded to the desk. The user could be at the train station, in a shopping mall or even in a different city. The role of computers in this environment has in a way become more like that of a personal assistant than, say, of a help-desk. This shift in interaction prompted us to look at how ubiquitous devices could assist users better by anticipating their preferences in a dynamic environment.

Currently, context-aware applications mostly rely on manually defined rules to determine application behavior for varying context. These rules can be predefined by application developers [1, 2] or, alternatively user-configured either by static preferences [3, 4, 5] or formed over time from user feedback [2]. Static rules are inflexible and difficult to customize for individuals, whereas the underlying learning process in the latter case has a long learning curve and can be tedious for users. More importantly, these systems are unable to predict a user's preference in an unseen situation.

T. Strang and C. Linnhoff-Popien (Eds.): LoCA 2005, LNCS 3479, pp. 244–253, 2005.

It is understandably difficult for a computer to judge the taste of a user, so in recent years, we have seen a trend towards recommendation systems that leverage the opinions of other users to make predictions for the user.

Collaborative Filtering (CF) is a technology that has emerged in e-Commerce applications to produce personalized recommendations for users [6]. It is based on the assumption that people who like the same things are likely to feel similarly towards other things. This has turned out to be a very effective way of identifying new products for customers. CF works by combining the opinions of people who have expressed inclinations similar to yours in the past to make a prediction on what may be of interest to you now. One well-known example of a CF system is Amazon.com.

To date, CF has mostly been applied to web applications for which the context is undefined, i.e., the content is static, and the recommendations do not change with the environment. In the dynamic environment of ubiquitous computing, a user's decision can be influenced by many things in the surrounding context. For example, when people travel on holiday, their preferred activities might largely depend on the weather. Existing CF systems could not model this complexity of context. They are as likely to recommend mountain routes for a person who likes hiking whether it rains or shines. Applications in ubiquitous computing exist that have used CF to give recommendations [2, 7]; however they did not utilize the context information in the environment, and hence did not break out of the boundaries of a normal CF application.

In this paper we propose a design for a context-aware CF system in which we leverage the pervasive context information such that a user's preference is not only predicted from opinions of similar users, but also from feedback of other users in a context similar to that the user currently is in. In contrast to existing methods that manually determine how each context will influence the desirability of an activity, we use CF to automatically predict the impact of context for an activity by leveraging past user experiences. In our system, context provides the hints necessary to explore different options, rather than just limiting the set of options.

The remainder of this paper is organized as follows: we begin by examining the existing process in CF in Section 2. We then introduce context in Section 3.1 and define the requirements for a context-aware CF system. Next we discuss how to model context data in a CF system in Section 3.2 and how to measure similarity between different contexts in Section 3.3, and give an algorithmic extension to the CF process to incorporate context in Section 3.4. Finally we discuss future work in Section 4, and conclude in Section 5.

2 Background: Collaborative Filtering Process

The task of collaborative filtering is to predict how well a user will like an item given a set of feedback made by like-minded users [8]. An active user provides the CF system with a list of items, and the CF system returns a list of predicted ratings for those items.

Various classes of algorithms have been used to solve this problem, including Bayesian networks, singular value decomposition, and inductive rule learning [9]. The prevalent class of algorithm used is the *neighborhood-based methods*. In these methods, a subset of users is chosen based on their similarity to the active user, and subsequently a weighted aggregate of their ratings is used to generate predictions for the active user.

In this section we will describe the stages of neighborhood-based methods to provide a general understanding of a CF system and give the reader a point of reference for the latter sections, when we extend this process to incorporate context.

2.1   Building a User Profile

The first stage of a CF process is to build user profiles from feedback (generally in the form of ratings) on items made over time. A user profile comprises these numerical ratings assigned to individual items. More formally, each user u has at most one rating $r_{u,i}$ for each item i.

2.2   Measuring User Similarity

The key in CF is to locate other users with profiles similar to that of the active user, commonly referred to as "neighbors". This is done by calculating the "weight" of the active user against every other user with respect to the similarity in their ratings given to the same items.

The *Pearson correlation coefficient*, which measures the degree of a linear relationship between two variables, is commonly used to weight user similarity [10]. The similarity weight between the active user a and neighbor u as defined by the Person correlation coefficient is

$$w_{a,u} = \frac{\sum_{i=1}^{m}(r_{a,i} - \bar{r}_a) \cdot (r_{u,i} - \bar{r}_u)}{\sigma_a \cdot \sigma_u}. \tag{1}$$

This equation combines the similarity between the relative ratings given by users a and u on the same item over all the items they have both rated. It gives a value between -1 and +1, where -1 means that these two users have the exact opposite taste, and +1 means they have the same taste.

2.3   Generating a Prediction

We can now combine all the neighbors' ratings into a prediction by computing a weighted average of the ratings, using the correlations as the weights [10]. The predicted rating of the active user a on item i can hence be formulated as

$$p_{a,i} = \bar{r}_a + k \sum_{u=1}^{n}(r_{u,i} - \bar{r}_u) \cdot w_{a,u}, \tag{2}$$

where n is the number of best neighbors chosen and k is a normalizing factor such that the absolute values of the weights sum to unity.

This covers the basic steps of the CF process in generating a prediction for a user. We now consider the issues involved when introducing pervasive context into the equation.

3   Incorporating Context into CF

3.1   Introducing Context: Concepts and Challenges

Context is a description of the situation and the environment a device or a user is in [11]. For each context a set of features is relevant. For each feature a range of values is determined by the context. From this model a hierarchy of context subspaces can be developed. Schmidt et al. [11] categorized context into six high-level subspaces. The first three relate to human factors: information about the user (e.g., habits, biophysiological conditions), social environment (e.g., social interaction, co-location with other users), and user's tasks (e.g., active tasks, general goals). The other three concern the physical environment: location, infrastructure (e.g., resources, communication), and physical conditions (e.g., noise, light, weather).

It is worthwhile noting that the first context identified on the list – knowledge about the habits of a user – is what a CF system currently models. CF uses this context to deduce any unknown habits of the user from habits of other similar users. What is lacking in CF is the knowledge about all the other contexts that help define a user's habits in different situations.

To model context in a CF system, we need to associate a user's choice or preferences with the context in which the user made that choice. This means that we need to capture the current context each time the user makes a choice. The same applies for the reciprocal: when a user asks for recommendations, we need to capture the current context and evaluate what others have chosen in a similar context in the past.

This poses two main problems: how do we manage context in the user profile in terms of data modeling and storage, and how do we measure similarities between contexts.

3.2   Context Modeling in CF

In a standard CF system, which we described earlier, a user's profile consists of a set of items with at most one rating assigned to each item. An item could be a product, a place or an action, and the rating represents the user's fondness of or preference towards that item. In a dynamic environment, a user's preference towards an item may change with the context. For example, Bob may want to visit a family diner instead of a posh restaurant when he is with his kids. To capture the different preferences towards an item in different contexts, a snapshot of the context need to be stored along with a user's rating for an item.

A snapshot of the context is a composite of different types of context data from various sources. This context can either be acquired from the embedded sensors in the mobile device itself or from an infrastructure placed in the smart environment which provides these data for the device. Consequently, various context data can be available or unavailable, depending on the infrastructure that is accessible in the current environment. This yields the requirement that different context types should be managed independently, and that their combined impact be calculated algorithmically.

In modeling context in CF, we took the approach of maintaining all the values and the structure within each context type. For example, the Location context object would maintain a hierarchy of all the different locations, such as "Czech Republic" and "Prague". Thereby we can minimize redundancy in the system and improve efficiency. When a user rates an item, the rating is associated with the current context value inside each available context, see Figure 1. In this example figure, Bob went to a spa when he was on holiday in Prague, he enjoyed it and rated this activity 5 out of 10. In the system a rating object is created that links Bob as the user and "Spa" as the item. There were two context data available to Bob's device at that time: the location and the temperature. To model this, the rating object also links to the value of "Prague" in the Location context and value of "5 degrees" in the Temperature context.

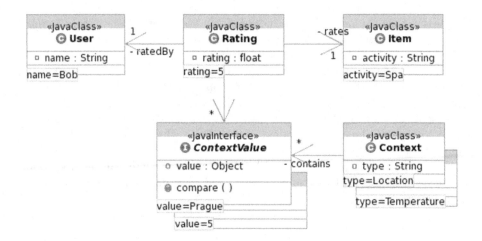

Fig. 1. Associating context with ratings

The links are bi-directional so that the system could easily traverse all the items rated for a given context and the entire context in which an item is rated. Context data is now available to the system, but to render it usable for making a prediction, we need to be able to compare the context of one user to that of another.

3.3   Context Similarity

The goal of calculating context similarity is to determine which ratings are more relevant for the current context. For instance, when Bob wants to go fishing in spring, ratings of fishing locations in spring would be more relevant than ratings of fishing locations in autumn. The similarity of the context in which an item is rated with the current context of the active user determines the relevance of this rating. Consequently, for each context type, there needs to be a *quantifiable measure* of the similarity between two context values.

Formally we define context here as a tuple of z different context types modeled in the system:

$$C = (C_1, C_2, \cdots, C_z), \qquad (3)$$

where C_t $(t \in 1..z)$ is a context type (e.g., Location, Temperature or Time). For each context type t, there exists a similarity function $sim_t(x, y)$, with $x, y \in C$, which returns a normalized value denoting the similarity between x and y with respect to C_t.

A general comparator can be defined for a context type which has one of the following properties:

- *Categorical*, where values in the same category are alike (e.g., transport).
- *Continuous*, where closer values are more similar (e.g., temperature).
- *Hierarchical*, where a more general context can be used when no ratings for a specific context are available (e.g., location).

Context types which do not have these characteristics may require a custom comparator to be defined for them.

Context types can be vary widely, and it would be difficult to manually define a similarity function for each context type, so we devised an automated method to compare the relevance of one context value to another for the same context type.

We make the assumption that if user preferences towards an item do not differ much in different contexts, then the ratings given in one context would also apply for the other. So if the ratings for an item are similar for two different context values, then these two values are very relevant to each other.

We use the *Pearson's correlation coefficient*, which was used to calculate the similarity weight for a user in Equation (1) to measure the correlation between two different context variables with respect to their ratings. We denote the rating given by the user u on item i in context $x \in C$ as $r_{u,i,x}$, and formulate the similarity weight for two different context variables, x and y, for item i as follows

$$rel_t(x, y, i) = \frac{\sum_{u=1}^{n} (r_{u,i,x_t} - \bar{r}_i) \cdot (r_{u,i,y_t} - \bar{r}_i)}{\sigma_{x_t} \cdot \sigma_{y_t}}, \qquad (4)$$

where $rel_t(x, y, i)$ returns the *relevance* of two context values in C_t over all the ratings users gave in these contexts. Compare this with $sim_t(x, y)$, which returns the *similarity* between two context values.

Either one or both could be used to weight each rating for the given context when making a prediction. The advantage of using the similarity function is that it can be better tailored to a particular context type, although it may be more restrictive in its comparison. Using the correlation function we were able to capture non-obvious relations between two context values, but each context value needs to be populated with many ratings for it to work.

Next we look at how this context is incorporated into the CF process to generate context-dependent predictions.

3.4   Context-Aware Extensions to the CF Process

Let us revisit the steps of a collaborative filtering process and redefine it to include context.

Building User Profile. In the context-aware CF system, a user's feedback needs to be put into context. This implies that the context needs to be recorded when the user selects or performs a recommendation. This poses a problem when the user wants to give explicit feedback on an item long after the action has passed and the context has changed.

Our advantage is that context data can also provide information on the user's current activity, which would enable the system to make implicit feedback for the user. Implicit feedback is when the system infers a user's rating for an item based on the user's behavior. Which behavior is monitored is application-dependent. For example, a tourist application could monitor how long a user stays in the location associated with a particular activity, whereas a shop application could look at items ordered, and a media-streaming application could see which items the user chooses to view or skip.

We could also use this implicit feedback to mitigate the problem of delayed feedback, by prompting the user to give feedback when the application senses the user has finished an activity (e.g. leaving a location or finished playing a clip). It could also allow the user to rate from a list of past activities the application has implicitly captured.

The feedback together with the available context would make up the user profile in the context-aware CF system as described in Section 3.2.

Measuring User and *Context* Similarity. In this step a normal CF system would "weigh" the active user against other users in the system with respect to similarity in their ratings to locate similar users. To incorporate context we would also "weigh" the current context of the active user against the context of each rating with respect to their relevance as described in Section 3.3 to locate ratings given in similar context.

Generating a Prediction. In this step a prediction is calculated by combining neighbors' ratings into a weighted average of the ratings, using the neighbors' correlations as the weights, see Equation (2).

In the context-aware CF system, each rating has an associated context. The similarity of the context in which an item was rated with the context of the active user determines how relevant this rating is in the current context, so we need to extend this to use the weighted rating with respect to the relevance of the rating's context to the current context.

We define $R_{u,i,c}$ as the weighted ratings for the user u on an item i in context c, where c is the current context of the active user, using context similarity as weights. The context is multi-dimensional so we assume linear independence and calculate the similarity for each dimension separately, i.e.,

$$R_{u,i,c} = k \sum_{x \in C} \sum_{t=1}^{z} r_{u,i,x} \cdot sim_t(c, x), \tag{5}$$

where k is a normalizing factor such that the absolute values of the weights sum to unity. It has nested sums: the inner loops over each dimension in context, e.g., Location, Weather; the outer loops over all the values in that dimension, e.g., "Zurich", "Prague", "Tokyo" for the Location context.

We now substitute $R_{u,i,c}$ for the rating of user u on item i without context $r_{u,i}$ in Equation (2). The predicted rating of the active user a on item i can hence be formulated as

$$p_{a,i,c} = \bar{r}_a + k \sum_{u=1}^{n} (R_{u,i,c} - \bar{r}_u) \cdot w_{a,u}. \tag{6}$$

This calculation combines all the weighted ratings, with respect to similarity in context, of all the neighbors, which is then further weighted with respect to the similarity of user, to give an overall prediction for the active user on an item in the current context.

4   Future Work

The next step is to evaluate the algorithms. The difficulty here is that we need real user data to validate whether the predictions match the user's actual decision. In addition, the collaborative nature of the system requires large user participation to generate good predictions. As we are the first to apply the CF technique to pervasive context, there is no readily available datasets to test it on. One possibility is to initialize the system with ratings from existing CF systems, such as VirtualTourist.com [12], adding any implicit context information available (e.g., location). This enables the system to provide general recommendations before more context data is collected to produce context-oriented recommendations.

We are currently developing a tourist application for mobile phones to demonstrate and collect data for the context-aware CF engine. We plan to test this application on a group of students in the laboratory, who will use the application for their weekend travels. The data collected from this deployment will allow us to evaluate and calibrate our algorithms.

On the modeling side, it would be interesting to look specifically into the social context and consider the influence of its complex interactions. Whether one should model relationships as a context (e.g., husband, girlfriend, children) or combine the profiles of individual participants.

Another important aspect to consider is the privacy issue. The system needs to keep usage statistics for each user in order to generate personalized recommendations. These statistics coupled with the context information can yield interpretations to potentially track everything from user movements to social behavior. Thus it is important that they be managed and protected properly.

5   Conclusion

We have designed a context-aware recommendation system that predicts a user's preference using past experiences of like-minded users. We used collaborative filtering to automatically predict the influence of a context on an activity. We defined the requirements of introducing context into CF and proposed a solution that addresses modeling context data in CF user profiles and measuring context similarity by applying CF techniques. Finally we gave an algorithmic extension to incorporate the impact context has on generating a prediction.

Much work still needs to be done to deploy the prediction engine in a real user environment, as we discussed in the section on future work. Predicting user behavior is an elusive art that requires iterative calibration to adapt to the user in a dynamic environment. This is why collaborative filtering could be very beneficial in solving this problem, because who could better to help us predict the users' preferences than the user themselves.

References

1. Xiao Hang Wang, Tao Gu, Da Qing Zhang, Hung Keng Pung: Ontology Based Context Modeling and Reasoning using OWL. In: Second IEEE Annual Conference on Pervasive Computing and Communications Workshops. (2004) 18–22
2. Mark van Setten, Stanislav Pokraev, Johan Koolwaaij: Context-Aware Recommendations in the Mobile Tourist Application COMPASS. In: Adaptive Hypermedia 2004. Volume 3137 of LNCS. (2004) 235–244
3. L. Ardissono, A. Goy, G.P.: INTRIGUE: Personalized recommendation of tourist attractions for desktop and handset devices. Applied Artificial Intelligence **17** (2003) 687–714
4. Keith Cheverst, Nigel Davies, Keith Mitchell, Adrian Friday, Christos Efstratiou: Developing a Context-aware Electronic Tourist Guide: Some Issues and Experiences. In: CHI. (2000) 17–24
5. Joseph F. McCarthy: Pocket RestaurantFinder: A Situated Recommender System for Groups. In: Workshop on Mobile Ad-Hoc Communication at the 2002 ACM Conference on Human Factors in Computer Systems. (2002)

6. J.B. Schafer, J. Konstan, J. Riedl: Recommender Systems in E-Commerce. In: Proceedings of the 1st ACM conference on Electronic commerce, New York, NY, USA, ACM Press (1999) 158–166
7. R.D. Lawrence, G.S. Almasi, V. Kotlyar, M.S. Viveros, S.S. Duri: Personalization of Supermarket Product Recommendations. Data Mining and Knowledge Discovery **5** (2001) 11–32
8. Jonathan L. Herlocker, Joseph A. Konstan, John Riedl: Explaining Collaborative Filtering Recommendations. In: Computer Supported Cooperative Work. (2000) 241–250
9. Jonathan L. Herlocker, Joseph A. Konstan, Al Borchers: An algorithmic framework for performing collaborative filtering. In: Proceedings of the 22nd Annual International ACM SIGIR Conference on Research and Development in Information Retrieval, New York, NY, USA, ACM Press (1999) 230–237
10. John S. Breese, David Heckerman, Carl Kadie: Empirical Analysis of Predictive Algorithms for Collaborative Filtering. In: Proceedings of the Fourteenth Conference on Uncertainty in Artificial Intelligence. (1998) 43–52
11. Albrecht Schmidt, Michael Beigl, Hans-W. Gellersen: There is more to context than location. Computers and Graphics **23** (1999) 893–901
12. VirtualTourist.com: (http://www.virtualtourist.com)

Mobile Context Inference Using Low-Cost Sensors

Evan Welbourne[1], Jonathan Lester[2], Anthony LaMarca[3], and Gaetano Borriello[1,3]

[1] Department of Computer Science & Engineering,
[2] Department of Electrical Engineering,
University of Washington,
Box 352350, Seattle, WA 98195 USA
{evanwel, jlester}@u.washington.edu
[3] Intel Research Seattle,
1100 NE 45th Street, Suite 600,
Seattle, WA 98105 USA
{anthony.lamarca, gaetano.borriello}@intel.com

Abstract. In this paper, we introduce a compact system for fusing location data with data from simple, low-cost, non-location sensors to infer a user's place and situational context. Specifically, the system senses location with a GSM cell phone and a WiFi-enabled mobile device (each running Place Lab), and collects additional sensor data using a 2" x 1" sensor board that contains a set of common sensors (e.g. accelerometers, barometric pressure sensors) and is attached to the mobile device. Our chief contribution is a multi-sensor system design that provides indoor-outdoor location information, and which models the capabilities and form factor of future cell phones. With two basic examples, we demonstrate that even using fairly primitive sensor processing and fusion algorithms we can leverage the synergy between our location and non-location sensors to unlock new possibilities for mobile context inference. We conclude by discussing directions for future work.

1 Introduction

The near ubiquitous deployment of cell phones and other ever more advanced mobile platforms in our daily lives has spurred an increased interest in mobile context inference. In the past few years, the decreasing cost of on-device location technology has led to a number of systems that use location traces to learn a person's significant places and his daily patterns of residence and transit among those places [1, 2, 3, 7, 17]. In other recent work, researchers have developed mobile systems that use simple, low-cost, non-location sensors to infer a user's (or a device's) situational context (e.g. activity, whether or not the device is being carried) [4, 5, 6, 19]. Some of these systems have used their sensors to infer high-level context about a user's place and situation, but only in simple ways and with limited flexibility. For example, Patterson et. al. used GPS logs and instantaneous velocity to recognize transit activities such as driving a car or riding a bus, but could only do so with a fairly regular GPS signal, and by using a large amount of a priori information (e.g. street maps and bus schedules).

T. Strang and C. Linnhoff-Popien (Eds.): LoCA 2005, LNCS 3479, pp. 254–263, 2005.

Our goal is to truly fuse location and non-location sensors and to leverage the synergy between them to enable a wider variety of high-level mobile context inference. In this paper, we introduce a system that is a significant next step in this direction. The system uses Place Lab [18] to provide indoor-outdoor location information with an average resolution of 20 - 30 meters in covered areas. To collect additional sensor data we use the 2" x 1" multimodal sensor board (MSB), which contains a variety of common non-location sensors (e.g. accelerometers, barometric pressure sensors). We illustrate the fundamental advantages of our system by combining location context with simple movement detection in two representative examples. In particular, we show that like previous systems, our system can:

- Classify a mode of transit (as driving, walking, or running) when the user is moving;
- Extract significant places within a user's daily movements.

However, unlike previous systems, our system can also:

- Classify mode of transit without GPS, precise velocity information, or learned knowledge of transit routines;
- Classify or highlight significant places based on the activity that occurs there.

Our principal contribution is to show how the problem of mobile context inference can be tackled using technologies that are or will be readily available in common computing platforms that are becoming truly ubiquitous, such as cell phones and wrist-watches. By using Place Lab as our location system we alleviate the need for costly infrastructure as well as for uncommon hardware, instead relying on existing infrastructure (cell phone towers and wireless access points) and commodity hardware (WiFi cards/chips and GSM receivers). We use sensors that vendors could easily incorporate into their mobile platforms; in fact, several of them have already done so. For example, Pantech's PH-S6500 phone is equipped with six motion sensors that are used to detect how fast and how far its owner has walked, while new IBM ThinkPads are equipped with accelerometers that measure forces so as to park the hard drive in the event that the laptop is dropped. With our centralized system design, we also collect and fuse location and non-location sensor data without requiring sensors or computation to be distributed around the body.

The remainder of this paper is organized as follows. In the next section we review recent work in mobile context inference. In section 3, we describe our system and its components in detail. Then, in section 4, we present two examples that demonstrate how our system can fuse location and non-location sensor data to expand the range of possibilities for mobile context inference. Before concluding in section 6, we discuss directions for future work.

2   Related Work

There has been a substantial amount of work in mobile context inference since the early 1990's. In the past five years, research has focused increasingly on systems that could be deployed ubiquitously in the near future by leveraging existing mobile infrastructure (e.g. commodity GPS devices, cell phones).   Specifically, recent projects have tended toward one of two areas: location-based user modeling or situational context awareness using simple, non-location sensors.

A number of location-based projects have used long-term GPS or GSM traces of a user's daily movements (on the order of weeks) to generate a probabilistic model of that user's residence and travel patterns [1, 3, 4, 17]. The resulting systems can help determine where a user is and where she is going, but fall short of being able to determine other kinds of equally important context (e.g. what she is doing at that place).   Moreover, these schemes are limited by the drawbacks of the location technology used: GPS-based systems have problems in urban canyons and indoors, while GSM cellular location data is often too coarse to allow distinction of neighboring places. Kang et. al. [10] circumvented these problems by using a WiFi-based location system (Place Lab) and were able to perform indoor place extraction with a location resolution of 20 - 30 meters [18]. We show how to augment Kang's method with basic movement detection using a common 3-axis accelerometer that costs about $10 US.

Most situational context-awareness projects have focused on human activity inference (e.g. walking, speaking, riding a bicycle) using multiple simple, body worn sensors [5, 6, 19]. This approach allows determination of what a user is doing at a particular point in the day, but falls short of being able to frame that activity in the context of his larger patterns of daily movement. Schmidt et. al. built a system with a multi-sensor board and GSM cell phone that is quite similar to ours, but focused much less on location.

A few systems [1, 4, 8] have combined place and situational context, but have been limited both in approach and in the technology used. Both Patterson and Liao [1, 7, 8] use incoming GPS data to infer mode of transit (e.g. foot, car, bus), but rely on a large corpus of learned travel patterns, and on prior knowledge of street maps and bus schedules. Marmasse and Schmandt process GPS and accelerometer data in real-time to distinguish transit activities such as walking and driving between places, but require that GPS signal be continually available and that sensors be distributed across the user's body rather than centralized on one device.

We overcome these limitations by using a single location system that works both indoors and outdoors, and by fusing location information with information from a collection of simple, non-location sensors. The use of these additional sensors can simplify the handling of location information by reducing or eliminating the need for a priori knowledge and extensive training. We also group all additional sensors on a centralized, worn device.

3   System Components and Sensors

Our system consists of three components: a GSM cell phone, a waist-worn, WiFi-enabled mobile device, and a multi-modal sensor board (MSB).   The cell phone provides GSM data for location and could also be used to take photos and video during experience sampling method (ESM) studies. The mobile device provides WiFi data for location and is also the central processing and storage unit.   The MSB captures a variety of sensor data and is discussed in more detail below. This design was developed with the foresight that the phone and mobile device functions would eventually be merged into a GSM PDA phone with WiFi.

The cell phone and the mobile device both run Place Lab software and communicate via Bluetooth; fusion and logging of location data is performed on the mobile device. The MSB is attached via USB or Compact Flash to the mobile device, which is responsible for both logging and processing its sensor data in real-time.

Our first implementation used a Nokia 6600 cell phone and a laptop with the MSB. After conducting our initial experiments (see section 4), we also assembled a more compact version of the system that uses a Nokia 6600, the MSB, and an iPaq hx4705. We plan to use the iPaq-based system in user studies because the iPaq hx4705 is smaller and less intrusive than most other mobile devices.

3.1   Multi-modal Sensor Board

The MSB (Figure 1) is a 2" x 1" sensor platform we built to allow us to experiment with different sensors. The MSB contains seven sensors (listed along with sampling rates in Table 1) and is capable of running independently, collecting data from all seven sensors, and then relaying that information to a handheld, laptop, or to Intel's iMote where it can be processed or stored. The iMote is an ARM7 Bluetooth sensor node which can be attached to the MSB [11, 12]. The MSB has an on-board ATmel ATMega128L microprocessor running at 7.3728 MHz which is capable of performing only simple data analysis before relaying the data to the host device.

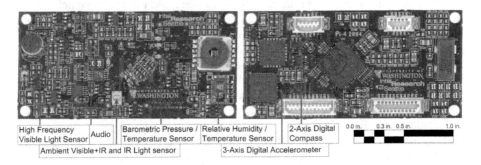

Fig. 1. Photograph of the top and bottom of the MSB

Table 1. The sensors available on the MSB and their sampling rates

Manufacturer	Part No.	Description	Sampling Rate
Panasonic	WM-61A	Analog Electric Microphone	15,360 Hz
Osram	SFH-3410	Analog Visible Light Phototransistor	549 Hz
STMicro	LIS3L02DS	3-Axis Digital Accelerometer	548 Hz
Honeywell	HMC6352	2-Axis Digital Compass	30 Hz
Intersema	MS5534AP	Digital Barometer / Temperature	14 Hz
TAOS	TSL2550	Digital Ambient (IR and Visible+IR) Light	5 Hz
Sensirion	SHT15	Digtial Humidity / Temperature	2 Hz

In our current setup we connect the MSB directly to a laptop or iPaq though either a USB or Compact Flash connection. The sensors available on the MSB were chosen to simplify future development; a majority of the sensors are simple digital components that could be easily integrated into devices such as cell phones. This allows us to experiment with a rich sensor set and to easily determine which sensors are appropriate for our goals.

4 Mobile Context Inference Examples

To demonstrate how our system can fuse location and non-location sensor data to infer high-level context and to enable new types of mobile context inference, we apply it to two common problems in location-based user modeling. Specifically, we show that like related systems [1, 2, 3, 4, 7, 8, 10, 17] our system can perform basic mode of transit inference and extract significant places from a user's daily movements. Further, we show that we can infer walking vs. driving without GPS, precise velocity information, or prior knowledge of transportation routines; we also present an example of sub-place identification and classification using a primitive form of activity recognition (movement detection). Both examples are presented and discussed below in more detail.

4.1 Mode of Transit

Using time, location, accelerometer data, and a very simple classification scheme, we can distinguish three modes of transit: walking, running, and riding in a vehicle. With an average location history of 25 to 120 seconds (shorter histories for areas with dense AP coverage, longer histories for sparser areas) we can classify a user as not in transit if she is settled within a geometric region or in transit if she is not settled within any geometric region. We can then use the location history to compute a rough estimate of the user's speed, which is classified as "vehicle speed" if she is in transit and consistently faster than 9 m/s (approximately 20 miles/hour). We can further distinguish the mode of transportation using the accelerometer data.

Previous work in bio-mechanics has shown that the frequencies of interest for walking motions are below 10Hz [13]. Using the accelerometer data from the MSB, we can perform a simple sum of the FFT coefficients from 0.5Hz to 3Hz (which contains the first harmonic of the walking and running motions) to get a rough idea of the movements being experienced. Since we only want a rough result and intend to perform this calculation on devices with limited power and computation, we use the Goertzel algorithm [14] which allows us to compute DFT coefficients very efficiently with a 2nd order filter. We store two seconds worth of acceleration magnitudes in a buffer and compute a binary movement/no-movement estimate three times a second with a simple threshold of the sum (a sum of greater than 10 indicates movement and a sum of less than 10 indicates no movement).

Figure 3, shows some sample data where we have the recorded acceleration, the binary movement/no-movement estimate, and the estimated speed for a user in an area densely populated with 802.11 access points (Seattle's university district). At the beginning of the recording the user was walking, switched to running around the 150 second mark, and then got into and drove a car down a busy street from the 225 second mark until the 450 second mark. While driving, the user stopped at traffic lights near the 260 second and 325 second marks, pulled into a parking lot around the 450 second mark, and finally parked near the 460 second mark.

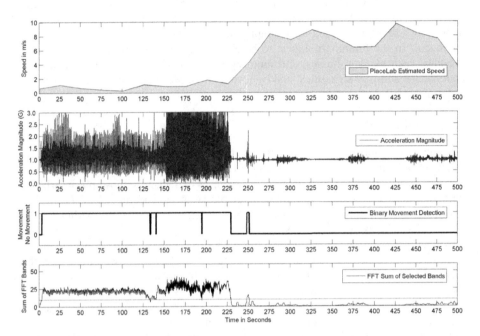

Fig. 3. The sample sensor trace with the user walking for ~150 seconds, then running until the ~225 second mark, then entering a car and driving for the remainder of the recording (with stops at the ~260 and ~325 second marks, and parking in a lot starting at the ~450 second mark). The summed frequency components of the acceleration are used along with acceleration magnitude and the "in transit" classification from the location history to determine whether the user is walking, running, or driving/riding in a vehicle. The top graph shows the PlaceLab estimated speed, calculated every 25 seconds. The 2nd graph shows the acceleration magnitude recorded from our sensor board, in G's. The 3rd graph shows the output of our binary movement (1) / no movement (0) threshold. The 4th graph shows the raw Goertzel sum calculated from the acceleration, the threshold at 10 is shown as a solid line across this graph. Points above the threshold are classified as 1, moving, and points below the line are classified as not moving, 0

Throughout this recording, the user was classified as in transit, and his estimated speed was calculated every 25 seconds. We can see that for almost the entire trace, the user's estimated speed was below 9 m/s so that estimated speed alone couldn't provide a reliable mode of transit classification. Thus, given that the user is in transit, we look at the binary movement/no-movement estimate and acceleration magnitude to distinguish mode of transit. We can see that for the walking and running phases, the binary movement/no-movement estimate (the black line in the 3rd graph in Figure 3,valued at 0 or 1) was almost always 1, while for the driving phase the movement/no-movement estimate was almost always 0. From this we an infer that the user was walking or running for the first 225 seconds, and in a vehicle for the remainder of the trace (because he was in transit but classified as exhibiting no movement). By looking at the acceleration magnitude for the first 225 seconds, it is also clear that the user was walking for the first 150 seconds, and running for the next 75.

While this overly simple method for inferring mode of transit inference may not be as accurate as other machine learning techniques (and could certainly be improved upon – by adding audio data for example), it shows the promise of our multi-sensor approach. In the future, we expect to be able to distinguish among different motor vehicles, such as cars, busses, and motorcycles, by their characteristic acceleration patterns and acoustic signatures; and by leveraging more advancing machine learning techniques [20].

4.2 Significant Places

We can also combine place extraction with activity recognition to perform more sophisticated inference about significant places. For example, we could "label" a place with the set of activities that occurs there or extract only those places where a particular activity occurs. Our simple example is an enhancement to Kang's place extraction method and uses information about the user's movements within a given place. Specifically, if a user has spent 5 or more minutes within a place and our movement-detection algorithm reports that he has not moved, then we decide that he is at a desk or some other significant, fixed station. In this case, we might label that place as one where a stationary activity occurs. Alternatively, we can "zoom-in" on that location by taking a WiFi fingerprint [15]; this provides us with a way to distinguish high-resolution (2 - 5 meter) sub-places that are currently finer-grained than Place Lab's average resolution (20 - 30 meter) allows.

To test our "zoom-in" enhancement to Kang's method, we recorded a day's worth of place and movement data for one student on a university campus. A visualization of the collected data (Figure 4) illustrates how our algorithm can extract multiple indoor places and sub-places. This basic example demonstrates how using an indoor-outdoor location technology in combination with a collection of simple, non-location sensors allows us to explore the relationship between activity and place.

5   Future Directions

The preceding examples have demonstrated how our system can fuse location and activity to solve common problems addressed by previous mobile context inference systems. We have also shown how our system could unlock new possibilities for mobile context inference. In this section we discuss possible directions for future work with our system.

5.1   Infrastructure Assist

One key benefit of fusing location and non-location sensors is that we can use the result to assist the underlying location infrastructure. For example, mode of transit could be used to select an appropriate motion model for a location particle filter

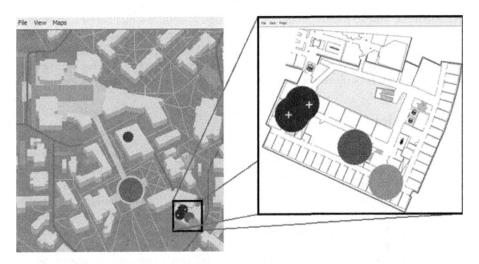

Fig. 4. The places and sub-places extracted from a day's worth of data collection using our movement-detector-enhanced place extraction algorithm. The zoomed view shows three places within the same building (left to right): a student area, a hardware lab, and a faculty office. In the student area, the student spent time sitting at a desk and on a couch - both of these positions were recorded as sub-places (denoted '+' and depicted at an arbitrary position inside the place) and stamped with a WiFi fingerprint. No sub-places were identified in the hardware lab because the student was walking back and forth between tables, and no sub-places were recorded in the faculty office because the student spent only a short time there

which would in turn improve Place Lab's accuracy. Collecting fingerprints of sub-places that are identified with activity (e.g. sitting) is another way to distinguish places at a higher resolution than what Place Lab currently offers. It is likely that other improvements can be made to Place Lab using data from more sensors. For example, digital compass data could be used to further augment a motion model by adding directional information.

5.2 Activity and Place

The meanings of place and activity are often closely tied. For example, it is often the case that a place is defined by the activities that occur there (e.g. a conference room is a place where meetings are held). Our system provides us with a rare opportunity to work with place and indoor activity – usually where most activities occur, and where the most time is spent. Some ideas in this area that may be worthy of investigation include the labeling of places with the activities that occur there, and work with the idea of a "mobile place" (e.g. a bus or train).

5.3 User Studies

We should be able to use our compact iPaq-based system in a number of user studies that rely on place and situational context. For example, we plan to use the iPaq-based system for a long term study in which data is collected from users during daily life, and annotated using the experience sampling method. We could also use this system to study the effects of place and activity on interruptability and prompting.

6 Conclusions

We have presented and described a new system for mobile context inference that fuses location and non-location sensors, and is designed for deployment on future computing platforms that will be truly ubiquitous (e.g. cell phones, PDAs). Our system has the primary advantage that it uses Place Lab for both indoor and outdoor location using existing infrastructure; and by using low-cost, simple sensors (the 3-axis accelerometer on the MSB costs about $10 US) that are already appearing on many emerging mobile platforms, we can widen the possibilities for mobile context inference and strengthen the robustness of our inference algorithms. We presented two examples of how our system can solve common problems (inferring mode of transit and extracting significant places), and unlock new possibilities unavailable to previous systems. Finally, we have discussed a number of promising areas for future work.

References

1. Patterson, D., Liao L., Fox, D., and Kautz, H., "Inferring High-Level Behavior from Low-Level Sensors", The Fifth International Conference on Ubiquitous Computing, 2003.
2. Ashbrook, D., and Starner, T., "Using GPS to Learn Significant Locations and Predict Movement Across Multiple Users", Personal and Ubiquitous Computing, v.7 n.5, pp. 275 – 286, 2003.
3. Marmasse, N., and Schmandt, C., "A User-Centered Location Model", Personal and Ubiquitous Computing, pp. 318 – 321, 2002.
4. Marmasse, N., Schmandt, C., and Spectre, D., "WatchMe: Communication and Awareness Between Members of a Closely-Knit Group", The Sixth International Conference on Ubiquitous Computing, , pp. 214 – 231, 2004.
5. Schmidt, A., Aidoo, K., Takaluoma, A., Tuomela, U., Laerhoven, K., and Velde, W. "Advanced Interaction in Context", Proceedings of the 1st international symposium on Handheld and Ubiquitous Computing, pp. 89 – 101, 1999.
6. Bao, L., and Intille, S., "Activity Recognition From User-Annotated Acceleration Data", Proceedings of the Second International Conference on Pervasive Computing, 2004.
7. Patterson., D., Liao, L., Gajos, K., Collier, M., Livic, N., Olson, K., Wang, S., Fox, D., and Kautz, H., "Opportunity Knocks: a System to Provide Cognitive Assistance with Transportation Services", The Sixth International Conference on Ubiquitous Computing, 2004.
8. Liao, L., Fox, D., and Kautz, H., "Learning and Inferring Transportation Routines", Proc. of the National Conference on Artificial Intelligence, 2004.
9. Stäger, M., Lukowicz, P., Perera, N., Büren, T., Tröster, G., and Starner, T., "SoundButton: Design of a Low Power Wearable Audio Classification System", Seventh IEEE International Symposium on Wearable Computers, pp. 12 – 17, 2003.

10. Kang, J., Welbourne, W., Stewart, B., and Borriello, G., "Extracting places from traces of locations", Proceedings of the 2nd ACM international workshop on Wireless mobile applications and services on WLAN hotspots, pp. 110 – 118, 2004.

11. Hill, J., et al., "The platforms enabling wireless sensor networks", Communications of the ACM, pp 41-46 , vol 47, no. 6, 2004.

12. Culler, D., and Mulder, H., "Smart Sensors to Network the World", Scientific American, pp 84-91, 2004.

13. Winter, D.: Biomechanics and Motor Control of Human Movement (2nd ed.). New York: Wiley (1990).

14. G. Goertzel, "An Algorithm for the Evaluation of Finite Trigonometric Series", Amer. Math. Month., vol. 65, pp. 34-35, Jan., 1958.

15. P. Bahl and V. N. Padmanabhan, "RADAR: An RF-Based In-Building User Location and Tracking System", Proc. IEEE Infocom, March 2000.

16. LaMarca A., Chawathe Y., Consolvo S., Hightower J, Smith I., Scott J., Sohn T., Howard J., Hughes J., Potter F., Tabert J., Powledge P., Borriello G., Schilit B. "Place Lab: Device Positioning Using Radio Beacons in the Wild", Intel Research Technical Report: IRS-TR-04-016.

17. Laasonen, K., Raento, M., and Toivonen, H., "Adaptive On-Device Location Recognition", In Proceedings of the 2nd International Conference on Pervasive Computting, April 2004.

18. LaMarca, A., et al., "Place Lab: Device Positioning Using Radio Beacons in the Wild." International Conference on Pervasive Computing (Pervasive 2005), Munich, Germany, May 2005 (to appear)

19. Siewiorek, D., et. al., "SenSay: A Context-Aware Mobile Phone", IEEE International Symposium on Wearable Computers (ISWC 2003), New York, New York 2003.

20. Brunette, W., et. Al., "Some Sensor Network Elements for Ubiquitous Comput-ing", The Fourth International Conference on Information Processing in Sensor Networks (IPSN 2005), Los Angeles, CA, 2005 (to appear).

Where am I: Recognizing On-body Positions of Wearable Sensors

Kai Kunze[1] Paul Lukowicz[1,2], Holger Junker[2], Gerhard Tröster[2]

[1]Institute for Computer Systems and Networks UMIT Hall i. Tirol, Austria
csn.umit.at
[2]Wearable Computing Lab ETH Zurich, Switzerland
www.wearable.ethz.ch

Abstract. The paper describes a method that allows us to derive the location of an acceleration sensor placed on the user's body solely based on the sensor's signal. The approach described here constitutes a first step in our work towards the use of sensors integrated in standard appliances and accessories carried by the user for complex context recognition. It is also motivated by the fact that device location is an important context (e.g. glasses being worn vs. glasses in a jacket pocket). Our method uses a (sensor) location and orientation invariant algorithm to identify time periods where the user is walking and then leverages the specific characteristics of walking motion to determine the location of the body-worn sensor.

In the paper we outline the relevance of sensor location recognition for appliance based context awareness and then describe the details of the method. Finally, we present the results of an experimental study with six subjects and 90 walking sections spread over several hours indicating that reliable recognition is feasible. The results are in the low nineties for frame by frame recognition and reach 100% for the more relevant event based case.

1 Introduction

A promising approach to context and activity recognition is the use of motion sensors (predominantly accelerometers) attached to different parts of the user's body. Various types of activities ranging from simple modes of locomotion analysis [14] to complex everyday [2] and assembly tasks[8] have been successfully recognized using such sensors. One thing that most of the work in this area has in common is that it relies on sensors being placed at specific locations on the body. Typically this includes the wrists, the arms, legs, hips, the chest and even the head. Once a subset of locations has been chosen, the system is trained on this specific subset and will not function properly if the sensors are placed at a different locations. This implies that the user either has to explicitly 'put on' the sensors each time he/she dresses up or the sensors have to be permanently integrated into the individual pieces of clothing.

T. Strang and C. Linnhoff-Popien (Eds.): LoCA 2005, LNCS 3479, pp. 264–275, 2005.

For wide spread and more near term applications of context recognition it would be more convenient to place sensors in appliances and accessories that most of us carry with us on a daily basis anyway. Such appliances include mobile phones, PDAs, key-chains, watches, hearing aids, headphones, badges and other smart cards and soon maybe even displays in glasses. Most of these devices are equipped with electronics and some already have sensors, including motion sensors (e.g Casio camera watch).

1.1   Context Recognition with Standard Appliance

While convenient for the user, the use of standard appliances and accessories for context recognition poses considerable difficulties. The main problem is that one can never be sure which devices the user carries with him and where on the body they are located. A mobile phone might be in a trousers pocket, in a holster on the hips, in a jacket or shirt pocket or in a backpack. Similar is true for a PDA, a wallet or a key-chain. Even such location specific devices as a watch, headphones or glasses might sometimes be carried in a pocket or a backpack.

As a consequence any system using standard appliances for context recognition must address the following issues:

1. Enough different body locations must be covered by sensor enabled appliances to provide sufficient information for the recognition task at hand.
2. The system is either able to deduce where on the body each device is located at any given moment or the recognition algorithm is location invariant.
3. The system must either be able to deduce the orientation of each device or the recognition algorithm is orientation invariant.
4. If the recognition algorithm is not location invariant, then it must be able to deal with different combinations of locations.

The work described in this paper constitutes the first step in our effort to facilitate the use of such appliances for complex context recognition. It focuses on the second point: the device location. We show that the most common human activity, namely walking, can be recognized in a location independent way. We then demonstrate how the information that the user is walking can be leveraged to determine device location.

The motivation for starting with device location is twofold. First we consider it to be the most critical issue. Experience shows that people usually have several accessories with them and mostly carry them at different locations. As an example in a typical scenario the user might carry a key-chain in his trousers pocket giving us the leg information, a watch on the wrist, a mobile phone in a holster on the hip and a smart card in a wallet in a jacket pocket. Thus while a systematic study will eventually be needed to determine the most common locations for different applications, at this stage we assume that there are relevant situations where point one is satisfied. Concerning device orientation, it has been shown by [10] and [5] that it can be derived from three axis acceleration sensors by looking at either time periods where the norm of the acceleration signal is 1 (no motion just gravity) or in an approximation by looking at the low pass filtered signal. Finally while there is considerable research potential in optimizing an adaptive classifier, simple solutions also exist proving that it is feasible for a system to deal with variable locations. In the worst case, a separate classifier could be trained for every relevant combination of locations. Since for the classifier it only matters that a sensor is at a certain location, not which device it is embedded in, the number of combinations is reasonable. In addition, only a limited number of all possible combinations will be relevant for a given recognition task.

The second motivation for looking at device location is the fact that location information itself is an interesting context. As an example, knowing if the glasses are worn or if they are in a pocket can be an important clue to the user's activity.

1.2   Related Work and Paper Contributions

The potential of body-worn sensors for context and activity recognition has been demonstrated in many scientific papers [3, 2, 11, 15, 14, 9, 16, 5, 6].

Irrespective of the signal processing schemes and recognition algorithms used, the variety of approaches presented differ in a number of ways. While some approaches rely on different types of sensors (e.g.[11, 15, 14, 5]), others solely use a single type of sensor such as accelerometers ([3, 2, 9, 16, 6]) for the recognition task. Furthermore, the approaches may differ with respect to the type of context targeted for classification ranging e.g. from the classification of walking behavior (e.g. [14]) to the recognition of complex everyday activities [2]. There also exist approaches targeting the same recognition task using the same sensors, but differ with respect to the placement of the sensors. One common example for this is the recognition of walking behavior (level walking, descending and ascending stairs) with accelerometers (e.g. [16, 12]). Despite the many differences, one thing that most of the approaches have in common is the fact that their recognition engines are unaware of the sensor locations and simply rely on individual sensors being placed at specific locations on the body. Consider the case, where e.g. two accelerometers A_1 and A_2 are used for a certain recognition task. The recognition system works properly when sensor A_1 is placed on the wrist, and sensor A_2 mounted on the thigh which is the configuration which the system has been trained for. If the two sensors would be exchanged after training, the system's recognition performance would without debt decrease, if it was not aware of this change. To allow recognition systems to adapt to such changes, it must be aware of the locations of the individual sensors on the body.

To our knowledge the method described in this paper is the first published attempt to facilitate this functionality. This paper describes the details of this novel method and presents an extensive experimental evaluation showing that it produces highly reliable results. The experimental evaluation elaborates not just the final results but also gives an insight into the effect of the different, individual optimizations steps contained in our method.

The work that comes closest to ours has been presented by Lester et. al [7]. They introduced a method to determine if two devices are carried by the same person, by analyzing walking data recorded by low-cost MEMS accelerometers using the coherence function, a measure of linear correlation in the frequency domain. In addition Gellersen et. al have shown how a group of devices can be 'coupled' by being shaken together. Furthermore there were a number of attempts detect whether an appliance such as a mobile phone is on a pocket in the hand or on the table [4].

2   Approach

2.1   General Considerations

Our approach is based on the obvious observation that different parts of the body tend to move in different ways. As an example, hand motions contain much more higher frequency components and larger amplitudes than hip or head motions. In addition, physiological constraints mean that certain types of motions are not permissible at all for some parts of the body (you can not turn your leg around the vertical axis in the knee or tilt your head more than 90 degrees). Also, some parts tend to be motionless for longer periods of time than the others. Thus, in theory, a statistical analysis of the motion patterns over a sufficient period of time should be able to provide information about the location of a sensor on the body. However, when implementing this idea in practice one has to deal with a number of issues. For one, the value of such a statistical analysis depends on the user activity during the analysis window. Little information will for example be gained if the user is sleeping during the whole time. In addition, the signal of a motion sensor placed on a given body part contains a superposition of the motion of this body part with the motion of the body as a whole. Thus while it is not possible to tilt the head more than 90 degrees, such a tilt will be registered when the user lies down. Finally, many of the motion characteristics that can be used to distinguish between body parts involve absolute orientation which is hard to detect, in particular if the orientation of the sensor is not known.

Our method deals with the above issues in two ways:

1. The analysis is constrained to the time during which the user is walking. This is motivated by two considerations. First, walking is a common activity that occurs fairly often in most settings. Thus, being able to detect the position of devices during walking phases should provide us with a sufficiently accurate overall picture of where the devices are located. In addition, once the location has been determined during a walking phase, this knowledge can be used to detect possible changes in placement.

   The second reason for focusing on walking is the fact that walking has such a distinct motion signature that it can be recognized without any assumptions about sensor location.

2. We base our analysis on the norm of the acceleration vector which is independent of the senor orientation.

2.2   Recognition Method

With the considerations described above our method can be summarized as follows.

Features Computation. Basic physical considerations confirmed by initial tests have lead us to use following features computed in a sliding window that is 1 sec long (overlapping 0.5 sec):

RMS: $\sqrt{\frac{1}{N} * \sum_i x_i^2}$, where N is the number of samples a sliding window contains, and x_i the i'th sample of the window.

75%Percentile: Given a signal s(t) the 75%percentile, also known as the third quartile, is the value that is greater than 75% percent of the values of s(t) .

InterQuartileRange: The interquartile range is defined as the difference between the 75th percentile and the 25th percentile.

Frequency Range Power: Computes the power of the discrete FFT components for a given frequency band.

Frequency Entropy: The frequency entropy is calculated according to the following formula: $H_{freq} = -\sum p(X_i) * log_2(p(X_i))$, where X_i are the frequency components of the windowed time-domain signal for a given frequency band and $p(X_i)$ the probability of X. Thus, the frequency entropy is the normalized information entropy of the discrete FFT component magnitudes of the windowed time-domain- signal and is a measure of the distribution of the frequency components in the frequency band. This feature has been used by [1].

SumsPowerWaveDetCoeff: describes the power of the detail signals at given levels that are derived from the discrete wavelet transformation of the windowed time-domain signal. This feature has successfully been used by [13] to classify walking patterns with acceleration sensors.

Training. Using a selection of relevant device positions walking recognition is trained. The recognition is trained in a location independent manner by putting the data from all locations into a single training set. As will be elaborated in the experiments section 3, best results were achieved with a C 4.5 classifier. However Naive Bayes, Naive Bayes Simple and Nearest Neighbor have also produced acceptable recognition rates. In the next phase, data collected during walking is used to train the location recognition.

Recognition. The recognition is performed separately by each sensor using the system trained according to the method described above. It consists of the following steps:

1. *Frame by Frame Walking Recognition* In this phase the features are computed in a sliding window of length 1s as described above and each window is classified as walking or non walking. The window length has been selected such that in a typical case it contains at least one step.

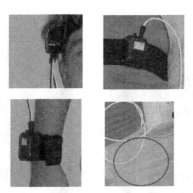

Fig. 1. Sensor placement

2. *Walking Recognition Smoothing* Using another jumping window of length 10 sec jumping by 5 sec the results of the frame by frame walking classification are then smoothed. The smoothing retains only those windows, where more then 70% of the frames were classified as walking. This ensures that the subsequent location classification is based only on 'clean' walking segments.
3. *Walking Segment Localization* The smoothed frame by frame recognition results are then used to localize walking segments that are long enough to allow reliable recognition. We define appropriate length to be at least a few tens of seconds and not longer than a 2 or 3 min. If a walking segment is longer than this boundary, it is automatically divided into several segments. The rationale behind this approach is that most devices are likely to remain in the same place for a few minutes. Changes on a smaller timescale must be considered as isolated events (e.g taking out a phone and rejecting an incoming call) and have to be detected separately by each device.
4. *Frame By Frame Location Recognition* A sliding window of the length of 1 sec. is then applied in each segment that has been identified as a relevant walking event. In each window the features for location recognition are computed and classification is performed.
5. *Event Based Location Recognition* For each segment a majority decision is performed on the frame by frame location classification.

3   Experimental Results

3.1   Experimental Setup

To evaluate the performance of our method an experiment was conducted with 6 subjects. For each subject, 3 experimental runs were recorded. Each run was between 12 and 15 min and consisted of the following set of activities:

1. Working on a desk (writing emails, surfing, browse through a book).
2. Walking along a corridor.

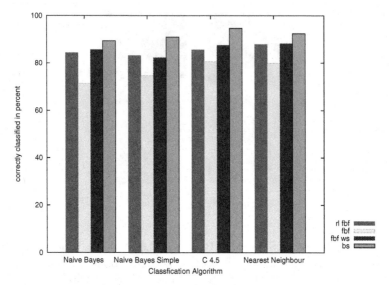

Fig. 2. Overview over the different classification algorithms and the varying approaches. The abbreviations have the following meaning: **rl fbf**=frame by frame using reference labeling (3.3), **fbf** = for frame by frame location recognition using frame by frame walking (3.3 and 2.2), **fbf ws** = frame by frame location recognition using smoothing over walking (2.2), **bs** = smoothing approach for both location and walking (2.2)

3. Making coffee, cleaning coffee kitchen, opening/closing drawers.
4. Walking along a corridor.
5. Giving a 'presentation'.
6. Walking.
7. Walking up and down a staircase.
8. (optional) working at desk.

Thus, there are 18 data sets containing a total of 90 walking segments (including the stairs).

All results presented below are based on a 10 fold-cross validation for evaluation on a subject by subject basis. For classification the WEKA [1] java package was used.

Hardware. The sensor system used for the experiment is the XBus Master System (XM-B) manufactured by XSens [2]. For communication with the XBus a Bluetooth module is used, thus data is transmitted wirelessly. Therefore, the test subjects just have to carry the XBus plus sensors and are not burdened with any additional load. The conductor of the experiment carries a Xybernaut to collect

---

[1] http://www.cs.waikato.ac.nz/ml/weka/
[2] http://www.xsens.com

the data and to supervise the experiment. Four of the XSens Motion Tracker sensors connected to the XBus are affixed on the test subject on four different body parts. The locations that have been chosen represent typical locations of appliances and accessories. Furthermore, they are relevant for the context recognition. Below is a short description of the locations used:

- Sensor 1: Wrist. This simulates a watch or a bracelet while worn.
- Sensor 2: Right side of the head above the eyes. This emulates glasses that are being worn.
- Sensor 3: Left trouser's pocket. This is a typical location for a variety of appliances such as key chains, mobile phones or even a watch that was taken off the wrist.
- Sensor 4: Left breast pocket. Again a typical location that would also include smart cards, glasses, (e.g. in a wallet).

3.2   Location Recognition on Segmented Data

As already mentioned earlier, the location recognition is only done during walking. Thus we begin our analysis by looking at the performance of the location recognition on hand picked walking segments. The results of the frame by frame recognition an all 90 segments contained in the experimental data is shown in figure 2. Using a majority decision on each segment leads to a 100% correct recognition (124 out of 124). The smallest segment is 1 minute long.

3.3   Continuous Location Recognition

Walking Recognition. The first step towards location recognition from a real life, continuous data stream is the detection of walking segments. As shown in Table 1 a frame by frame walking recognition (walking vs. not walking) showed an accuracy between 69% and 95% (mean 82%). However, for our purpose the mere accuracy is not the main concern. Instead we are interested in minimizing the number of false positives, as the subsequent location recognition works correctly only if applied to walking data. Here a mean of 18%(over all experiments) it is definitely to high.

As a consequence a false positive penalty has been added to the classification algorithms. Tests (see Figure 3) have lead to a minimal false positive rate considering a misclassification of 'Not Walking' four times worse than a misclassification of 'Walking'. While the overall correct rates goes down to between 61% and 85% (mean 76%), the percentage of false positives for 'Walking' is reduced to an average of 4% (between 0.5% and 7%).

The best results for the walking recognition is provided by the C4.5 tree algorithm with a mean of 82%, the worst by the Naive Bayes Simple with a mean of 65%.

In the next step the effect jumping window smoothing was investigated showing an average false positive rate of 2.17% with 84% of the windows being correctly recognized.

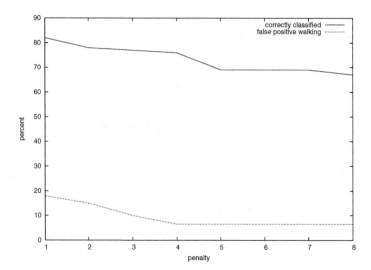

Fig. 3. Relation between correctly classified and false positives for walking

Table 1. Overview Classification for Walking in Percent

		Person1	P2	P3	P4	P5	P6	Mean
Frame by Frame	Correctly Classified	95	69	87	78	82	85	82.67
	False Positives for Walking	14	8	26	10	34	18	18.33
Frame by Frame with penalty	Correctly Classified	83	61	78	75	79	81	76.17
	False Positives for Walking	3	5	0.5	8	6	6	4.75
Frame by Frame penalty, jumping window	Correctly Classified	93	72	89	85	78	92	84,83
	False Positives for Walking	2	3	1	2	2	3	2.17

Walking Segments Location. In the last walking recognition step the walking segment location was applied to the smoothed frame by frame results. This has lead to 124 segments being located, none of which was located in a non-walking section. As shown for an example data set in figure 4 the only deviations from the ground truth was the splitting of single segments and the fact that the detected segments were in general shorter then the ground truth segments. However in terms of suitability for location recognition this is not relevant.

Frame By Frame Location Recognition. With the walking segments detected the frame by frame location recognition was applied. The results are shown in 2. They were later improved using the jumping window smoothing method which has lead to the results shown in 2 and 4.

The confusion matrices depicted in Table 3 indicate that the sensors attached to Head and Breast, as well as, Trousers and Wrist are most often confused. Especially, the confusion between Hand and Trousers is significant in size. One possible reason is that the movement pattern of Hand and Leg is similar while walking, particularly if the test subjects swings with the hand.

Table 2. Mean of C4.5 over all data sets for pre-labeled frame-by-frame (89,81 % correctly classified)

a	b	c	d	← classified as
856	2	87	5	a = Head
21	804	0	12	b = Trousers
101	32	765	4	c = Breast
0	103	5	819	d = Wrist

Table 3. Mean of C4.5 over all data sets for frame-by-frame using frame-by-frame walking recognition (80 % correctly classified)

a	b	c	d	← classified as
567	4	94	4	a = Head
3	431	3	178	b = Trousers
83	32	678	10	c= Breast
12	155	24	754	d =Wrist

Table 4. Mean of C4.5 over all data sets for both smoothed walking and location (94 % correctly classified)

a	b	c	d	← classified as
965	2	31	2	a = Head
0	847	4	49	b = Trousers
42	0	883	1	c= Breast
17	68	10	921	d =Wrist

Event Based Location Recognition. In final step majority decision was performed in each segment leading to an event based recognition. Just like in the hand segmented case *the recognition rate was 100 %*.

4   Conclusion and Future Work

The work described in this paper constitutes a first step towards the use of sensors integrated in standard appliances and accessories carried by the user for complex context recognition. It is also motivated by the relevance of device location for general user context.

We have introduced a method that allows us to recognize where on the user's body an acceleration sensor is located. The experimental results presented above indicate that the method produces surprisingly reliable results. The method has found all walking segments in each experiment and has produced perfect event based recognition. Note that for practical use such event based recognition and not the less accurate frame by frame results that are relevant.

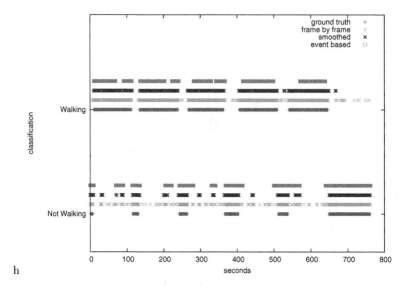

Fig. 4. Sample set containing different approaches for recognizing the walking segments

Despite this encouraging results it is clear that much work still remains to be done. The main issue that needs to be addressed is the detection of location changes that happen when the user is not walking and short duration location changes occurring during walking (e.g. taking out a mobile phone events during walking). Here, we see communication and cooperation between different devices as the key. Thus, for example, if all devices but the one located in a breast pocket detect walking then it must be assumed that something is happening to this device. Similar conclusions can be drawn if devices known to be in the trousers side pockets detect the user sitting and all but one devices report no or little motion while a single devices detects intensive movement. How such device cooperation can be used in real life situations and how reliable results it can produce is the subject of the next stage of our investigation.

Another interesting issue is the study of how well the recognition method can distinguish between locations that are close to each other on the same body segment. Finally we will need to see how well the actual context recognition works with standard appliances given only approximate location and not a sensor tightly fixed to a specific body part.

References

1. L. Bao. Physical activity recognition from acceleration data under semi-naturalistic conditions. Master's thesis, MIT, 2003.
2. L. Bao and S.S. Intille. Activity recognition from user-annotated acceleration data. In F. Mattern, editor, *Pervasive Computing*, 2004.

3. Ozan Cakmakci, Joelle Coutaz, Kristof Van Laerhoven, and Hans-Werner Gellersen. Context awareness in systems with limited resources.

4. Hans W. Gellersen, Albercht Schmidt, and Michael Beigl. Multi-sensor context-awareness in mobile devices and smart artifacts. *Mob. Netw. Appl.*, 7(5):341–351, 2002.

5. N. Kern, B. Schiele, H. Junker, P. Lukowicz, and G. Tröster. Wearable sensing to annotate meeting recordings. In *Proceedings Sixth International Symposium on Wearable Computers ISWC 2002*, 2002.

6. N. Kern, B. Schiele, and A. Schmidt. Multi-sensor activity context detection for wearable computing. 2003. European Symposium on Ambient Intelligence.

7. J. Lester, B. Hannaford, and G. Borriello. "are you with me?" - using accelerometers to determine if two devi ces are carried by the same person. In A. Ferscha and F. Mattern, editors, *Pervasive Computing*, 2004.

8. P. Lukowicz, J. Ward, H. Junker, M. Staeger, G. Troester, A. Atrash, and S. Starner. Recognizing workshop activity using body worn microphones and accelerometers. In *Pervasive Computing*, 2004.

9. J. Mantyjarvi, J. Himberg, and T. Seppanen. Recognizing human motion with multiple acceleration sensors. In *2001 IEEE International Conference on Systems, Man and Cybernetics*, volume 3494, pages 747–752, 2001.

10. D. Mizell. Using gravity to estimate accelerometer orientation. In *Proceedings Seventh International Symposium on Wearable Computers IS WC 2002*, 2003.

11. A. Schmidt, K. A. Aidoo, A. Takaluoma, U. Tuomela, K. Van Laerhoven, and W. Van de Velde. Advanced interaction in context. *Lecture Notes in Computer Science*, 1707:89–93, 1999.

12. M. Sekine, T. Tamura, T. Fujimoto, and Y. Fukui. Classification of walking pattern using acceleration waveform in elderly people. In D. Enderle, J., editor, *Proceedings of the 22nd Annual International Conference of the IEEE Engineering in Medicine and Biology Society Cat*, volume 2, 2000.

13. M. Sekine, T. Tamura, T. Fujimoto, and Y. Fukui. Classification of walking pattern using acceleration waveform in elderly people. *Engineering in Medicine and Biology Society*, 2:1356 – 1359, Jul 2000.

14. L. Seon-Woo and K. Mase. Recognition of walking behaviors for pedestrian navigation. In *Proceedings of the 2001 IEEE International Conference on Control Applications (CCA'01) (Cat*, pages 1152–1155, 2001.

15. K. Van-Laerhoven and O. Cakmakçi. What shall we teach our pants? In *Digest of Papers. Fourth International Symposium on Wearable Computers.*, pages 77–83, 2000.

16. P. H. Veltink, H. B. J. Bussmann, W. de Vries, W. L. J. Martens, and R. C. Van-Lummel. Detection of static and dynamic activities using uniaxial accelerometers. *IEEE Transactions on Rehabilitation Engineering*, 4(4):375–385, Dec. 1996.

Context Obfuscation for Privacy via Ontological Descriptions*

Ryan Wishart[1], Karen Henricksen[2], and Jadwiga Indulska[1]

[1] School of Information Technology and Electrical Engineering,
The University of Queensland
{wishart, jaga}@itee.uq.edu.au
[2] CRC for Enterprise Distributed Systems Technology
kmh@dstc.edu.au

Abstract. Context information is used by pervasive networking and context-aware programs to adapt intelligently to different environments and user tasks. As the context information is potentially sensitive, it is often necessary to provide privacy protection mechanisms for users. These mechanisms are intended to prevent breaches of user privacy through unauthorised context disclosure. To be effective, such mechanisms should not only support user specified context disclosure rules, but also the disclosure of context at different granularities. In this paper we describe a new obfuscation mechanism that can adjust the granularity of different types of context information to meet disclosure requirements stated by the owner of the context information. These requirements are specified using a preference model we developed previously and have since extended to provide granularity control. The obfuscation process is supported by our novel use of ontological descriptions that capture the granularity relationship between instances of an object type.

1   Introduction

The use of context information has been widely explored in the fields of context-aware applications and pervasive computing to enable the system to react dynamically, and intelligently, to changes in the environment and user tasks. The context information is obtained from networks of hardware and software sensors, user profile information, device profile information, as well as derived from context information already in the system. Once collected, the context information is usually managed by a context management system (CMS) in a context

---

* The work reported in this paper has been funded in part by the Co-operative Research Centre for Enterprise Distributed Systems Technology (DSTC) through the Australian Federal Government's CRC Programme (Department of Education, Science, and Training).

T. Strang and C. Linnhoff-Popien (Eds.): LoCA 2005, LNCS 3479, pp. 276–288, 2005.

knowledge base as a collection of context facts. The pervasiveness of the context collection means that the context facts provide ample fodder for malicious entities to stage attacks on the context owner. These attacks may range from annoying targetted advertisements to life threatening stalking. As the information was collected to be used by the pervasive network and context-aware applications, total prevention of context information disclosure is counter productive. Thus, privacy mechanisms are needed to protect the context owner's privacy by applying access control to queries on the context information. These privacy mechanisms either authorise or prohibit context information disclosure based on a set of preferences established by the context owner.

Preferences that only allow or deny access to the context information are very course-grained, and prevent context owners from releasing less detailed, though still correct, context information. In this paper we present a preference mechanism that gives context owners fine-grained control over the use and disclosure of their context information. This would enable a context owner to provide detailed location information to family members, and low granularity information, e.g., the current city they are located in, to everyone else.

The mechanism we present to support this operates over different context types, and provides multiple levels of granularity for disclosed context information. The obfuscation procedure it uses is supported by detailed ontological descriptions that express the granularity of object type instances. Owner preferences for disclosure and obfuscation are expressed using an extended form of a preference model we previously developed for context-aware applications [7].

The remainder of the paper is structured as follows. Section 2 provides an overview of related work in obfuscation of context information. In Section 3 we present an example context model which we use as a vehicle for later examples. The paper then continues, in Section 4, with ontologies of object instances and our novel approach to using them to capture granularity levels for the purpose of context obfuscation. Our preference language for controlling obfuscation and expressing privacy constraints is discussed in Section 5, while the evaluation of these preferences is discussed in Section 6. An example demonstrating the operation of the obfuscation mechanism is then presented in Section 7. Finally, our concluding remarks and discussion of future work are provided in Section 8.

2   Related Work

Granular control over the release of context information was established by Hong and Landay [10] as a desirable feature for privacy protection. The notion of granular control is explored by Johnson et al. [11] who refer to it as "blurring". However, the technique is only mentioned with respect to location and spatial context administered by a Location Management System presented in the paper. Restricting the "blurring" functionality to a limited subset of the context information handled by the CMS is undesirable, as the CMS supports many types of context information, all of which should be potentially available at different levels of obfuscation.

The P3P [4] (Platform for Privacy Preferences) standard is intended for specifying information usage policies for online interactions. The supporting APPEL [3] language can be used to express user privacy preferences regarding the P3P usage policies. Although P3P can provide data at different levels of granularity, it does not engage in obfuscation in the sense that specific data is modified to meet disclosure limits on granularity. In addition, as the preference rules either evaluate to true or false, requests on user context are either permitted or rejected. There is no concept of plausible deniability, where the system can dynamically provide "white lies" for the user or give a vague answer such as "unknown", with the interpretation left to the recipient.

The obfuscation of context information to different levels of precision is mentioned by Lederer et al. [12], who propose four different levels of context information: precise; approximate; vague; and undisclosed. This approach of using only a set number of levels hampers the user's control over the disclosure of the context. For example, assume exact location, the building the user is currently in, the suburb, and the city in which the user is located are provided as the four levels of obfuscation. A user not wanting to provide her exact location, but wanting to give a more specific value for her location than the building, such as a floor number, cannot do so. A better approach would be to support an arbitrary number of levels, with the number of levels set according to the type of context information.

A different approach is pursued by Chen et al., who discuss in [2] the development of a privacy enforcement system for the EasyMeeting project. Obfuscation of location information is supported as part of the privacy mechanism. The obfuscation relies on a predefined location ontology, which defines how different locations link together spatially. Obfuscation of other context types is not supported in their system.

A privacy preference language is presented by Gandon and Sadeh [5] that can express release constraints on context information as well as provide obfuscation of context. Obfuscation requirements are expressed as part of privacy preferences. This means that a commonly used obfuscation has to be respecified on each privacy preference. Furthermore, if obfuscation is used, the value to which the context should be obfuscated is required. As a result, this makes the obfuscation static. If a user specified that her location should be obfuscated to the city in which she is currently located, the name of that city would be specified as the obfuscation value. If she then moved to another city, the preference would become invalid. A more flexible technique is needed that looks for the current city, and uses that value rather than using a hard-coded value in the preference.

In summary, an obfuscation mechanism should be applicable to a wide range of different context types, not just a single type, like location. In addition to this, the mechanism should provide an arbitrary number of levels of obfuscation. The exact number of levels should depend on the type of context information being obfuscated. The mechanism should also be able to overcome missing context information in the context knowledge base. If a value required for obfuscation is unavailable, the system should choose the next most general value that still meets the limitations specified by the context owner. Furthermore, granularity preferences that are applicable to multiple privacy preferences should only have to be specified once. This can be achieved by separating the granularity preferences from the privacy preferences.

In the remainder of the paper we present an obfuscation method that can meet these challenges. We use ontologies to describe how different instances of an object type relate to one another with respect to granularity. This information is used to find lower granularity instances of an object type during obfuscation. Context owners can control the obfuscation procedure using our preference language for specifying constraints on granularity. In the event that obfuscation is not possible, the release of a value of "unknown" is used to provide the system with plausible deniability.

3   The Sample Context Model

In this section we present a sample context model that is used later to demonstrate the functionality of our obfuscation mechanism and preference model extensions. While we use the model extensively for demonstrating our approach, our approach is not tied to the model, and is general enough to be used to provide obfuscation in other context management systems.

The sample model was constructed using the Context Modelling Language (CML). CML was developed as an extension of the Object Role Modelling method (ORM) [6], with enhancements to meet the requirements of pervasive computing environments. A more rigorous discussion of CML is provided in [7].

The sample context model is graphically represented in Figure 1. It models four object types: Activities, Devices, Persons, and Places. These are depicted as ellipses, with the relationships between them shown as boxes. Each box is labelled with the name of the relationship, and annotated with any contraints on that relationship. With reference to terminology, instances of object types are referred to as objects. Relationships between object types are captured as fact types, and relationships between objects are captured as facts.

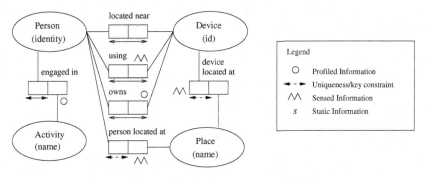

Fig. 1. A sample context model

According to our example model, a Person can be modelled as being at a particular Place through the locatedAt fact type. Similarly, a Person may interact with a Device which they may own, use and be located near. The respective fact types for these are own, using and locatedNear. Two other fact types, engagedIn and locatedAt, model the current Activity of a Person and a Person's location, respectively.

Using a previously developed technique, the context model is mapped onto a context knowledge base [8] such that fact types are represented as relations. Context information in the database is stored as context facts or tuples, which express assertions about one or more objects. The context information can be accessed by directly querying facts, or through the use of situations. These situations are a high-level abstraction supported by the modelling technique we use from [7]. Situations are defined using a variant of predicate logic and can easily be combined using the logical connectives *and*, *or* and *not*, as well as special forms of the existential and universal quantifiers. The example situation below, EngagedInTheSameActivity, takes two people as parameters and returns true if they are working on the same activity.

EngagedInTheSameActivity(person1, person2):
$\exists activity,$

- engagedIn[person1, activity],
- engagedIn[person2, activity].

Our context model captures ownership of context information to enable the CMS to determine whose preferences should be applied. The ownership definitions we use for this also enable the resolution of overlapping ownership claims that arise in pervasive computing. These can occur when multiple parties claim jurisdiction over context information. Our scheme, presented in [9], models ownership at both the fact and situation levels, such that facts and situations can have zero, one or multiple owners. Facts and situations with no owners are considered public, and are always visible. Facts and situations that have owners are considered non-public, are always visible to their owners, but are only disclosed in accordance with the owner's privacy preferences to third parties. The specification of ownership directly on situations is necessary for efficiency, as situations can potentially operate on context owned by multiple parties. Performing ownership tests on each context fact is extremely inefficient, particularly when the situation involves large numbers of context facts, all with different owners.

The following section discusses our novel use of ontologies to provide obfuscation of context information.

4   Ontologies for Obfuscation

In our new approach to obfuscation, each object type in the context model is described with an ontology. These ontologies are constructed from what we refer to as "ontological values", which are arranged in a hierarchy. The hierarchy represents the relative granularity between the values, with parent ontological values being more general than their child values. To determine an object's granularity, the obfuscation procedure matches it with an ontological value. Objects that match to the ancestors of this ontological value are of lower granularity than the object. Conversely, objects which match to the children of the ontological value are considered to be of higher granularity. Ontological values at the same level in the hierarchy are considered to be of the same granularity. Example ontologies for Activity, Person and Place are provided in Figure 2.

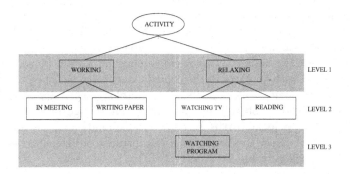

(a) Activity Ontology

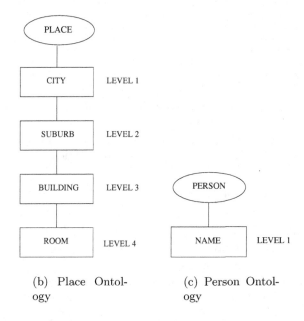

(b) Place Ontology

(c) Person Ontology

Fig. 2. Ontologies for obfuscation

When context facts are obfuscated, the CMS decomposes them into their component objects. These objects are then matched with ontological values from their respective object type's ontology. The behaviour of the system then branches, depending on the nature of the ontology being used. We have identified two different classes of ontology, which we refer to as Class I and Class II.

The first class of ontology is constructed from ontology values which are themselves valid instances of the object type. As such, the ontology values can be used as the result of obfuscation. These ontologies we refer to as Class I ontologies. In the second class of ontology, referred to as Class II ontologies, this is not the case. The ontology value must be matched with an actual instance from the context knowledge base.

The ontology for Activity is an example of a Class I ontology. Thus, the ancestor value of "reading", i.e "relaxing", is a valid Activity object. A potential obfuscation of engagedIn[Alice, reading] is then engagedIn[Alice, relaxing].

Place and Person are examples of object types with Class II ontologies. With respect to Person, the "name" ontological value is not a valid Person object, and must be matched with an entry in the context knowledge base before being used. Likewise, the ontological values in the Place ontology, e.g. "city" and "suburb", need to be matched to objects in the context knowledge base. Obfuscating a person's location to locatedAt[Bob, city], is meaningless. A meaningful result would be locatedAt[Bob, Brisbane], where an instance of Place has been matched to the "city" ontology value.

The approach presented in this paper considers both the Class I and Class II ontologies to be system-wide and largely static. We are currently working on extending the system to enable per user customisation of the ontologies.

5   Specification of Preferences

In our approach the privacy policy of a context owner is captured as a set of preferences. The preference language we have developed to facilitate this supports two kinds of preferences for controlling the release of context information: the *privacy preference*, which limits the disclosure of context information to third parties; and the *granularity preference* which the context owner can use to control the level of detail of disclosed context information.

5.1   Privacy Preferences

Privacy preferences enable context owners to authorise or deny disclosure of context information based on a set of activation conditions. To express these preferences we use an extended version of a preference model we developed previously for context-aware applications [7].

In our privacy preference model, each preference is given a unique name. The activation conditions for the preference are then listed as part of a scope statement, which is declared beginning with a "when". These activation conditions can be specified over fact types, situations and conditions on parameters like *re-*

quester or *purpose*. If all the access conditions are met, the preference activates and either ratifies or prohibits the disclosure. These two options are represented as ratings of oblige or prohibit, respectively. An example privacy preference for an entity called Alice is given below:

```
privacy_pref = when locatedAt[Alice, Home] AND
                    requester = Bob AND
                    purpose = Advertising AND
                    factType = engagedIn
               rate prohibit.
```

This preference activates when Alice is located at home, and Bob requests the activity that Alice is currently engaged in, which he intends to use for advertising purposes. As this is an example of a negative preference, when all the activation conditions are met, the preference prohibits the disclosure of the queried information. Not all of the four conditions in the example preference need to be present. By removing conditions, the preference is made more general, e.g, removing "purpose" and "factType" will cause the preference to block all requests from Bob when Alice is at home.

The example preference can be converted into a positive preference by replacing the prohibit rating with oblige. This flexibility to specify both positive and negative preferences is lacking in other preference languages such as that used by Gandon and Sadeh [5]. Both positive and negative are desirable, as they make preference specification considerably easier for users [1].

5.2   Granularity Preferences

In our novel approach to obfuscation, context owner control over the obfuscation procedure is provided with the granularity preference. This new preference is separated from the privacy preference as it operates on instances of objects rather than on facts. As a result of this, one granularity preference can cover all assertions in which instances of the protected object type participate. This is an improvement over other obfuscation techniques, like [5], where an obfuscation rule only operates on one assertion.

With regard to structure, the granularity preference definition begins with the preference name, and then lists the activation conditions in the scope statement. As the format for the granularity and privacy preference scope statement is the same, it will not be repeated here.

An example granularity preference is given below:

```
granularity_pref = when requester = Bob
                   on Activity
                   limit level 1
```

When the activation conditions specified in the scope are met, the specified granularity limit becomes active on the object type. While the granularity preference is active, context facts containing instances of the protected object type are only disclosed if the protected instances are less than or equal to the granularity limit. Instances with a granularity higher than the limit are obfuscated to meet the granularity limit.

The limit can be specified in two ways: as a cut-off; or as a granularity level number. The cut-off method uses a value in the ontology below which disclosure is not permitted. Consider the Activity ontology. If a cut-off of "watching TV" was specified on Activity, any objects matching "watching program" could not be disclosed in response to queries, while both "watching TV" and "relaxing" could be disclosed freely.

The second means states the limit as a granularity level number taken from the relevant ontology. The level numbers start from 1 and increase up to the highest level of granularity supported in the ontology for that object type.

For a comparison between the two approaches, consider Figure 2. A granularity level of 1 would allow the release of objects which match with the "working" and the "relaxing" ontological values. Access to other Activity objects would be blocked as all other Activity objects have a granularity in excess of level 1. In contrast, a cut-off of "watching TV" prevents access to "watching program" only. If the "watching TV" cut-off granularity preference were active, and the user were engaged in a work-related activity, (i.e, not relaxing), the granularity preference would have no effect.

6   Preference Evaluation

A context information query on the context knowledge base issued by a third party causes the CMS to evaluate the current context. If the requested context information is available, the CMS determines whether or not the context is owned by checking the context ownership definitions, which are described in more detail in an earlier paper [9]. If the information is public, meaning there are no owners, the request is granted. If the request was submitted by an entity with ownership rights to the context, the request is permitted. If the requester is not an owner, the owner's preferences are retrieved and evaluated by the CMS.

6.1   Evaluation of Privacy Preferences

The privacy preferences of the owners are evaluated using the current context information in the CMS. If any of the preferences specified by the owner prohibit the disclosure, the evaluation is aborted, and an "unknown" value is returned. If the collective privacy preferences of all the owners permit the disclosure, any granularity preferences specified by the context owners are retrieved and evaluated.

6.2   Evaluation of Granularity Preferences

Provided that disclosure is authorised, the granularity preferences of the owners are then evaluated. If any of the granularity preferences are activated by the current context, the release of all context facts containing an instance of the object type on which the granularity preference operates become subject to obfuscation. The most restrictive granularity preferences are used to obfuscate the context information.

To obfuscate a context fact, it is necessary to obfuscate its component objects. To achieve this, the objects must be extracted from the fact and matched to ontological values in their object type's ontology. Obfuscation then involves selecting an ancestor of this value from the hierarchy that satisfies any disclosure limitations specified by the context owner(s). The process then diverges, depending whether the ontology is Class I or Class II.

6.3   The Affect of Obfuscation on Situations

When evaluating situations it may be necessary to access context facts which refer to specific objects. If the object is protected from disclosure by granularity preferences, then the situation cannot be evaluated. An example of this can be seen by considering the situation InBrooklynWithAlice below. This situation takes a person object as its parameter and returns true if the person is in the suburb of Brooklyn at the same time as Alice.

InBrooklynWithAlice(person):
locatedAt[Alice, Brooklyn] and
locatedAt[person, Brooklyn].

This situation requires specific locatedAt context facts from both Alice and the parameter *person*. If either of these people have granularity preferences which prevent release of Place instances at the suburb level (i.e., Brooklyn), the situation cannot be evaluated, and will return *unknown*.

6.4   Failure of the Obfuscation Process

The two classes of ontology defined for obfuscation behave quite differently during the obfuscation process. With Class I ontologies the ontology values are themselves valid instances of the object type, and so obfuscation will never fail to find a return value. However, with Class II ontologies, it is possible to have ontological values for which there is no match in the context knowledge base. To overcome this, the obfuscation procedure attempts to obtain a more general value from the hierarchy by recursively testing to see if any of the ancestor values are present. Obfuscation fails if the root of the hierarchy is reached and a match has not been made. Should this occur, the *unknown* value is returned.

7   Example Scenario

In this section we present an example to demonstrate the operation of the obfuscation mechanism. The granularity preferences in the example make use of the Activity ontology depicted in Figure 2.

A person, Alice, defines a set of preferences for the disclosure of her context information. The preferences relevant to this example are given below.

```
p1  = when requester = Bob          g1 = when requester = Bob
      rate oblige                        on Activity, Place
                                         limit level 1
```

Through privacy preference p1, Alice allows Bob access to her context. However, she places constraints on the level of detail available. If Bob issued a query on Alice's current activity, the query would be processed by the CMS which would consult the ownership definitions to determine that Alice is the owner of the context information. The CMS would then apply Alice's privacy preferences, which in this case permit the release of information to Bob. Any granularity preferences specified by Alice would then be evaluated by the CMS. In this case, preference g1 limits Bob's access to instances of Activity and Place so that he may only access them up to granularity level 1. If Alice is currently engagedIn[Alice, Watching_Program], the CMS would extract the object instances and obfuscate those of type Activity. In this instance Watching_Program is of granularity level 4. By examining the ontology for Activity, the value would be obfuscated to granularity level 1, Relaxing. The result returned to Bob would then be engagedIn[Alice, Relaxing].

Not to be thwarted, Bob knows that Alice always carries around her PDA device. From the location of her device, he could possibly infer more detail about her activity. To obtain this information he queries the location of Alice's PDA. Again, privacy preference p1 permits the access. However, granularity preference g1 activates and obfuscates the location of the PDA to granularity level 1. All that Bob learns is the city in which Alice's PDA is located, i.e., locatedAt[Alice_PDA, Brisbane]. This information is of no help to Bob.

The power of the approach presented in this paper lies in the separation between privacy preferences and granularity preferences which operate on object types. The context owner defines their granularity requirements for object types. These granularity requirements are then applied to all instances of the object types over which the owner has jurisdiction. Ownership rights, or jurisdiction, are then determined by the CMS based on the ownership definitions provided in the context model. In our example, Alice did not have to specifically specify limitations on Bob accessing her device location, as all location information belonging to Alice was protected by preference g2.

8   Conclusions and Future Work

In this paper we presented an obfuscation mechanism to enhance user privacy. The mechanism is capable of modifying the granularity of context information to meet limits specified by the owners of the context information regarding its disclosure. Unlike other obfuscation mechanisms, ours supports variable levels of obfuscation for arbitrary types of context information. The obfuscation procedure is supported by ontological descriptions of object types. These descriptions capture the relative granularities of object type instances in a hierarchical fashion. Users control the obfuscation procedure by specifying granularity preferences which control the extent to which context information is obfuscated before being released to those who request it. Unlike the solutions presented in [5] and [2], the granularity restriction is not tied to a particular privacy preference, but rather is specified on an object type from the context model. This means that a general granularity preference can be reused across many different privacy preferences.

We are currently exploring ways to provide per user customisation of ontologies. This will enable users to more closely tailor the obfuscation mechanism to their needs. We are also developing a preference specification tool to conduct usability testing on our approach.

References

1. R. Agrawal, J. Kiernan, R. Srikant, and Y. Xu. An XPath-based preference language for P3P. In *Proceedings of the twelfth international conference on World Wide Web*, pages 629–639. ACM Press, 2003.
2. H. Chen, T. Finin, and A. Joshi. A Pervasive Computing Ontology for User Privacy Protection in the Context Broker Architecture. In *Technical Report TR-CS-04-08*, Baltimore County, Maryland, USA, 2004. University of Maryland.
3. L. Cranor, M. Langheinrich, and M. Marchiori. *A P3P preference exchange language 1.0 (APPEL1.0)*, April 2002. http://www.w3.org/TR/P3P-preferences; last accessed 4/11/2004.
4. L. Cranor, M. Langheinrich, M. Marchiori, M. Presler-Marshall, and J. Reagle. *The Platform for Privacy Preferences 1.0 (P3P1.0) Specification*, 2001. http://www.w3.org/TR/P3P/; last accessed 4/11/2004.
5. F.L. Gandon and N.M. Sadeh. A Semantic e-Wallet to Reconcile Privacy and Context Awareness. In *The 2nd International Semantic Web Conference (ISWC2003)*, pages 385–401, Sanivel Island, Florida, USA, 2003. Springer-Verlag Heidelberg.
6. T.A. Halpin. *Information Modeling and Relational Databases: From Conceptual Analysis to Logical Design*. Morgan Kaufman, San Francisco, 2001.
7. K. Henricksen and J. Indulska. A software engineering framework for context-aware pervasive computing. In *2nd IEEE Conference on Pervasive Computing and Communications, PerCom'04*, Orlando, March 2004.
8. K. Henricksen, J. Indulska, and A. Rakatonirainy. Generating Context Management Infrastructure from Context Models. In *Mobile Data Management 2003 (MDM2003)*, 2003.
9. K. Henricksen, R. Wishart, T. McFadden, and J. Indulska. Extending context models for privacy in pervasive computing environments. In *Proceedings of Co-MoRea'05 (to appear)*, 2005.
10. J.I. Hong and J.A. Landay. An architecture for privacy-sensitive ubiquitous computing. In *Proceedings of the 2nd international conference on Mobile systems, applications, and services*, pages 177–189. ACM Press, 2004.

11. C. Johnson, D. Carmichael, J. Kay, B. Kummerfeld, and R. Hexel. Context Evidence and Location Authority: the disciplined management of sensor data into context models. In *Proceedings of the first International Workshop on Context Modelling, Reasoning and Management at UbiComp2004*, pages 74–79, September 2004.
12. S. Lederer, C. Beckmann, A. Dey, and J. Mankoff. Managing Personal Information Disclosure in Ubiquitous Computing Environments. Technical Report IRB-TR-03-015, Intel Research Berkley, 2003.

Share the Secret: Enabling Location Privacy in Ubiquitous Environments

C. Delakouridis, L. Kazatzopoulos, G. F. Marias, and P. Georgiadis

Dept. of Informatics and Telecommunications, University of Athens, Greece, TK15784
{delak, kazatz, georgiad}@di.uoa.gr, marias@mm.di.uoa.gr,

Abstract. Anonymity and location privacy in mobile and pervasive environments has been receiving increasing attention during the last few years, and several mechanisms and architectures have been proposed to prevent "big brother" phenomena. In this paper we present a novel architecture to shield the location of a mobile user and preserve the anonymity on the service delivery. This architecture, called "Share the Secret - STS", uses totally un-trusted entities to distribute portions of anonymous location information, and authorizes other entities to combine these portions and derive the location of a user. STS simply divides the secret, and as a lightweight scheme it can be applied to network of nodes illustrating low processing and computational power, such as nodes of an ad-hoc network, smart gadgets and sensors.

1 Introduction

Ubiquitous computing and communications are inducing the way of living, fostering smart environments that enable seamless interactions. Smart spaces, enriched with sensors, tags, actuators and embedded devices, offer anytime – anywhere computing capabilities, deliver integrated services to users and allow seamless interactivity with the context, beyond any location frontiers. Location is considered as an essential component for the development and delivery of context-aware services to nomadic users [1-3]. Several approaches have been proposed to gather the location of a user, device, or asset. They differ in several semantics, such as scale (e.g., worldwide, indoor, outdoor), accuracy, location measurement units involved, time-to-first-fix and costs (e.g., computational, or investment). These approaches include the worldwide GPS, outdoor cellular-based methods, and indoor positioning mechanisms based on wireless LANs. Recently, positioning systems include RF-tags and specialized sensors that provide higher accuracy, enabling location aware applications that require the calculation of the exact position. The aforementioned solutions provide the means for easily location gathering, monitoring, and management, operations that considered essential for the elaboration of pervasive computing and the exploitation of Location Based Services (LBS). On the other hand, the networking technologies (e.g., GSM/GPRS) and the web offer the infrastructure for advertisement of the location information, and, thus, potential eavesdropping and unauthorized use (or misuse) of it. An adversary can potentially derive location information at different layers of the network stack, from the physical to the network layer [4]. Furthermore, data

T. Strang and C. Linnhoff-Popien (Eds.): LoCA 2005, LNCS 3479, pp. 289–305, 2005.

collection and mining techniques might produce historical location data and movement patterns [4], which they are subject to unauthorized use, as well. Minch presents and discusses several risks that are associated with the unauthorized disclosure, collection, retention, and usage, of location information [5]. Additionally, the Location Privacy Protection Act, announced on 2001 in the United States, addresses the necessity and identifies several risks related to the privacy of the location information [6]. Location privacy is considered of high importance, since individuals should be able and free to explicitly identify what location information will be disclosed, when this can happen, how this information will be communicated and to whom this information will be revealed. Even through anonymization, as defined in [7], personal data collection is combined with privacy, the disclosure of the personal identity might be useful for the delivery of personalized, pervasive or location-aware, services, especially when accounting and charging is a requirement.

This paper introduces an innovative architecture, called STS, to address several aspects of the location privacy issue. Firstly, it elaborates on the control of privacy that an individual should have over his/her location. It enables individuals to define different levels or rules of privacy (i.e. secrecy) over different portion of location data, depending on the operation environment (e.g., hostile) and the service that indents or asks to use location data. This is achieved through data and secret sharing techniques, which are discussed in a later section. Additionally, it gives users, or location targets, the capability to explicitly identify services or pervasive applications that might collect, store, and use location information. Finally, it provides anonymity on the distribution of the location information. STS architecture is designed to operate within an un-trusted environment. Portions of a target's location data are distributed to public available nodes, which act as intermediate and temporal location servers. Targets authorize, on-demand, the pervasive or LBS services to access and combine the distributed location information. This is accomplished implicitly, through the disclosure of the mapping between a pseudonym and user's or target's location data. In the last few years, several methods have been proposed which deal with the location privacy issue on the lower layers of the communication stack (e.g., the IP layer). STS address location privacy in the application layer, enabling end-user to disclosure location information to authenticated pervasive applications and location-based services.

The STS architecture is applicable to environments that consist of user devices with several idiosyncrasies, such as luck of energy autonomy, memory and processing power. To provide secrecy it avoids complex computational tasks, such as producing key pairs, hashing and message authenticated codes. Instead, it deals with the division of location data into pieces and the distribution of these pieces to serving entities. Thus, it is applicable to smart spaces featuring ubiquitous services through the involvement of small gadgets, RF-tags, sensors and actuators that activated to infer the location of a user or target.

The remainder of this paper is structured as follows. First we discuss several approaches that address the location privacy issue. We also briefly describe STS and its advantages compared to relative approaches. In Section 3 we provide the assumptions that STS considers for its operation. In Section 4 we present the STS architecture and the entities that it consists of. It introduces the secret share mechanism that enables the anonymity and privacy of location information. Subsequently, in Section 5, we discuss the procedures followed by the entities of the architecture. In Section 6, we evaluate the robustness of the STS architecture against attacks. In Section 7, we illustrate the applicability of the STS to a real case scenario. Finally, we provide some concluding remarks and directions for further research.

2 Related Work

Several approaches that enable location privacy have been proposed in the literature. Some of them focus on the secrecy of Medium Access Control (MAC) identifiers or IP-layer address. On the MAC layer, Gruteser and Grunwald discuss the problem of interface identifiers [8] that uniquely identify each client, allowing tracking of his/her location over time. They introduce privacy through the frequent disposal of a client's interface identifier. In the IP layer, Mobile IP [9] address location privacy by associating two different IPs to the same subject; the static one, related to the user's home network, and the dynamic, related to his current network. As the user roams through different networks, his/her mobile device registers the current location to a home agent located at the user's home network. The Non-Disclosure Method [10] considers the existence of independent, security, agents that are distributed on the IP network. Each security agent holds a pair of keys and forwards the packets in an encrypted format. The sender routes a message to a receiver through a path that consists of security agents. Mist System [11-12] handles the problem of routing a message though a network while keeping the sender's location private from intermediate routers, the receiver and potential eavesdroppers. The system consists of a number of routers, called Mist routers, ordered in a hierarchical structure. According to Mist, special routers, called "Portals", are aware of the user's location, without knowing the corresponding identity, whilst "Lighthouse" routers are aware of the user's identity without knowing his/her exact location. The aforementioned methods deal with location privacy in the MAC and IP-layer, and propose solutions that apply to MAC authentication, IP mobility and IP routing methods. On the other hand, the STS scheme addresses location secrecy on the application layer. Thus, some of these methods (e.g., the Mist) can be considered as complementary, or underlying, secure infrastructures for the deployment of STS.

Beyond the solutions that apply to the lower layers of the OSI protocol stack, two main approaches have been proposed to address the location issue on the application layer. The first is the policy-based approach, which incorporates a Service Provider as the basic element for privacy building. The Service Provider introduces privacy policies that identify rules concerning data collection (type of data that can be collected, purpose of collection) and data dissemination. The subjects (i.e., target of location) communicate with the Service Provider in order to evaluate and select a privacy policy. The second is the anonymity-based approach that aims to unlink the location information from the subject, before the collection of the location information. In the context of location privacy in the application layer, several architectures introduce the idea of using of pseudonyms instead of users' identities. Leonhardt and Magee [13] propose the replacement of the identity with a sequence of chained idempotent filters governed by a rich formal policy language. Kesdogan et. al [14], propose the use of temporary pseudonymous identities to protect the identity of users in the GSM system.

Furthermore, the IETF Geopriv Workgroup [15] addresses the location privacy issue by introducing the usage of a Location Server (LS) which registers the position of the subject (i.e., a target of location). The LS discloses this information using specific rules. These rules are generated by a Rule Maker (e.g., the subject itself) and managed by a service called Rule Holder. Each subject can explicitly specify the desired level of privacy, through the Rule Holder service.

The majority of the aforementioned schemas are based on the existence of a Trusted Third Party (TTP, e.g., the Geopriv's LS). The LS entity maintains access to the data related to both user's identity and location. As a result it could jeopardize the user's location privacy by revealing this information to unauthorized users. Furthermore, due to the fact that LS is the central storage of the location information, the privacy architecture becomes sensitive to eavesdropping and attacks. The former might reveal location information through traffic analysis, whilst the latter might produce availability risks (e.g., denial of service attacks and flooding), or derogation of the privacy service (e.g., though impersonation and spoofing). On the other hand, non-TTPs based systems, such as the Mist, enable location privacy using rerouting and hard encryption, and, therefore introduce heavy processing tasks, which influence their applicability in real scenarios. Finally, in several mobile and wireless networks, such as ad-hoc networks, TTPs servers are ephemerally present, due to the mobility of the nodes. Thus, centralized location privacy architectures that rely on a trust server might be inapplicable.

The proposed STS architecture enables location privacy without relying on the existence of trusted parties (e.g., TTPs, LS). The main idea is to divide the location information into pieces and distribute it to multiple servers, called STS Servers. These are no-trustworthy, intermediate, entities, assigned to store, erase and provide segments of location information that anonymous users register or update. Third party services, such as LBS or pervasive applications, access multiple STS servers to determine user's location data through the combination of the distributed pieces. This is achieved after a well-defined authentication process motivated by the corresponding user. Furthermore, due to the decentralized STS architecture, a possible collusion between STS servers, each of one maintaining only partial knowledge of the user location, is fruitless. The secret i.e. (location of a user) is not distributed to specific STS Servers, and, furthermore, each user dynamically chooses different servers to distribute the location information segments, according to the requested service (e.g., location-based friend-find service) and to policy rules. Additionally, even if multiple STS Servers collude successfully, the actual location information is not revealed, since the segments are stored through pseudonyms. The main characteristics of the STS are:

- Location privacy is offered as a service to users who desire to hide their location information.
- Different privacy levels can be defined, based on users' profiles, policies or end-applications that the users are registered to (e.g., segmentation of location information into different segments).

- The user maintains full control over the location data. The position is calculated on the user's device, the user defines which STS servers will have partial knowledge of it, and which third-party services are authorized to access and combine it.
- The location privacy is achieved within rational or distrusted environments; it does not employ trustworthy or reliable entities for data storage.

3 Definitions and Assumptions

Before describing in details the STS architecture it is essential to define what we consider as location data and which are the possible technologies, currently available, for position retrieval. As Location Information *(LocInfo)* we use a variation of the definition introduced in [16], i.e., we use the triple of the following form: *{Position, Time, ID}*. Thus, *LocInfo* is defined as a combination of the *"Position"* that the entity with identifier *"ID"* maintained at time *"Time"*, within a given coordinate system. This triple is adopted in order to emphasize that, within the STS architecture, time is considered as a critical parameter in terms of location identification, as well. Friday et. al in [16], provide further details on the semantics of this definition. The STS scheme incorporates multiple user profiles to provide different precision of *LocInfo* i.e. a drive-navigation service needs high precision location data to provide exact instructions, whilst a Point Of Interest service requires data of lower precision. A user could use different pseudonyms per service and choose a discrete profile to denote the required level of the precision on the *LocInfo*. We assume that each STS server keeps a table called "Location Information Table" *(LIT)*. Each record of the *LIT* table keeps a pair of values *(ID, σ(LocInfo))*, where *σ(LocInfo)* is a random segment of the *LocInfo* of the corresponding user, produced from the SSA algorithm, as will be discussed later.

As previously mentioned, position technologies differ in terms of scale, accuracy, time-to-first-fix, installation and operation costs. Additionally, the *Position* of an object can be either physical (e.g., at 38014′49"N by 23031′94"W) or symbolic (e.g. in the office, in Athens). Another classification uses the absolute or relative *Position* definitions. Absolute position is depicted on a common coordinates system, and the *Position* for an object is the same and unique for all the observers. Unlike, the relative position represents the *Position* of a located object in reference to the observer. *LocInfo* is defined to incorporate any of the four aforementioned *Position* semantics. In terms of scale, GPS is a technology that focuses on outdoor, wide range, location identification, providing high accurate results (less than 30 feet) using a trilateration positioning method in the three-dimensional space [17]. Location determination can also be achieved through terrestrial, infrastructure-based, positioning methods. These positioning methods rely on cellular networks equipment to infer an estimation of a mobile users' position. In [18] a general brokerage architecture that gathers the location information independently of the positioning technology, such as Time Of Arrival, Enhanced Observed Time Difference, and Global Cell ID, is proposed. Location retrieval can rely on the infrastructure of the wireless LAN communication network, such as the IEEE 802.11, to provide higher accuracy, especially in indoor environments. Positioning systems like RADAR [19] and Nibble [20] operate using the IEEE 802.11 data network to gather the location of a user or object. For context-aware services that require higher accuracy, specialized positioning infrastructure needs to be established (e.g., sensors or RF-tags). Among the positioning systems that have been proposed to provide finer granularity to the pervasive applications is the *Active Badge,* a proximity system that uses infrared emissions emitted by small infrared badges, carried by objects of interest [21]. *Active Bat* resembles the Active Badge using an ultrasound time-of-flight lateration technique for higher accuracy [3]. *Cricket* relies on beacons, which transmit RF signals and ultrasound waves, and on receivers attached to the objects [22]. *SpotON* measures the signal strength emitted by RF tags on the objects of interest and perceived by RF base stations [23]. *Pseudolites* are devices that are installed inside buildings and emulate the operation of the GPS satellites constellation, [24]. *MotionStar* incorporates electromagnetic sensing techniques to provide position-tracking. Additionally, according to [25], there are several other types of tags and sensors that provide user's location with high accuracy, depending the requirements of the ubiquitous service. *Easy Living* uses computer vision to recognize and locate the objects in a 3D environment. *Smart Floor* utilizes pressure sensors to capture footfalls in order to track the position of pedestrians. *Pin Point 3D-iD* uses the time-of-flight lateration technique for RF signals emitted and received by proprietary hardware. Positioning technologies are further divided into two main categories: centralized and distributed. Technologies of the former category rely on a centralized, dedicated, server, responsible for calculating the position of the registered users. Methods of the latter category are based on specialized software of hardware, installed on the subject's device, which performs advanced calculations to estimate user's current position. The STS framework takes advantage of the distributed approach; the entity *ID* determines the *Position,* either autonomously or using contributed measurements, based on calculations performed at a given *Time.*

4   STS Architecture and Algorithms

4.1 Architecture

The STS architecture does not rely on the existence of a single and trusted third party. Multiple STS servers are geographically distributed. A subset of them is assigned to store portions of the location information of a target. The segments of the location information are constructed through the "Secret Share Algorithm" (SSA), a novel threshold scheme that is proposed here to cope with location privacy. The fig. 1 illustrates the entities of the proposed STS architecture. STS servers might offered by a Wireless Internet Service Provider (WISP), an Internet feed provider, a VAS and content provider, or offered by a mobile operator. In the ad-hoc networking scenario, in the presence of STS servers the targets might use these as location information storage areas. If the STS servers are unreachable, ad-hoc nodes, RF tags and sensors might be used as local caches to store the *LocInfo* segments.

The Value Added Service (VAS) provider offers location based and pervasive services to subscribers (registered users) or to ephemeral users (pay per view model). VAS might offer a wide range of services, such as indoor/outdoor navigation, proximity services (e.g., find-friend), positioning and point-of-interest, tracking person's location (e.g., children, elderly), localized advertisement and content delivery (e.g., city sightseeing), and emergency (e.g., E911). Hereafter we use the term *Service* to denote value added or pervasive applications that require the location information of a user, asset or device to operate and provide a location service. Each *Service* requires the location information, maintained in intermediate STS servers, to provide the required level of service to an individual.

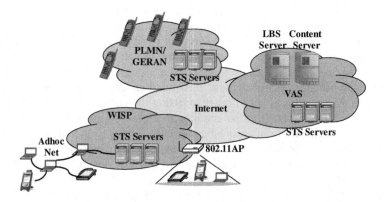

Fig. 1. Entities of the STS architecture

4.2 Secret Share Algorithm

Several variations of the SSA have been proposed [26]. The basic algorithm is the "Trivial Secret Sharing". According to this basic form, the secret is divided into n pieces and all the pieces are required to restore it. An improved SSA, called "Perfect Scheme", assumes that the secret is shared among the n out of m available entities ($n<m$) and any set of at most $n-1$ entities cannot rebuild the secret. The SSA we have adopted for the STS architecture is the perfect sharing scheme of Shamir, referenced to as *threshold scheme* [27]. Shamir's algorithm is based on the fact that to compute the equation of a polynomial of degree n, one must know at least $n+1$ points that it lies on. For example, in order to determine the equation of a line (i.e., $n=1$) it is essential to know at least two points that it lies on. Let assume that the secret is some data L, which is (or can be easily made) a number. According to Shamir's (n,m) threshold scheme, to divide L into pieces L_i one can pick a random, $n-1$ degree, polynomial

$$f(x) = S_0 + S_1 x^1 + S_2 x^2 + ... + S_{n-1} x^{n-1}$$

in which $f(0)=S_0=L$ and evaluate $L_1=f(1)$, ..., $L_i=f(i)$, ..., $L_n=f(n)$. Given any subset of n of these Li values, we can determine the coefficients of $f(x)$, and rebuild L, since $L=f(0)$. On the other hand, knowledge of at most $n-1$ of these values is not sufficient to determine L. In the STS architecture, L is the *LocInfo* triple.

5 STS Procedures

Using real life identifiers, such as user names, in each *LIT* entry enables STS servers to collude and, eventually, reconstitute the location information of a user, based on his/her username. To prevent collusions and the disclosure of *LocInfo*, we replace the username of the *LIT* entries with a pseudonym. This approach does not prevent hackers from obtaining the location of an unknown entity, by combining entries with identical pseudonyms. Instead, it discourages them from obtaining the location of a specific user. Only the entities that have a prior knowledge about the mapping between a pseudonym and a real identity can combine segments to produce useful conclusions. To increase the security level, the pseudonym is divided into parts, as well. Each segment of the *LocInfo* is sent to the STS server along with a discrete part of the related pseudonym. As a result, a potential eavesdropper should map multiple pseudonym's segment (say p_i) in order to obtain the correct record $(p_i$, $\sigma_i(LocInfo))$ from all the STS servers for the monitored user. If a perfect SSA is used, the pseudonym is divided into *m* parts where *m* corresponds to the "*n* out of *m*" perfect scheme. Regulating the value of *m*, one can increase the complexity, and the privacy level. Note that for each *ID* only one *LIT* record is stored on a STS server. We assume that a software random integer generator is running on user's device and performs the generation of the pseudonym. If there are *m* STS servers and at least *n* (*m>n*) can rebuild the *LocInfo*, the random integer generator generates a random *m*k*-bits number (i.e., the pseudonym). This number is then divided into *m* segments (*1...k* bits represent the first segment, *k+1...2k* represents the second segment etc.) where each segment acts as user's pseudonym.

The robustness of the aforementioned scheme could be enhanced if the secret is not shared to *m* STS servers but to *x,* randomly selected, servers, where *x≤m*. In this approach the secret can be recovered from *y* segments, *y<x* (i.e., a "*y* out of *x*" perfect scheme). This variation increases the complexity required by an eavesdropper to identify which STS servers keep the location of a specific user. To provide a paradigm let assume that a farm of 15 STS servers are available and the secret is being shared among *x,* randomly selected, servers, where *8≤x≤12*. An attacker should first determine the value of *x* and then calculate all the possible, C_1, combinations of the STS servers in order to combine the pseudonym

$$C_1 = \sum_{i=m-A}^{m} \binom{15}{i}$$

Assume *m* STS servers, where *x* of them have the secret, and *y* (*y<x*) is the minimum number of servers required to obtain the *LocInfo*. Furthermore, let each pseudonym constitutes of *k*x* bits. The random pseudonym generator splits the pseudonym into *x* segments (p_x), each of *k*-bit length. The variable *x* takes values from (*m-A, m*) where *A* (*A≥0*) is an integer that affects the desired security level of the STS architecture. More specifically, if *A* is small, an attacker has to gain access to a greater number of STS servers. However, small *A* reduces the number of different combinations of STS servers keeping the secret's segments, and, thus, the security level of the system is diminished. On the other hand, for a large *A*, the number of different combinations of STS servers sharing the secret is increased. However, large *A* reduces the number of servers that an attacker has to gain access to reveal the

secret. Regarding the maximum value of variable y, this is set to x-t, where t ($t>0$) is an integer that depends on the desired robustness of the STS architecture. If t is set close to zero and B out of x STS servers go down at the same time, where $B>t$, the system eventually runs out of service. The variables A and t may vary per *Service*, based on e.g., charging policy or the user profile. Furthermore the value of t is a function of x, ($t=f(x)$), where as x increases t increases, as well. The steps followed by the target during the creation of its pseudonym, the calculation of the updated *LIT entries* and the dissemination of these to the STS servers (location update procedure) are as follows:

step 1. Find a random x, such that m-$A<x<m$ step 2. Generate a random $x*k$– bit length number step 3. Split that number into x segments of k – bit length each step 4. Register to the x STS servers	Pseudonym and STS registration procedure
step 5. Apply the SSA at the *LocInfo* to derive the σi pieces step 6. Produce the pairs $(p_x, \sigma_x(LocInfo))$ step 7. Send to each one of the x STS servers the *LIT* pair $(p_x, \sigma_x(LocInfo))$	*LocInfo* segmentation and update procedure
step 8. If *LocInfo* is changed go to step 4	

Steps 1 – 4 are followed once, per user and per *Service* that requires the *LocInfo*. Each time the location information is altered, the target performs the steps 5 – 7 to inform STS servers and update the *LIT* entries.

Registration to the STS Service. The random generators that reside in different user's device might produce identical pseudonyms. In such a case, one STS server might receive two location segments (i.e., σ_x) that belong to the *LocInfo* of two discrete users, but associated with the same pseudonym. As a result, the STS server might forward an inaccurate segment to the *Service* that requested the user's *LocInfo*.

To avoid such a conflict, the STS scheme requires from each user to perform a registration procedure (step 4), and to enter his/her pseudonym's segments to the x STS servers. This procedure requires a handshake between the user and each STS server. Upon the arrival of a registration request from the user, the I-th STS server searches the *LIT* table to determine whether the p_i already exists. If so, a failure message is send back to the user. Upon receiving this message, the user generates a new pseudonym p', splits this into x segments and restarts the registration (step 1 to step 4). Alternatively, instead of generating a $k*x$ bits length number and split this into x segments, the user might generate x k-length bit numbers, one for each σ_x. According to the former approach if one of the x STS servers sends a failure message a new p has to be generated and the registration procedure has to be restarted. In the latter approach only one k-length number has to be generated and send to the specific STS server without affecting the registration procedure on the remaining x-1STS servers.

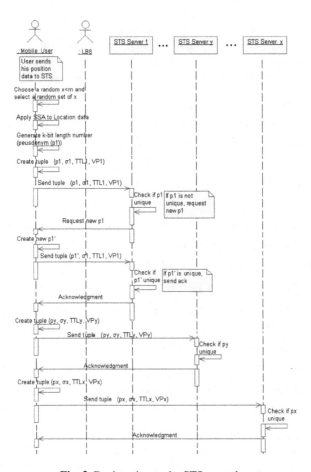

Fig. 2. Registration to the STS procedure

Location Retrieval Procedure. When the *Service* requires user's location, it sends a request to each one of the *x* STS servers. When the *x* segments are received, the *Service* rebuilds user's location. During the location retrieval some *LIT* entries might not be synchronized among the STS servers. More specifically, if the location update and the location retrieval are performed simultaneously, there is the possibility that some of the location segments, which held on the STS servers, to index a previous *LocInfo* of a target. To avoid these phenomena each *LIT* entry is tagged with a sequential number (*SEQ*). On the location retrieval, the *Service* checks whether at least *y* out of the *x* received segments are tagged with the same highest *SEQ* (Updated LIT, or *ULIT*). If *ULIT≥y*, the *Service* reconstructs target's *LocInfo*, otherwise (i.e., *ULIT<y*) it waits for an interval and then sends requests only to the *x-UTIL* servers from which it retrieved aged *SEQ* numbers.

LIT Update Procedures. On the other hand, when an STS server receives an update request (step 7) it searches the existing *LIT* records to determine whether the p_x

already exists. If so, it updates the corresponding *LIT* entry, importing the $(p_x,$ $\sigma_x(LocInfo))$ record.

Registration to a Service. Beyond the registration procedure to the STS servers, user has to provide to the *Service* provider the details of how to obtain his/her location information. Therefore, the user provides the following information during the registration procedure to the *Service*:

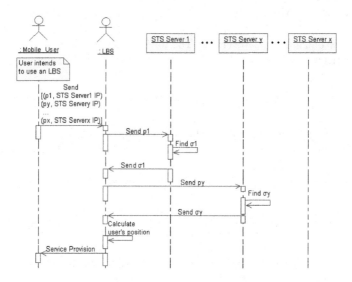

Fig. 3. Registration to the STS procedure

- The *x* STS servers that store his/her location information (e.g., IP address or URL)
- His/her pseudonym's segments, created by the user for the specific *Service*
- The mapping between pseudonym's segments and STS servers (i.e., to which STS server each segment corresponds)

The aforementioned information is communicated to the *Service*, though a public key scheme, ensuring authentication, confidentiality and integrity.

LIT Entries Remove Policy. Normally, the end-user authorizes a *Service* to access his/her location information for a given period of time, (e.g. a month), through a selection of a predefined policy. Consequently, *LIT* records of each STS server should be deleted after that period. This can be performed explicitly or implicitly. In the former case the user sends a specialized message requiring the deletion of the *LIT* entry from a specific STS server. This message can be sent when the user desires to prohibit a previously authorized *Service* from accessing his location information. A possible multicast message can inform the entire set of the *x* STS servers to perform a deletion. In the latter case, an extra field on the $(p_i, \sigma_i(LocInfo))$ record is added to enable end-user to define the time-to-live (*TTL*) value for the record. The STS server checks this field, and in the case it expires, the record is deleted.

Location Information Validity Period. In ubiquitous environments users are characterized by high mobility. Thus, the location information that is stored on the STS servers must be associated with a Validity Period (*VP*). VP is an additional field on the $(p_i, \sigma_i(LocInfo), TTL_i)$ record. The *VP* period depends on several parameters, such as:

- The nature of the *Service* provided: A navigation *Service* might require higher location-update rates to provide exact instructions, whilst a tourist guide *Service* needs periodical updates.
- User's velocity. For a driver the location is continuously altered, while for a walker the location changes at a slower rate. The former requires lower *VP*, whilst a higher *VP* is sufficient for the latter.

6   Threats Against STS

Steps 1 – 7 encourage a brute-force attacker to combine *LocInfo* entries in order to reveal the location information. The STS scheme produces *C* possible combinations, where

$$C = \sum_{i=m-A}^{m} \left[\binom{m}{i} * 2^{k*i} \right] \tag{1}$$

Assigning appropriate values to parameters k, A, t and m the reliability of the STS architecture can be enhanced, because the C is increased, and, thus, the STS robustness is increased. A brute force attack might be inefficient, because C combinations should be examined. On the other hand, an attacker might determine that the pseudonym he/she sends to the first STS server does not exist, and, thus, he might prune the possible combinations using a pseudonym that is unknown to the first STS server. The term 2^{k*x} can be replaced with the following expression:

$$2^{k*x} = \overbrace{2^k * ... * 2^k}^{x} \Rightarrow$$

$$[2^k * P(Z_1)]*[2^k * P(Z_2)]*...*[2^k * P(Z_x)] = 2^{k*x} * P(Z_1) * P(Z_2) *...* P(Z_x) \tag{2}$$

where $P(Z_j)$ refers to the probability that the generated pseudonym already exists at the STS server$_j$. Combining (1) and (2) we have:

$$\sum_{i=m-A}^{m} \left[\binom{m}{i} * 2^{k*i} * P(Z_1) * P(Z_2) *...* P(Z_i) \right]$$

Assuming that $P(Z_1) = P(Z_2) = ... = P(Z_x) = P(Z)$ (i.e., all the servers maintain the same number of *LIT* entries) and due to the fact that the Z_is are independent then:

$$\sum_{i=m-A}^{m} \left[\binom{m}{i} * (2^k * P(Z))^i \right]$$

Another threat occurs when an attacker gain access simultaneously into a random number of z STS servers and obtains their records. In such as case, to identify whether some user is registered to those z servers, he has to calculate R^z possible combinations, where R is the entire *LIT* records that each STS server maintains (all servers maintain the same number of records). Furthermore, the probability an attacker, who broke z servers, to reconstitute a user location, is equal to the probability of the potential event E, defined as "The min number of servers that a user selects to reconstitute his location is y or $y+1$ or ...or z, where $y<=z$". $P(E)$ is defined as,

$$P(E) = \frac{\sum_{i=0}^{z-y}\left[\binom{z}{z-i}*\binom{m-z}{i}\right]}{\binom{m}{z}}$$

where y is minimum number of STS servers needed to rebuild location information, z is the STS servers under attack and m is the total number of STS servers. For example, if $m=20$, $z=10$ and $y=7$, $P(E)=0,00156$. This low value should not be interpreted as feasible, and a potential attack should not be considered as effective, since for the previous estimation we assumed that the $z=10$ servers were under attack simultaneously. Actually, the attacker must try an average of

$$M = \frac{\sum_{i=1}^{T} i}{T} = \frac{\frac{T}{2}(1+T)}{T} = \frac{1+T}{2}, \quad where\ T = R^z$$

possible combinations in order to find out if the location segments obtained from the z servers are sufficient to rebuild the location information of any (and not specific) user. Assuming that $z=10$ and $R=1000$, then M is approximately equal to $5*10^{29}$. Even if the attacker is able to try 10^{12} (THz) combinations per second using e.g., Grids, he/she needs an average of 10^{10} years in order to determine if the compromised set of z STS servers can rebuild the location of a random user, and this can happen with probability $P(E)=0,00156$.

7 A Real Case Scenario

To further elaborate on STS architecture, we describe here a paradigm of usage. Assume that Phevos intents to visit Athens as a spectator of "Athens Olympic Games 2004". His new mobile device is equipped with an embedded GPS receiver. A pervasive *Service*, called "Olympic Navigator", that takes advantage of the recently deployed STS servers, is offered by the Cellular Operator to the mobile users. This *Service* covers the urban areas of Athens and each sport complex, providing navigation services into Olympic facilities and rich multimedia content that is related to the event the spectator is attending. Phevos registers to the "Olympic Navigator" from its home PC, using the Internet and some Web forms (in that case a pre-pay method has to be introduced for the service provision), or through a service advertisement that is offered on specific hotspots e.g., the airport (in such as case we assume that the charging policy might incorporate a pay-as-you-use logic). In the latter case, upon Phevos' arrival at the new Athens International Airport, he receives a message to his mobile phone to inform him about the available service. The typical registration procedure and the charging policy are out of the scope of the current work. Regarding schedule, Phevos arrives by plane and intents to stay in Athens for five days. During the registration procedure, the service informs Phevos that he will be traced every T seconds, unless Phevos explicitly asks for a complementary navigation service, which is charged on a per-use policy. Phevos uses the "Olympic Navigator" and, at the same time, keeps his position private from any other unauthorized entity. In order to do so, a software module is downloaded at Phevos' device, to run the location update procedure, to produce the pseudonym of Phevos and to communicate segments of Phevos' *LocInfo*, as gathered by the GPS, to STS servers. The latter is performed every T seconds, and this is achieved transparently to the end-user. Additionally, the module creates the pseudonym of Phevos, and sends this information to the provider of the "Olympic Navigator" service. This communication is performed exactly once, and it might be associated with public key cryptography to provide strong authentication, confidentiality and integrity. This step enforces the service provider to rapidly resolve the *LocInfo* of Phevos, combining the pieces from the *LIT* entries that correspond its real identity.

For five days the "Olympic Navigation" service is authorized to obtain the location information of Phevos. During the registration procedure Phevos chooses the "spectator" profile. This profile corresponds to a specific location privacy policy, and defines specific values of several parameters that correspond to the number of the STS server that will be exploited on behalf of Phevos (i.e., A, t and m), as well as the parameters *TTL* and *VP*. Different registration profiles might be offered to athletes, members of Olympic Committees, sponsors, or journalists. After five days of "unforgettable and dream games"[1], Phevos departs; his is automatically unregistered from the "Olympic Navigator" and the STS service, and his location information is permanently deleted from the STS system.

8   Remarks and Future Work

This paper introduces an innovative architecture to provide location privacy in insecure, wireless and pervasive environments. The architecture relies on the Shamir's threshold algorithm [27] to divide the location information into small pieces and distribute those in a numerous, public available, entities, called STS servers. The scheme is enriched with pseudonyms, and, thus, guarantees that only the authorized entities, such as location-based services and pervasive applications, have the knowledge to rebuild the secrets based on the distributed segments. We further elaborate on the proposed architecture, evaluating the potential threats and its robustness against brute-force and sophisticated attacks. We argue that the STS architecture is stealth against such attacks, since we proved that it is hard to be compromised.

---

[1] Dr. J. Rogge, IOC president, Athens Olympics 2004 Closing Ceremony, Aug. 29, 2004, Athens.

To increase the reliability, to prevent DoS attacks, and to solve synchronization issues, we extend the perfect sharing scheme, using additional entities to store the location information segments. This policy will be further investigated through simulations, which are under development. Additionally, we plan to investigate load-balancing issues when distributing location information to the STS servers, and the tradeoff between the usage of false records that enhance the scheme's robustness and the required storage overheads. Since STS architecture does not utilize hard computational and processing tasks, such as cryptosystems, it is applicable to ad-hoc as well as spontaneous networks and systems. We plan to evaluate the architecture, in terms of energy consumption and communication overheads. In the ad-hoc and spontaneous scenario we plan to further investigate the usage of caching schemes, taking into account relative works (i.e., [28-29]) and the applicability of reciprocity principles, where two or more nodes have the mutual intensive to share their location information segments, as in [30]. Finally, the under-process development, will tune the architecture in terms of time thresholds that are associated with the location information, providing ideal refresh and update time intervals.

The STS architecture addresses location privacy in the application layer. Existing solutions in the lower layers of OSI model that provide secrecy and authentication (i.e., tunneling, IPSec, SSL, see [31]), can be used as underlying assets to enhance privacy and prevent attacks, such as traffic analysis. Such heavyweight schemes can be applied to the information exchange between the STS Servers and the Service Provider, since this part of communication involves stationary and powerful processing units, fixed networks and high capacity lines (e.g., broadband links).

References

1. M. Satyanarayanan, "Pervasive Computing: Vision and Challenges," IEEE Personal Communication, Aug. 2001, pp. 10–17
2. D. Saha, and A. Mukherjee, "Pervasive Computing: A Paradigm for the 21st Century", IEEE Computer Mag., Mar. 2003, pp. 25–31
3. A. Harter, A. Hopper, P. Steggles, A. Ward, and P. Webster, "The anatomy of a context-aware application", in Proc. 5th ACM/IEEE International Conference on Mobile Computing and Networking, Aug. 1999
4. M. Gruteser and Dirk Grunwald, "A Methodological Assessment of Location Privacy Risks in Wireless Hotspot Networks", in Proc. Security in Pervasive Computing, Mar. 2003
5. R. P. Minch, "Privacy Issues in Location-Aware Mobile Devices", in Proc. 37th Hawaii International Conference on System Sciences, Jan. 2004
6. U.S. Location Privacy Protection Act of 2001, Bill Number S.1164, Introduced July 11, 2001, US Congress, available at
   http://www.techlawjournal.com/cong107/privacy/location/s1164is.asp
7. A. Pfitzmann, and M. Koehntopp, "Anonymity, unobservability, and pseudonymity – a proposal for terminology", in Proc. Workshop on Design Issues in Anonymity and Unobservability, Jul. 2000

8. M. Gruteser, and D. Grunwald, "Enhancing Location Privacy in Wireless LAN Through Disposable Interface Identifiers: A Quantitative Analysis", in Proc. WMASH'03, Sept 2003

9. A. Fasbender, D. Kesdogan, and O. Kubitz, "Variable and Scalable Security: Protecting of Location Information in Mobile IP", in Proc. 46th IEEE VTC, Mar. 1996

10. A. Fasbender, D. Kesdogan, and O. Kubitz "Analysis of Security and Privacy in Mobile IP", in Proc. 4th International Conference on Telecommunication Systems, Modelling and Analysis, Mar. 1996

11. J. Al-Muhtadi, R. Campbell, A. Kapadia, D. Mickunas, and S. Yi, "Routing Through the Mist: Privacy Preserving Communication in Ubiquitous Computing Environments", in Proc. International Conference of Distributed Computing Systems, Jul. 2002

12. J. Al-Muhtadi, R. Campbell, A. Kapadia, D. Mickunas, and S. Yi, "Routing through the Mist: Design and Implementation", Technical Report UIUCDCS-R-2002-2267, Mar. 2002, available at ciae.cs.uiuc.edu/mist

13. U. Leonhardt, and J. Magee, "Security Considerations for a Distributed Location Service", Journal of Network and Systems Management, Vol. 6, No. 1, Sept. 1998

14. D. Kesdogan, P. Reichl, and K. Junghärtchen, "Distributed Temporary Pseudonyms: A New Approach for Protecting Location Information in Mobile Communication Networks", In Proc. 5th European Symposium on Research in Computer Security, Sept. 1998

15. Available at http://www.ietf.org/html.charters/geopriv-charter.html

16. A. Friday, H. Muller, T. Rodden, and A. Dix, "A Lightweight Approach to Managing Privacy in Location-Based Services", in Proc. Equator Annual Conference, Oct. 2002

17. R. Bajaj, S. L. Ranaweera, and D. P. Agrawal, "GPS: Location- Tracking Technology", IEEE Computer, Apr. 2002, pp. 92–94

18. G. F. Marias, N. Prigouris, G. Papazafeiropoulos, S. Hadjiefthymiades, and L. Merakos, "Brokering Positioning Data From Heterogeneous Infrastructures", Wireless Personal Communications Vol. 30 (Issue: 2-4), Sept. 2004, pp. 233–245

19. P. Bahl, and V. N. Padmanabhan, "RADAR: An In-Building RF-based User Location and Tracking System", in Proc. of IEEE INFOCOM2000, Mar. 2000

20. P. Castro, P. Chiu, Ted K. and R. Muntz, "A Probabilistic Room Location Service for Wireless Networked Environments", in Proc. Ubicom2001, Sept. 2001

21. R. Want, A. Hopper, V. Falcao, and J. Gibbons, "The active badge location system", ACM Transactions on Information Systems, 10(1) Jan. 1992, pp. 91-102

22. N. B. Priyantha, A. Chakraborty, and H. Balakrishnan, "The cricket location-support system", in Proc. 6th ACM MOBICOM, Aug. 2000

23. J. Hightower, G. Borriello and Roy Want, "SpotON: An Indoor 3D Location Sensing Technology Based on RF Signal Strength", University of Washington, Technical Report #2000-02-02, Feb. 2000

24. C. Kee, et.al, "Development of Indoor Navigation System using Asynchronous Pseudolites", in Proc. ION GPS -2000, Sept. 2000

25. J. Hightower and G. Borriello, "Location Systems for Ubiquitous Computing", IEEE Computer, 34(8), Aug. 2001, pp. 57–66

26. B C. Hoffman, et. al, "Secret Sharing Schemes, Project Specification" COSC 645, Spring 2004

27. A. Shamir, "How to share a secret", Comm. of the ACM 22 (1979), pp. 612–613

28. F. Y-C Hu and D. B. Johnson, "Caching Strategies in On-Demand Routing Protocols for Wireless Ad Hoc Networks", In Proc. 6th Mobile Computing and Networking, Aug. 2000

29. R. Beraldi, and R. Baldoni, "A caching scheme for routing in mobile ad-hoc networks and its application to ZRP", IEEE Trans. on Comp, Vol. 52, No.8, 2003
30. L. Buttyan, and J. P. Hubaux, "Nuglets: a Virtual Currency to Stimulate Cooperation in Self-Organized Mobile Ad Hoc Networks", Technical Report No. DSC/2001/001, EPFL, Jan. 2001
31. W. Staling, "Network Security Essentials", Prentice Hall, ISBN 0-13-016093-8, 2000

Filtering Location-Based Information Using Visibility

Ashweeni Beeharee and Anthony Steed

Department of Computer Science, University College London, Gower Street, London
WC1E 6BT, United Kingdom
{A.Beeharee, A.Steed}@cs.ucl.ac.uk

Abstract. In this paper we present an approach for exploiting knowledge about features in the real world in order to compute visibility of buildings. This is performed with the awareness of the inconsistencies and lack of accuracy in both mapping technology and GPS positioning in urban spaces. Electronic tourist guide systems typically recommend locations and sometimes provide navigation information. We have augmented this system to exploit visibility knowledge about neighbouring physical features

.

1 Introduction

One reason that mobile, ubiquitous and mixed-reality systems are attracting a lot of attention recently is that they can present information about the real world that surrounds the system and its users. Along with increased power of portable systems, and accessible tracking technology such as GPS, this has enabled a new class of highly interactive 'guiding' that can give information and instructions to the user as they move through their environment [10]. This is a very broad class of application, but typical applications include presenting route knowledge from road databases [4] or location-specific information such as web pages about buildings and spaces. There are many configurations of such systems depending on the technology available from WAP-based phone, through to wearable personal computers. They rely mostly on retrieving some form of location data about the user, either by explicit input or passive tracking, and matching this against a database of knowledge about the world.

In this paper we start by claiming that such systems can suffer from two problems. The first is that the forms of information that are presented are often quite abstract, and thus hard to relate to the real world. For example, route knowledge might be presented as a series of instructions such as 'Turn right after 100m', that can be difficult to interpret accurately. This is because those instructions require judgements that users might find it difficult to make such as relative distance. The second problem is that the information often ignores critical parts of the user's perception of their own context: what they can or can't see from their current location.

We present a guiding application for urban environments that exploits knowledge about the physical structure of the real world. The guiding application has

T. Strang and C. Linnhoff-Popien (Eds.): LoCA 2005, LNCS 3479, pp. 306–315, 2005.

fairly standard functionality for a guiding application: as the user moves around the real world, their location is plotted on a map and icons which link to geo-tagged and context sensitive information are presented to the user. The location is either sensed by a GPS receiver, or explicitly input if a GPS unit is not available or a usable GPS signal is not available. As they move, a location-based recommendation system presents a series of media resources and links that might be relevant. Primary amongst these is photographs of the real world.

Our contributions are in exploiting knowledge about the real world. Specifically we exploit building geometry from a geographic information system (GIS). We use this knowledge in two ways. Firstly, information can be attached to the buildings themselves rather than simply to coordinates in space. This could be done explicitly, but our first technical contribution is in presenting a method for taking photographs and associating them to the buildings that are actually in the photograph. Secondly, when retrieving information we filter it based on whether it is likely to be visible from the current location. Our contribution here is in showing how this can be implemented even if the user's location is known only imprecisely.

By using building geometry as the primary means to index information we get around some of the problems of the imprecision and inflexibility of just associating information to positions in space. By retrieving information based on an estimate of visibility, we can present data that the user can more easily relate to their actual experience.

In this paper we will first discuss some related work in more detail. We then present an overview of the system encompassing the user interface devices (a laptop or PDA) and the back-end database services. In the following three section we then present the implementation: user interface for editing recommendations, geo-indexed photograph database and process for generating and filtering recommendations. Finally we discuss the deployment of the system and some preliminary experiences with the system.

2   Related Work

There are many examples of electronic consumer guiding systems available. For example, see [3] for a review of the capabilities and issues of vehicle GPS navigation systems, and [12, 14] for example, location-based PDA guide systems. In-car systems often utilise a GPS system to give location information. Indeed moving vehicles are a situation where GPS performs well: some consumer GPS units exploit velocity information and models of vehicle dynamics for improving tracking accuracy. Also most roads are uncovered and it is often possible to get a good view of the sky because roads are built away from buildings. On a vehicle, a good GPS aerial position can be chosen and the equipment can be powered reliably.

GPS-based systems perform less well in urban environments and in situations where they are carried by pedestrians. There are several reasons including, the occlusion of the sky by buildings and trees, less stable aerial placement, presence of an occluding body near the device (the user) and less certain movement behaviour of the carrier. Thus mobile city guides use a much wider variety of location technologies, including 802.11 wavelan [15] and explicit input [6, 13]. For the purpose of this paper it is worth noting that all these technologies provide different qualities of error. GPS varies in accuracy over time [16], and can often fail to provide a position. However, when it does provide a position, it can be accurate to within a few meters, and this is expected to improve as technical improvements are made to the supporting technology. For phone cell, and other region-based location technologies it can often be difficult to describe the region that is sensed. For example, see [8] for a description of tracking accuracy using a 802.11-based system. For our work, we will be able to use any approximation to the location of the user since we use a sampling approach to visibility analysis

3   System Overview

The George Square system (subsequently referred to simply as GSSystem) provides a guiding application for urban spaces [7]. A user has a PDA or tablet PC running User Viewer software. The user viewer software shows a map of the local area. The map is generated automatically from the MasterMap data [2] using the mapserver software[1]. As the user's movements around the real world are tracked by GPS or as they move an icon representing them on a map, the GSSystem pushes recommendations on to the user interface. A recommendation is a link to a resource such as a web page, or a photograph. The recommendations have a focus that is a polygonal region of space on the map to which they are attached. A single focus can have multiple recommendations attached to it. The system includes the ability to take a photograph and attach it to an existing focus. The focus closest to the user is used.

The GSSystem is novel in that recommendations are not generated purely by proximity to the user; rather they are based on likely contextual relevance based on analysis of previous use of the recommendations [9]. That is, a recommendation will more likely to be made in the past, another user who shared some previous recommendations, then went on to find subsequent recommendations useful. The service that performs this, Recer Service, is a form of collaborative filtering, but note that the symbols over which matches are made are not explicitly defined, rather activity in the system is recorded verbatim as text strings. The GSSystem is also novel in that it is peer to peer in nature. This means that although recer is an independent service, any number of such services can be run. Recer services communicate through a shared data space system, EQUIP [5]. The recer service records behaviour of all user viewer clients that connect to this data space and can make recommendations to all peers on that space. It provides no other services and, in particular, peers automatically set up collaboration between themselves as required.

The original GSSystem thus comprises the elements User Viewer, Shared Data Space, Recer Service and Focus Database from Figure 1. In this paper we discuss the Enhanced George Square System (EGSSytem), which also includes the Visibility Filer, Map Data and Map Database elements from Figure 1.

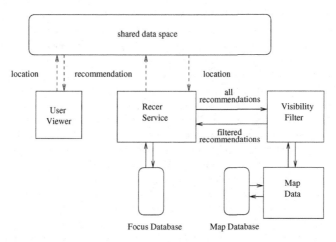

Fig. 1. Overview of the Enhanced George Square System in Run-Time Mode

4   Photo Visibility Computation

The main contribution of this paper is in the use of visibility information to filter recommendations. A recommendation is a list of resources. In the original GSSystem, these were purely generated based on prior activity [9] and our enhancement is to also provide a filter based on probability of visibility from current location.

In the original EGS, new photographs were simply attached to the closest focus. We provided a simple editor such that more specific information about the photographs of the buildings can be stored and used to compute visibility of real buildings within those photographs. The images need not be attached to any particular existing building but can exist at any location on the map. By using the editor, the user can also specify the view volume for each photograph by drawing on the map.

In [11] Cohen-Or et al. present an overview of visibility algorithms from a computer graphics research point of view. Although analytic solutions for visibility exist, because our system is probability based, we need to estimate how much of a photograph a building covers. This, and ease of implementation suggests that a simpler, sampling-based approach is more suitable. We have used a raycasting algorithm to compute the visibility of foci boundaries from a photograph's view volume. This algorithm has the advantage of being easy to implement, and the from region visibility we will use later, is a simple variant.

The visibilities are calculated using the following steps. Firstly, a set F is populated with all features (such as buildings) which intersect with the view volume of the image. About hundred rays, with the view volume's origin as their initial point, are created according to the expanse of the view volume. These rays are cast for each of the buildings in the set F.

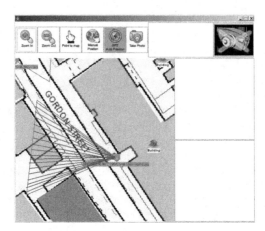

Fig. 2. Example of raycasting approach for associating foci boundaries to photographs. The lines originating from the photograph icon indicate some example rays that are cast from the photographs location. Note that because of the direction the photograph is taken in, the photograph is not associated to the closest building which is 'behind' the camera

If a feature is hit, a check is performed to ensure that there is nothing between the feature and the viewpoint where the picture was taken, that has been hit by the same ray. This is done by comparing the distances of the intersection points of the features to the view volume's origin. If the feature under inspection is the nearest one to be intersected by the ray, it's counted as a hit. Otherwise it moves on to the next ray. We repeat the same operations until the set F is exhausted.

5   Filtering Recommendations

The second contribution is in taking recommendations and filtering them based on visibility to the user's current location. A recommendation consists of a list of resources, each of which has an associated foci. We use these foci and test if they are visible from the user's location.

5.1   Visibility Filter

Because the user's location is known only inaccurately, we use a from-region visibility algorithm [11]. We assume that some form of probability distribution is given for the user's location. Once again, a sampling-based method is appropriate because it can estimate likelihood of visibility no matter the shape of the location probability distribution.

The algorithm used to compute the visibility of the recommendations is as follows. A set P is populated with all polygons to be considered in the visibil-

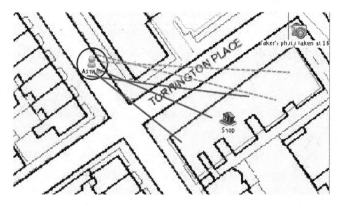

Fig. 3. Visibility computation for the filter - A random point is taken in the region around the location (the circle). Another random point is taken on the target building, and ray is defined between them. The thicker rays hit the target while the thinner rays are obstructed by buildings in their path

ity computation. A view polygon around the view point is considered. This is representative of the error in the GPS positioning.

Another set of polygons, R, is created corresponding to the recommended features. R need not be, but usually is a subset of P. The following steps are repeated for all polygons in R, referred to as target polygon.

1. From the set P, a new subset is defined as the polygons which lie within a rectangular region subtended by the viewpoint and the centroid of the target polygon. This subset of polygons is the set of potentially relevant polygons.
2. Select n number of random points within the view polygon and the target polygon.
3. For each ray from a point in the view polygon to the target polygon, find intersections with the set of relevant polygons. Search for intersection (for a line) is halted as soon as it intersects ANY polygon. The visibility confidence of a given target polygon is given by 1- (number of intersection found/number of rays considered).

This process is illustrated in 3, where five rays are shot, but only three hit the building, giving a confidence of 60using the set of focus polygons found in the recer database or by using the set of closed polygons from the Ordnance Survey map database.

5.2   Presenting Recommendations to the User

Figures 4 show simple example of filtering. a) Un-filtered recommendations b) Filtered recommendations. Note the building at the top of the figure has been removed. Also note that the photograph on the bottom left has not been removed, because the building it is a photograph of is still visible.

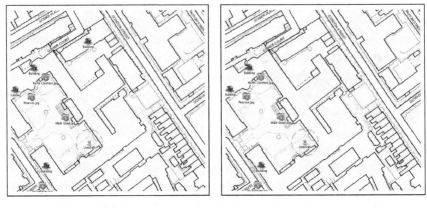

(a) Unfiltered Recommendation   (b) Filtered Recommendation

Fig. 4. Maps showing recommendations from Recer Service

5.3   Highlighting Recommendations at Run-Time

The building which is highlighted upon selection has a tool-tip which says 1) from which photograph it can be seen 2) how much space in the photograph it occupies.

5.4   Small Form Factor Versions

The previous sections have used screenshots from a version of the User Viewer software designed to run on a laptop or tablet PC. For mobile use we also created a version of the user viewer that runs on a PDA.

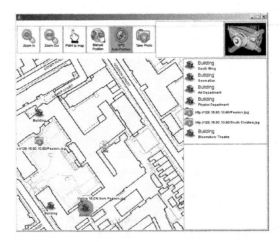

Fig. 5. Highlight recommendation to the user

This small form factor is a challenge to any mapping application, and filtering of recommendations on visibility can serve as an aid to removing clutter on the display. For this version, the recommendation service has been implemented as a web service. Figure 6 in the next section shows examples using the PDA version.

6   Experience on the Streets

The original GSSystem takes its name from the area of its original deployment in Glasgow, Scotland. Some minor technical changes within the EGSSytem have made it deployable anywhere in the UK given availability of the OS MasterMap information. A test system has been deployed in the Bloomsbury area of London, England. Thirty photographs were taken in the local area and published on a website. These were then added to the map and associated to buildings using the EGSSystem. The User Viewer software running on a laptop was used to test the system with GPS tracking on and explicit location input. Finally we tested with the PDA version. For this purpose we used publicly available 802.11b networks, though other forms of networking would be possible. It is possible to run the plain GSSystem standalone, but the EGSSystem requires access to a database.

Figure 6 shows a picture of the PDA version on the street. A photograph has been recommended, and it is obvious from the picture that the building in this photograph is visible from the user's location, though the actual photograph was taken from a very different angle. This is the core functionality that we wish to support in the application: presentation of information filtered based on likelihood of visibility. Note that the EGSSystem can be used in two modes: tracking by GPS, or explicit input. In the former mode, the photographs are useful for the user to orient themselves to the map and the other recommendations. In the latter mode, it is useful to be able to 'preview' an area to see what it will look like when you get there.

Informal observations of use in the street suggest that recommending photographs based on visibility is quite intuitive to understand. One concern was that allowing recommendation of photograph that were taken from locations that

Fig. 6. Recommending a picture to the user

are invisible but of buildings that are visible would be confusing for the user.
However, we realised that buildings are quite recognisable from different directions, so it would usually be obvious which building was meant. If this proved
a problem, it would be possible to extend the system to check which parts of
buildings were visible. A second concern was that visibility filtering would mean
results would change state frequently. However the sampling approach means
that visibility changes fairly slowly. Because a threshold is used, buildings that
are only marginally visible are not allowed through the filter.

A fuller user evaluation is currently being designed.

7   Conclusion

We presented a system, the Enhanced George Square System, a mobile guide
application that exploits the geometry of the real world. It uses GIS information
about buildings as a basis for filtering information based on whether it is likely
to be visible to the user in the real world. In this paper we have presented technical contributions on how to implement this filtering based on raycast sampling
methods. This method has the advantage of being simple to implement and fast.
We have also presented a system and application that exploit this visibility information. These should be seen as early examples of the potential of exploiting
visibility information. Furthermore, we would claim that this application is one
of the first to start investigating the exploitation of knowledge about the geometry of the world as a primary determinant of user context. There are several
avenues for further work. The algorithms we have presented are simple, and we
have already indicated that there is a lot of other work from the 3D graphics field
that could be incorporated to improve and speed up these algorithms. However,
the most fruitful areas for further work are on other methods or paradigms for
exploiting knowledge about local context.

Acknowledgements

This work was supported by the UK projects, EQUATOR Interdisciplinary
Research Collaboration (EPSRC Grant GR/N15986/01) and Advanced Grid
Interfaces for Environmental e-science in the Lab and in the Field (EPSRC
Grant GR/R81985/01). The vector data used was supplied by the UK Ordnance
Survey.

References

1. Mapserver. Web Resource - verified 15-02, 2005. http://mapserver.gis.umn.edu.
2. Uk ordinance survey.   Web Resource - verified 15-02, 2005.   http://www.
   ordnancesurvey.co.uk/oswebsite.
3. ABBOT, E., AND POWELL, D. Land-vehicle navigation using gps. *Proceedings of
   the IEEE 87*, 1 (1999), 145–162.

4. AGRAWALA, M., AND STOLTE, C.   Rendering effective route maps: Improving usability through generalization. In *SIGGRAPH 2001, Computer Graphics Proceedings* (2001), E. Fiume, Ed., ACM Press / ACM SIGGRAPH, pp. 241–250.

5. BENFORD, S., BOWERS, J., CHANDLER, P., CIOLFI, L., FLINTHAM, M., FRASER, M., GREENHALGH, C., HALL, T., HELLSTROM, S.-O., IZADI, S., RODDEN, T., SCHNADELBACH, H., AND TAYLOR, I.   Unearthing virtual history: Using diverse interfaces to reveal hidden virtual worlds. *Ubicomp 2001* (2001), 225–231.

6. BENFORD, S., SEAGAR, W., FLINTHAM, M., ANASTASI, R., ROWLAND, D., HUMBLE, J., STANTON, D., BOWERS, J., TANDAVANITJ, N., ADAMS, M., FARR, J. R., OLDROYD, A., AND SUTTON, J.   The error of our ways: The experience of self-reported position in a location-based game. *Ubicomp 2004* (2004).

7. BROWN, B., CHALMERS, M., BELL, M., MACCOLL, I., HALL, M., AND RUDMAN, P.   Sharing the square: collaborative visiting in the city streets. *CHI 2005* (2005), under submission.

8. CASTRO, P., CHIU, P., KREMENEK, T., AND MUNTZ, R.   A probabilistic location service for wireless network environments. In *UbiComp 2001* (2001), S. Verlag, Ed., pp. 18–24.

9. CHALMERS, M.   Cookies aren't enough: Tracking and enriching web activity with recer. *Preferred Placement: Knowledge Politics on the Web* (2000), 99–102.

10. CHEVERST, K., DAVIES, N., MITCHELL, K., FRIDAY, A., AND EFSTRATIOU, C.   Developing a context-aware electronic tourist guide: Some issues and experiences. In *CHI 2000* (2000), ACM Press, pp. 17–24.

11. COHEN-OR, D., CHRYSANTHOU, Y., SILVA, C., AND DURANT, F.   A survey of visibility for walk-through applications. *IEEE Transactions on Visualization and Computer Graphics 9*, 3 (2003), 412–431.

12. ESPINOZA, F., PERSSON, P., SANDIN, A., NYSTRM, H., CACCIATORE, E., AND BYLUND, M.   Social and navigational aspects of location-based information systems. In *UbiComp 2001* (2001), B. . S. Abowd, Ed., Springer, pp. 2–17.

13. KRAY, C., AND KORTUEM, G.   Interactive positioning based on object visibility. *Mobile HCI 04* (2004), 276–287.

14. LANE, G.   Urban tapestries: Wireless networking, public authoring and social knowledge. *Personal and Ubiquitous Computing* (2003), 169–175.

15. SCHILIT, B., LAMARCA, A., BORRIELLO, G., GRISWOLD, W., MCDONALD, D., LAZOWSKA, E., BALACHANDRAN, A., HONG, J., AND IVERSON, V.   Challenge: Ubiquitous location-aware computing and the place lab initiative. In *The First ACM International Workshop on Wireless Mobile Applications and Services on WLAN (WMASH 2003)* (2003), pp. 29–35.

16. STEED, A.   Supporting mobile applications with real-time visualisation of gps availability. *MobileHCI 2004* (2004), 373–377.

Introducing Context-Aware Features into Everyday Mobile Applications

Mikko Perttunen and Jukka Riekki

Department of Electrical and Information Engineering and Infotech Oulu,
P.O.BOX 4500, 90014 University of Oulu, Finland
{Mikko.Perttunen, Jukka.Riekki}@ee.oulu.fi

Abstract. We describe our approach of introducing context-awareness into everyday applications to make them more easy-to-use. The approach aims in shortening both the learning curve when introducing new technology to end-users and prototype development time, as well as results in more reliable prototypes. Moreover, we expect that the approach yields better quality user test results. To demonstrate the approach, we have employed context-based availability inference to automatically update the availability of IBM Lotus Sametime Everyplace users. This is likely to result in more reliable availability information and to make the application easier to use. Context inference is done using information from Lotus Notes Calendar and WLAN positioning technology.

1 Introduction

Context-aware applications adapt to the environment of the user, aiming in calmly supporting the tasks of the user [1, 2]. We have seen a plethora of applications being built based on and relaying on the concept of context-awareness. We have witnessed software architectures for context-aware services been designed and various visions of their ubiquitous existence in the future been presented.

Even if these visions of ubiquitous calm technology would not be realized, context-awareness will have a tremendous effect on our lives; it will change the applications we use everyday and the way we use them. But the change will not happen overnight. Context-awareness is generally seen as enabling technology in distributed systems [3]. Moreover, Abowd has noted that although context-awareness is crucial in ubiquitous computing because of frequent context changes, it is still important in context-sensitive desktop systems as well [4].

It is improbable that the users will desert the familiar applications and obtain totally new ones just because of the context-aware functionality. A more likely scenario is that context-aware functionality will be added gradually into the current applications, both in mobile computing and in distributed computing with fixed environments. For example, things like instant messaging applications automatically updating the availability of the user to unavailable, when she has not touched the keyboard for a while, can be seen as simple context-aware functions. Context-awareness will be just a feature among others that the users weigh when they are

T. Strang and C. Linnhoff-Popien (Eds.): LoCA 2005, LNCS 3479, pp. 316–327, 2005.

selecting the application to be purchased or maybe even an invisible, embedded, feature enhancing usability.

People don't want to suffer from long learning periods when taking new software into use. Naturally, such is the case for context-aware software as well. Adoption-centric research aims in finding technologies that might be useful for end-users and the adoption of which is likely to take place (e.g. [5]). Many of the context-aware applications for ubiquitous computing have been research prototypes built from scratch for demonstration purposes. Moreover, they often utilize proprietary user interfaces due to fast proof-of-concept development.

Aligned with the suggestion of adoption-centric research, we propose that prototypes based on well-established applications and application frameworks should be used to the extent possible in researching context-awareness. Furthermore, systems and frameworks providing plug in mechanisms or exposing application programming interfaces for modifying their features are suitable candidates for this approach, similarly to what is suggested with software engineering tools in [6]. Naturally, open source projects provide the greatest potential in this respect.

The above suggestions aim in shortening the time end-users spend on learning new technology, but also in shortening the prototype development time. Using well-established applications should result in more reliable prototypes as well. Moreover, we expect that the approach yields better quality user test results and may boost the development of realistic product concepts in the field of context-awareness. It does not, however, mean that self-standing prototypes for testing innovative ideas were not important; it rather suggests that context-aware software should be developed along these parallel tracks. Since it is hard, or even waste of resources, to implement a really ubiquitous computing environment in a single research project, we need also to follow the path of building reusable toolkits described by Abowd [4].

In this paper, we describe integrating context-awareness into mundane instant messaging software of office workers. Namely, we deal with our efforts in introducing context-aware features to IBM Lotus Sametime and its mobile extension, Sametime Everyplace (STEP). We describe a purely server-side solution; the client-side software comprises of commercial of-the-shelf products, and stays intact. Our system combines the usage of IBM Lotus Notes calendar as well, to get information on scheduled meetings. We describe our fully implemented solution and discuss our findings in general in the last section.

The rest of the paper is organized as follows. In section 2 we list related work. In section 3 we describe preliminary guidelines for introducing context-awareness to existing applications. In section 4 we review context-based availability inference in instant messaging. Section 5 presents our prototype system. It is followed by a discussion of the findings in section 6.

2   Related Work

Abowd lists the criteria for ubiquitous computing (ubicomp) research to include: a motivating application, a notion of scale, everyday use, and evaluation [4]. He engraves evaluation in everyday life and states that the system should be robust enough so that long testing causes the novelty of the system to vanish. Considering transparent interfaces, context-awareness, and automated capture as the general features of ubicomp applications, and rapid prototyping of systems as the way to develop them, he unveils the need for toolkit design, software structuring for separation of concerns, and component integration.

In [6] the authors describe building add-on software design tools on top of Lotus Notes. Their hypothesis is that new tools are easier to adopt, if they are "compatible with both existing users and existing tools". This hypothesis is supported by the familiar notion of "learning curve" from software development [7].

Walenstein [5] points out the generic problem in developing "research prototypes"; they are often ad-hoc implementations of the new innovations, so that user testing with real end-users is difficult. He suggests that by using an existing framework it is easier to find a user base that is familiar with the framework and thus, much of the prototype. In the paper, he describes a framework to enhance reengineering software systems that is based on theory of *cognitive support* [8], which means, shortly, that the computing environment helps the users in perceiving, understanding, and remembering things necessary for their tasks.

In [9] Dey et al describe a conceptual framework for context-aware application design and an implementation of it as a toolkit. Their methodology consists of the following steps (details not repeated here): identifying entities, identifying context attributes, identifying quality of service requirements, choosing sensors, deriving a software design. They also provide examples of context-aware applications developed using the toolkit, one of the applications being an existing application that is updated by introducing context-awareness. Although this is similar to what we are suggesting here, it differs in one major way; in our case we are targeting environments that provide plug in, or extension mechanisms. That is, the environments are not changed at the source code level, whereas in the example of Dey et al the sources of the existing application are available and modified.

We have summarized some of the related work on instant messaging in [10] and revisit a few of them next. In a recent study of introducing presence information into telephone usage so as to provide a "live address book" it was found that users generally do not always remember to manually update the presence information, which may lead to others not trusting the information [11]. Based on this study it can be assumed that also IM applications would be easier to use, if the updating of presence information were automatic.

MyVine [12] is a recent research prototype that integrates email, phone and instant messaging into one client that shows availability information of other users. Using the application, the users can initiate any of the above-mentioned communication methods; the real act of communication goes through the ordinary applications, i.e. the availability application provides an additional user interface. The system automatically infers the availability of the users from recognized context, which is formed of location information, speech detection, computer activity and calendar information. The MyVine client represents the availability of a user as four different levels of gray of a figure of the user, allowing others to see more detailed context information by moving mouse over the figure. The users mostly respected the inferred availability when the values were at the end of the scale i.e. highly unavailable or highly available. As the most important finding of their study we see that while the users easily saw that a person was moderately unavailable, they often still started to communicate via IM, if the detailed context information showed that the person was present in the building or in office. That is, physical presence seemed to be in these situations more important than the inferred availability.

In our previous work, we have developed a context-based method that aims in reducing unnecessary interruptions by automatically updating the presence, or availability, of the IM users [10]. The main idea was to use context relation, that is, in brief, to use the contexts of both the communication initiator and the receiver in inferring availability of the receiver. The method can be exploited in a context-aware ubiquitous system. In related work, only the context of the receiver had been employed.

3   Preliminary Guidelines for Introducing Context-Awareness

In this section, we describe the tasks comprising our simplistic guidelines of introducing context-awareness into everyday, existing applications. We propose the following four tasks: Application Analysis and Context Recognition, Context Sensing and Representations, Prototype Building, and User Experience Evaluation. The guidelines reflect typical interactive system design lifecycles (e.g. chapter 6 of [13]). The idea is that by guiding our research with these high-level and comprehensible tasks, we form a framework that allows us to perform loose comparisons of prototypes designed along these guidelines. The guidelines do by no means constitute a complete process.

Application Analysis and Context Recognition means studying how the applications used at the moment could be improved, particularly by utilizing context information. The analysis of requirements for a ubicomp application can start by observing and analyzing the real-life situations of probable end-users [14]. Next, data collection for the identified and meaningful contexts can be done using sensors and context recognition can be attempted offline. This phase is similar to our previous work on context recognition and data mining, where data mining methods are used to test what contexts can be recognized from given data [15]. Moreover, the application deployment environment is studied for the feasibility of introducing new software modules or features into it. This phase results in scenarios describing the context-aware features and provides some insight into the design possibilities.

Context Sensing and Representations includes developing the context recognition methods to be suitable for use in the application prototypes as well as designing required context representation methods for the prototypes. This phase is analogous to phases in many design guidelines for context-aware systems e.g. [9, 16]. In general, it can be considered as improving the toolkits or architecture providing the basis for context information. After this phase, the infrastructure is ready for the application development.

In Prototype Building various application prototypes introducing context-aware features are built. The main goal is a prototype that is reliable enough for real user testing. Introducing context-aware features sparingly may be a better approach than changing the application behavior too much at once [5]. The features introduced first are selected by weighing the ones that rely on the contexts that can be most reliably recognized and the ones that are expected to provide the most value for users.

Existing knowledge (though not well-established) from the field of context-aware application design is utilized. In particular, experience from previous prototypes of the same system, or similar systems, is considered.

When performing User Experience Evaluation, the target is in evaluating the usability of the application and the suitability of the underlying supporting technologies, for example context sensing methods. The relation of reliability of context recognition to user experience can be studied, for instance, since the required reliability may vary between features. As mentioned in [5], it should be easy enough to find suitable test users from the existing user base.

4   Instant Messaging and Context-Based Availability Inference

In instant messaging, the users log in to an instant messaging service that enables them to see if other users are available and to exchange messages with the other users in almost real time. The messages and the presence information are delivered through the IM service in the Internet. Similarly to RFC 2778 [17], we call the receivers of presence information *watchers* of the users the presence of which they receive. The users offering their presence information are called *presentities* (presence entity).

Generally, messages can be sent in IM when the recipient is logged in to the system and her availability is *available*. In many systems, however, messages can be sent also in some other states, but the sender is aware that the messages might not be read immediately (e.g. just opened on screen, when user is away from the desktop computer). The state available is interpreted by the sender as willingness of the recipient to receive messages and to respond to them in a timely manner. To avoid disturbance, a user can manually select to be *unavailable*. Because availability plays such a crucial role in IM, it is important that the information is reliable. The correctness of availability reduces unsuccessful communication attempts.

Context-aware systems adapt to the situation of the user in various ways. The availability of IM users can be updated automatically, using context information of the users [10, 12, 18]. This is important due to two reasons: First, the users don't always remember update their availability, which causes the information to be unreliable [11]. Unreliable information may cause unwanted communication when the user really wishes not to be disturbed. It may also be a waste of time for the ones trying to reach someone who is available according system, but doesn't want to communicate. Second, correct and timely automatic updating of availability is likely to ease the use of the application and let the users concentrate on more productive tasks. We expect that the influence of these factors to the user experience of instant messaging software increase along with the volume of IM communication.

To be easy to use, the IM application should be able to update the availability of its user without the user's intervention. This also increases the reliability of the availability information, taking into account that the user might sometimes forget to update it. In our work one major design guideline is that the receiver is always in control of her availability. Availability updating should be based on automatic context recognition and function according to rules accepted by the user. For example, the user might select a rule "When my activity is *conference*, I'm *unavailable*".

5   System Prototype

5.1   Capnet System

The Capnet (Context-Aware and Pervasive Networks) program focuses on context-aware mobile technologies for ubiquitous computing [10, 19]. One Capnet engine, shown in Figure 1, handles context in this prototype.

5.2   IBM Lotus Notes/Domino and Lotus Sametime

As mentioned in the introduction, we developed the context-aware automatic availability inference on to the Lotus Sametime collaboration system and used Lotus Notes as a source of calendar information.

Sametime is a real-time collaboration environment. To list some of the features, it enables the user to send chat messages and transfer files in real time, audio and video communications, shared file editing, presence-awareness, and organizing online meetings. The features can be used through a variety of client applications that can be run on devices ranging from STEP's Java MIDP devices to desktop PCs [20].

Sametime environment can be interfaced with other software through the Sametime client toolkits. The Sametime Links Toolkit enables web page developers to build presence-awareness into the web pages. In addition, there is Sametime COM Toolkit, Sametime C++ toolkit and Sametime Java toolkit. The Sametime Java toolkit provides the richest set of features [21].

Lotus Notes is widely used client-server document handling, email and calendar software. Domino web server that provides web access to many of the Notes applications accompanies the Notes database on the server. The web server contains also a servlet engine and some of the Sametime services are provided by servlets, for example.

5.3   From Application Analysis to Prototype Building

In our previous work [10], we started by analyzing the usage of current IM applications. Soon, we came up with the idea of automatic availability inference, and found out that others had already suggested it as well (see section 2). That is, in this case we already had the basic application usage scenario and we did not perform data collection or data mining.

Since we had the possibility of utilizing Sametime, we analyzed the extendibility potential of it, and found out the services provided through the toolkits. We expected the Notes calendar to most likely be in daily use by the users of Sametime, making Notes a reliable source of schedule information.

In this first prototype, we decided to begin in the simplest possible way; the prototype considers only the context of the user himself and employs a generic availability rule for all users. The rule defines the availability of a user to be *unavailable* when she is in a room where she has a scheduled appointment at the current time. This matches the basic scenario; automatically updating the availability of the user based on her context.

Ideas that are discarded for expected usability problems (e.g. too different from current application features, unreliable context recognition) are considered for following prototypes. For example, allowing the users to manage the availability rules could be done through a Notes application, or a web page. The users could associate rules to groups in their contact list to be available for a subset of people in certain contexts. Moreover, we decided not to include the context-relations [10] in availability inference, since introducing too many features at once could lead to difficulties in the adoption of the application and could introduce the need for prolonged user test periods. We will incorporate context relations in the following prototype versions.

5.3.1 Server-Side Software

The prototype consists of three major, originally independent systems: Sametime, Notes Calendar, and the Capnet system. The users' password and user name to Notes and Sametime are stored into the Capnet system to enable it to use these systems on behalf them. The Capnet system tracks the context of the users and infers their availability. The availability information is delivered to the Sametime system using the Sametime Java toolkit. Below, we will describe the interfacing systems briefly, without going into details of implementation.

The IM system is composed of the components shown in Figure 1. The positioning system is based on the Ekahau [22] Wireless LAN positioning engine that utilizes WLAN signal strengths measured in the devices and provides positioning information for the Capnet system.

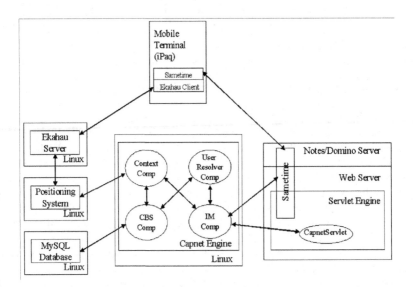

Fig. 1. Software components of the prototype IM system

The calendar information is accessed through a servlet executing in the servlet engine of the Domino web server. The servlet accesses the Notes database through

Domino Java API. The database resides on the local machine, which makes the access relatively efficient.

During startup, the system generates a Capnet IM session for all the Sametime users into the IM component. The IM component, in turn, performs a separate login for each Sametime user. If a user sets her availability manually, the Sametime client reports the value to the Sametime server, which delivers the information to the *watchers*. The instant messages travel through the Sametime server from client to client.

It's important to note, that the Capnet system acts as an add-on to the Sametime system; if the Capnet system crashes, it doesn't affect Sametime's normal functions. Also, the Capnet system can be started and stopped independently – only the availability inference is affected. Unique email-like identifiers identify users in Capnet, and an identity is associated with a user when she logs into the Capnet system.

In our current implementation, automatic availability inference is performed by the IM component, which receives position information from the context component. In the next version, we plan to update the context component to provide availability information as well. The inference is done using hard-coded rules common for all users; explicit representation of these rules is one challenge for further research. When the IM component receives a 'room changed'-event, it checks the current event in the calendar of the user and whether it is scheduled to take place in the new room. If a meeting or an appointment is scheduled to be currently going on, her activity is set to *meeting* and her availability is set to *away*. The availability values used in Sametime are *online*, *offline*, *away*, and *do not disturb*. *Away* is used by the Capnet system, because the STEP Connect application shows the availability message of the user only when the availability of the receiver is *away*.

5.3.2   Client-Side Software

The mobile devices, Compaq iPAQ PDAs, are equipped with WLAN cards and located with the Ekahau positioning engine running on a network server. Only the Sametime and Ekahau clients are running on the mobile device. That is, only off-the-shelf software is executed in the client terminal. This was one of the key ideas in developing the prototype, along with the fact that, this way, users can continue on using the applications they are familiar with. Because the client application will not be modified, the settings of the user and, for instance, the contact list with its associated privacy settings are preserved. These facts should greatly improve adoptability.

However, because there is no toolkit for modifying the STEP Connect application, we loose some of the adaptability possibilities. In testing the availability inference, it does not hurt too much. We expect that in general many new features can be tested without big modifications to the user interfaces. If this would become necessary, we would need to implement a new client by utilizing the Sametime toolkits.

In our current configuration, the users can access their calendars by using desktop installations of Notes clients or through iNotes (Notes web access) with a full-blooded web browser. That is, they can't access the calendar from the PDA. To provide better user experience, we will consider taking Domino Everyplace into use in the following user tests.

5.4   Test Environment and Testing

The university's premises covered by WLAN were used as the test environment. The meeting rooms of our laboratory's premises were configured into the Notes system, so that when users organized meetings or appointments through Notes' calendar, they could reserve a room in the ordinary manner.

Figure 2 shows the UI of Sametime Connect client on the PDA screen with a context menu (left) opened by tapping a user in the contact list and the user's availability message shown (right). From the screen shown on the left in Figure 2, one-to-one messaging can be started by clicking 'Chat' from pop-up menu.

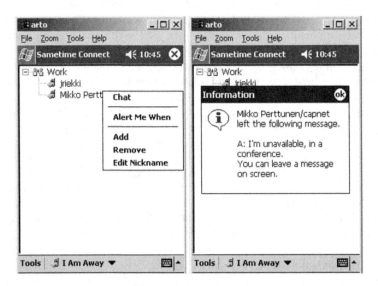

Fig. 2. Views from the Sametime Connect application on PDA screen. Left: Context menu. Right: Availability message

When a user is in a meeting, her availability is set to *away* by the Capnet system. If another user tries to start chatting with the user while she is in the meeting, the STEP Connect application on the PDA shows the availability message, as on right in Figure 2. Also this message is generated by the Capnet system as a result of the availability inference. The Sametime Connect client for PDAs suggests the user to leave a message on the screen of the other user if her availability is *away*.

We demonstrated the system with two users according to the scenario in section 5.3. While mobile, the users used the PDAs described section 5.3.2. We plan to organize real user testing soon. When a user entered a meeting room with a scheduled meeting, the Capnet system recognized the context change and inferred a new availability for the user. Naturally, this works also when one user sends a meeting request to another one who accepts it.

6   Discussion

We introduced context-awareness into the off-the-shelf Sametime instant messaging environment. The key idea was to demonstrate our approach of introducing context-awareness into everyday applications by implementing a context-aware IM system that does not require proprietary software in the client terminal. We presented an IM prototype that utilizes context to update Sametime users' availability information automatically. Automatic updating aims in enabling user-friendly IM applications with reliable availability updating. It is likely to reduce disruptions by minimizing the number of communication attempts by others when the user is not willing to communicate.

According to our approach, we aim in performing user tests with our prototypes, but not introduce too many features at once, as the difficulties in the adoption of them could take too big role, which could mean that user test periods would need to be prolonged. This would also ruin the idea of introducing context-awareness into mundane applications that users are readily able to use. We hope that this cautious approach results in more reliable user experience evaluation. Such evaluations may provide important information on the underlying technologies that the context-aware features rely on. For example, our current positioning system is based on WLAN, the capabilities and constraints of which could emerge in the user test results, thus providing valuable information of the suitability of current the technology for commercial applications. This is hard to reach with prototypes implemented from the scratch.

Furthermore, the Capnet system has been gradually developed alongside various context-aware prototypes. The underlying architecture for handling context representation, reasoning, and sensor fusion (for example) is necessarily the component that currently cannot be obtained off-the-shelf. For this reason, it is desirable for it to be mature and reliable enough to neither affect the evaluation of the technologies it relies on, nor the evaluation of the applications it supports.

The current results demonstrate that our goal is right. It is possible to analyze existing applications and to find usability issues for which context-awareness may be a remedy. Using more existing technology also facilitates the prototype development. However, before organizing and evaluating user tests, we are not yet able to fully prove the concept. Moreover, we recognize it as future work to analyze the relation of the simplistic guidelines to general software engineering processes.

Through our experience with the Sametime and Notes systems, we envision that the Lotus Notes application framework provides many other interesting possibilities to augment existing applications by using context information. This is supported by the fact, that Notes is often used as a platform for collaborative applications in organizations [23]. Although we lack comparative experience with other large application frameworks, we consider the applicability of our approach generally significant. We feel the approach is a necessary step towards ubiquitous context-aware applications.

Acknowledgments

This work was funded by the National Technology Agency of Finland. The authors would like to thank all the personnel in the Capnet program, and the participating companies.

References

[1] Schilit,B.N. & Theimer,M.M., (1994) Disseminating active map information to mobile hosts. IEEE Network, Vol. 8, 5, pp. 22-32.

[2] Weiser,M. & Brown,J.S., (1997) The Coming Age of Calm Technology. In: Denning,P.J. & Metcalfe,R.M., (edited) Beyond Calculation: The Next Fifty Years of Computing. Springer-Verlag, New York, pp. 66-75.

[3] Satyanarayanan,M. (2001) Pervasive computing: vision and challenges. IEEE Pers. Commun., Vol. 8, 4, pp. 10-17.

[4] Abowd,G.D. (1999) Software engineering issues for ubiquitous computing. Proceedings of the 21st international conference on Software engineering, Los Angeles, California, United States, pp. 75-84.

[5] Walenstein,A.Improving Adoptability by Preserving, Leveraging, and Adding Cognitive Support To Existing Tools and Environments. Proceedings of the 3rd International Workshop on Adoption-Centric Software Engineering, pp. 36-41.

[6] Ma,J., Kienle,H.M., Kaminski,P., Weber,A. & Litoiu,M., (2003) Customizing lotus notes to build software engineering tools. Proceedings of the 2003 conference of the Centre for Advanced Studies conference on Collaborative research, Toronto, Ontario, Canada, pp. 211-222.

[7] Raccoon,L.B.S. (1996) A learning curve primer for software engineers. SIGSOFT Softw.Eng.Notes, Vol. 21, 1, pp. 77-86.

[8] Walenstein A. (2002) Cognitive Support in Software Engineering Tools: A Distributed Cognition Framework. PhD thesis, School of Computing Science, Simon Fraser University.

[9] Dey,A., Abowd,G.D. & Salber,D., (2001) A conceptual framework and a toolkit for supporting the rapid prototyping of context-aware applications. Human-Computer Interaction, Vol. 16, pp. 97-166.

[10] Perttunen,M. & Riekki,J., (2004) Inferring Presence in a Context-Aware Instant Messaging System. The 2004 IFIP International Conference on Intelligence in Communication Systems (INTELLCOMM 04), November 23-26, Bangkok, Thailand, pp. 160-174.

[11] Milewski,A.E. & Smith,T.M., (2000) Providing presence cues to telephone users. ACM 2000 Conference on Computer Supported Cooperative Work, Dec 2-6, Philadelphia, Pensylvania, USA, pp. 89-96, New York, NY, USA.

[12] [12]   Fogarty,J., Lai,J. & Christensen,J., (2004) Presence versus availability: The design and evaluation of a context-aware communication client. International Journal of Human Computer Studies, Vol. 61, 3, pp. 299-317.

[13] J. Preece, Y. Rogers and H. Sharp, (2002) Interaction Design. John Wiley & Sons, New York, USA.

[14] Tamminen,S., Oulasvirta,A., Toiskallio,K. & Kankainen,A., (2004) Understanding mobile contexts. Personal Ubiquitous Comput., Vol. 8, 2, pp. 135-143.

[15] Pirttikangas,S., Riekki,J., Kaartinen,J. & Röning,J., (2003) Context-Recognition and Data Mining Methods for a Health Club Application. In proceedings of 12th Int. Conf. Intelligent and Adaptive Systems and Software Engineering (IASSE 2003), July 9-11, San Francisco, California, USA, pp. 79-84.

[16] Barkhuus,L. (2003) Context Information vs. Sensor Information: A Model for Categorizing Context in Context-Aware Mobile Computing. Symposium on Collaborative Technologies and Systems, pp. 127-133.

[17] Day,M., Rosenberg,J. & Sugano,H., (2000) RFC 2778, A Model for Presence And Instant Messaging. IETF.

[18] Begole,J., Matsakis,N.E. & Tang,J.C., (2004) Lilsys: inferring unavailability using sensors. Proceedings of the 2004 ACM conference on Computer supported cooperative work, Chicago, Illinois, USA, pp. 511-514.

[19] Davidyuk,O., Riekki,J., Rautio,V.-. & Sun,J., (2004) Context-Aware Middleware for Mobile Multimedia Applications. In Proc 3rd International conference on Mobile and Ubiquitous multimedia (MUM2004), October 27-29, College Park, Maryland, USA, pp. 213-220.

[20] IBM Lotus Sametime 3.1 User Guide.

[21] White Paper: Introducing the Sametime Client Toolkits (2002).

[22] Ekahau, Inc. URL: http://www.ekahau.com/.

[23] Karsten,H. (1999) Collaboration and collaborative information technologies: a review of the evidence. SIGMIS Database, Vol. 30, 2, pp. 44-65.

Predicting Location-Dependent QoS for Wireless Networks

Robert A. Malaney[1,2], Ernesto Exposito[1], Xun Wei[1], and Dao Trong Nghia[1,2]

[1] National ICT Australia, Bay 15, Locomotive Workshop,
Australian Technology Park, Eveleigh, NSW 1430, Australia
{robert.malaney, ernesto.exposito, xun.wei, dao.trong-nghia}@nicta.com.au
http://nicta.com.au/
[2] School of Electrical Engineering and Telecommunications,
University Of New South Wales, Sydney, Australia

Abstract. In wireless networks the quality of service (QoS) delivered to an end user will be a complex function of location and time. "QoS Seeker" is a new system which informs a user what location in the wireless network he should move to in order deliver the required QoS for his real-time applications. At the heart of QoS Seeker is a construct known as a "QoS Map" - which is the value of a specific QoS metric as a function of space. If any significant temporal trends are present in the wireless network, then the current QoS Map will be statistically different from a future QoS Map. In this report we investigate the use of adaptive linear filters as a means to predict future QoS Maps from historical QoS Maps. By using the received signal strength (RSS) as the QoS metric, we show that local adaptive filters can deliver very significant performance gains relative to last-measure and moving-average predictors. We also show how global adaptive filters can produce performance gains, albeit at a lower level. These results show that adaptive prediction techniques have a significant role to play in the QoS Map construction.

1 Introduction

GPS availability on small mobile wireless devices is rapidly becoming ubiquitous. Indeed, GPS chipsets directly embedded in hand-held devices are now commercially available [4]. In a recent work [1] we outlined a new system, termed "QoS Seeker", which uses such GPS capability in mobile communication devices, in order to optimize the quality of the services offered by a wireless network.

In QoS Seeker information of application QoS metrics and device location, which would otherwise remain unused, are utilized in order to make the user aware of the availability of better QoS within his vicinity. With QoS Seeker deployed an end user no longer has to randomly move in order to find a physical position where the QoS of his application is at an acceptable level - instead the user is informed of where the nearest position for acceptable QoS is. QoS Seeker is easily deployed and is cost-effective. It is purely a software based solution, with

T. Strang and C. Linnhoff-Popien (Eds.): LoCA 2005, LNCS 3479, pp. 328–340, 2005.

no network modifications necessary, and no additional network infrastructure is required.

In our previous report [1] various experimental and simulated tests of a VoIP QoS Seeker system were presented, as was detailed information of the system architecture and protocols. These tests showed that such a system is indeed viable and practical, and that when deployed, users of VoIP over 802.11b wireless local area networks (WLANS) could more readily find locations where their VoIP connection was at an acceptable quality.

At the heart of QoS Seeker is a construct known as a QoS Map. This map represents the best estimate of the expected QoS that an application can receive in a particular physical zone. This could be the QoS estimate at the present time, or the QoS estimate at some future time. QoS Map construction occurs at a server which collects all position and QoS information from all users currently in the network. The server uses adaptive learning algorithms to construct from historical QoS Maps, what a future QoS Map would look like. This is required since QoS Seeker may inform a user to move to a zone a few minutes travel time away. As such, it must have made a reasonable determination on what the QoS in that zone - a few minutes into the future - will actually be.

Once the future QoS Map is created within the server, it can be periodically transferred back to the users of the system so that they can be informed of the predicted future QoS in their own local vicinities. In our initial set-up, a graphical representation of the QoS Map is presented to the user via a graphical interface on his device. Note that QoS Maps can find uses in many wireless settings, such as WLANs, sensor networks, or even ad hoc networks. Indeed, the prediction algorithms applied to QoS Maps may be of particular value in the more challenging and dynamic QoS environment of an ad hoc network.

In many cases, such as a stable radio environment with constant noise levels, a prediction algorithm can do no better than setting the future QoS metric equal to the last measured value. Indeed, in many of our experimental tests we find this to be the case. However, in radio environments which possess evolving and/or periodic trends, prediction algorithms applied to QoS Maps are likely to have significant performance gains. Evolving trends in QoS metrics could arise from many factors, such as movement of users, and periodic interference sources. We do not wish to model possible trends here, but to simply assume that they could exist, and therefore could affect QoS Map functionality.

Our previous work [1] outlined some simple tests of the prediction algorithms embedded at the QoS Map server. Here we wish to explore this issue in much greater detail. Our original contributions in this work are two fold. First, we explore the performance of adaptive linear filter algorithms applied to QoS Maps for a wider range of simulated conditions than previously reported. Such work is necessary so to quantify the performance gains obtainable by adaptive filters, relative to simple last-measure predictors, and moving-average predictors. Secondly, we explore the performance gains achieved by global filters (the same filter algorithm applied to all physical zones of a QoS Map), compared to local adaptive filters (different filter settings applied at different zones). The issue of global *vs.* local filters has important practical considerations, since global filters are much easier to incorporate in the system architecture.

2   QoS Maps

In actual implementation the QoS Map can be considered in abstract terms as a matrix Q of dimensions i by j. The element Q_{ij} corresponds to a QoS metric value in a particular physical zone of the QoS Map. Without loss of generality we will assume all elements of the matrix represent a physical zone of equal area. Consider a vector of QoS metrics q. The elements of this vector will contain any information obtained by the receiver which could pertain to the QoS of the application, such as packet delays, packet losses, jitter, RSS and reported link speed (data throughput). In all our maps we translate the QoS metric into the range $0 - 1$, with zero meaning zero QoS. Note that for most metrics the actual algorithm that maps the measured QoS metric to the range $0 - 1$ is slightly dependent on the platform and application software used. This is a consequence of the fact that the perceived QoS of an application is marginally different using different software and hardware platforms.

For the large outdoor areas we study here, the QoS Maps possess individual zones with physical dimensions of order 20m square. Construction of the current QoS Map is straight forward. Considering sequentially each zone of the QoS Map (identified by i and j), the database is filtered so that only QoS measurements (within the last time bin) associated with devices in that specific zone are selected. Some algorithm (such as the average, or the most recent value) is then applied to this data in order to form the value Q_{ij}. Repeating this process for each zone will provide the most up-to-date QoS Map.

2.1   Predicting QoS Maps

In dynamic situations the current QoS Map may not be the optimal indicator of what is likely to happen in the near future. As stated earlier, wireless networks, and the QoS Maps describing them, are complex and there are many reasons why evolution in the QoS Map could occur. In order to improve upon the usefulness of QoS Maps we must attempt to model dynamic behavior of QoS Maps in some coherent fashion.

In terms of our abstraction of the QoS Map, the question we face is given the data currently in the server: what is the best estimate of the value of Q_{ij} (the QoS value in a specific zone of the QoS Map) at some future epoch? Analyzing each zone will collectively lead to the predicted QoS Map for the next epoch.

For our calculations here we will focus on the RSS. The RSS from an access point is by itself an important element of the QoS vector. There are several reasons for this. First, it is one of the metrics that will be device independent. Other QoS metrics, measured at the application layer are to some extent influenced by the processing load currently active on a device. Secondly, even when a device is not running an application, it can still be recording RSS measurements within the QoS Map area. Thirdly, historical correlations with other QoS metrics and the RSS can be used to predict future metrics in a zone where no previous QoS metrics have been measured.

In QoS Seeker the prediction of the next QoS Map is based on neural network architectures. Artificial neural networks have in fact been used before in order to predict maps of RSS [2] [3] [5]. However, in these works, propagation models, terrain models, and known access point placements are used to predict RSS coverage over the wireless area. Test measurements are then used to train the neural network in order to find the appropriate RSS over the entire area. This is important for QoS Seeker, as it is likely that many areas of the QoS Map do not possess any measurement data, particularly at the onset of QoS Map construction. Although we will encompass such wider RSS prediction techniques into our neural network design, we are more interested in this specific report in looking at predictions of RSS based solely on historical data for that zone. As such, we assume measured data already exists for a particular zone, and that there are no gaps in the data. Our aim is to train the neural network to predict the future RSS based on the historical measurements of the RSS in a particular zone.

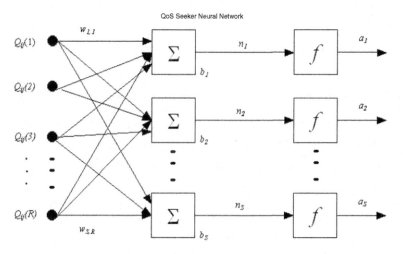

Fig. 1. Neural network architecture deployed in QoS Seeker

Determination of the appropriate network architecture (number of layers, number of neurons, type of transfer functions) to use for a given QoS Map requires experimentation. A specific architecture is unlikely to be optimal for all QoS metrics at all times. In the QoS Map server the architecture of the neural networks is based on Multi-Layer Perceptron (MLP) models. A generic architecture of such models is shown in Fig. (1). Here the inputs $Q_{ij}(1), \ldots Q_{ij}(R)$ are time delayed inputs representing historical values of the QoS Map metrics

in a specific zone. With this set-up, each QoS metric at each zone forms a specific Time Delay Neural Network (TDNN). A TDNN network can be made to
self-adapt to ongoing data collection (re-training) - a characteristic that makes
them ideal for the potentially complex and dynamic behavior that underpins
QoS Maps.

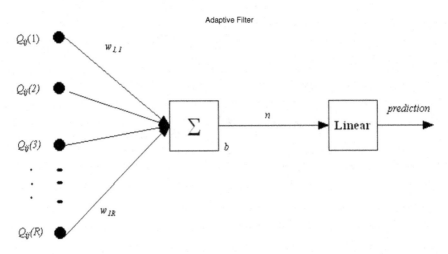

Fig. 2. Neural network architecture deployed for this work

In Fig. (1), the input vector of length R is used as input to a hidden layer of
S neurons. Each element of the R input vector $\boldsymbol{Q}_{ij}$ is connected to each neuron
input through a matrix of weights w_{SR}. Each neuron sums the weighted inputs
and a bias term b to form its own scalar output n_i which is then collectively acted
upon by a transfer function to produce the output vector output $\boldsymbol{a}$. This transfer
function can take various forms, with linear and log-sigmoid functions the most
common. The output $\boldsymbol{a}$ can be taken as input into another (hidden) layer of
neurons, and the process repeated. Finally, at the output layer the final inputs
are combined in a linear way to give the final prediction for the value of the next
epoch Q_{ij}. The number of neurons at each layer, the number of layers, and the
adopted transfer functions collectively describe the neural network architecture.
In the QoS Map server the actual architecture adopted could be different for
different QoS metrics. A model based on Fig. (1) with only one neuron and a
linear transfer function will lead to an optimized adaptive linear filter, and in
many cases such a simple architecture will suffice. Note, that as the adaptation
of the filter will try to minimize the squared error difference between prediction
and actual value, this type of filter is equivalent to a finite impulse response
(FIR) filter. Only in the presence of significant non-linear trends, and/or the
inclusion of additional RSS prediction techniques (such as those reported in
[2]), will multi-neuron systems be required. Although multi-neuron calculations
form part of our ongoing work, in this report we will only investigate a single

neuron system - which we henceforth refer to as the adaptive filter. The simplified architecture for this filter is shown in Fig. (2).

A key parameter is the learning rate of the network - the performance of the algorithm is very sensitive to the proper setting of this value. The learning rate is related to the step size the algorithm takes as it tries to find the minimum point in the QoS Map cost function. If the learning rate is set too high, the algorithm may oscillate and become unstable (see later). The optimal learning rate is difficult to predict *a priori*, and in fact the optimal value is likely to change as the algorithm adapts to new arriving data. We have carried out various tests on the data in order to pick the best learning rate for our simulated data.

In the simulations we assume an average RSS of -60 dBm, with a normal noise component of 7dBm. This noise component was determined from our experimental measurements of the variation of the RSS reported in [1]. We assume ten of these measurements are taken, and the average found is then adopted as the reported RSS for that particular zone of the QoS Map. We normalize the value of the RSS to one (as the actual RSS is in negative dBm the smallest value of the RSS present in the simulation is equal to 1). A periodic trend in the normalized RSS was added to mimic QoS Map evolution. We set the period and amplitude of the evolution to $P = 25$ time units and $A = 5$ dBm, respectively, and study the effect of relative changes to these values. Unless stated otherwise, the learning rate of the adaptive algorithm was set to 0.06 per time step, and the order of the filter was set at $R = 8$. Note that the time units for the simulation are arbitrary. Although we could attempt to make prediction several time units ahead, in this work we focus on the more reliable next step-ahead prediction.

2.2   A QoS Map Zone

Let us consider a particular zone of the QoS Map.

The performance of the adaptive prediction algorithm is shown in Fig. (3) for an evolution of period P and amplitude A. The top diagram of Fig. (3) represents the training epoch of the neural network. The solid line shows the actual value of the RSS, the dotted curve shows the adaptive filter value predicted for the RSS, and the dashed curve shows the moving-average predictor value. The moving-average predictor is based on the average of the last R measurements (R is also the order of the adaptive filter). Here the first 40 samples of the simulated RSS data are used to train the network.

In the middle diagram of Fig. (3) we show how the neural network adaptively learns as it evolves. This allows the network to re-adjust to the evolving trend of the measurements. Perhaps more illuminating is the bottom diagram of Fig. (3), where the squared error is shown. Here, the solid curve shows the square error if the previous RSS value is simply adopted as the predicted next RSS value (last-measure predictor), the dotted curve represents the squared error between the adaptive filter prediction and the actual value, and the dashed curve represents the squared error between the moving-average prediction and the actual value. The adaptive filter predictor is seen to show most performance gain. More quantitatively, we find for this simulation the mean squared error of

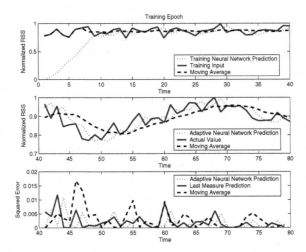

Fig. 3. Performance of the different prediction schemes with evolution period P and amplitude A

the adaptive filter predictor is 20% more efficient than that of the last-measure predictor, and 70% more efficient than the moving-average predictor. We note that in this simulation the moving-average predictor is a poorer estimator than the last-measure predictor. Tests where we reduced the order of the moving-average predictor showed improved results - with the moving-average predictor approaching the last-measure predictor performance as the order reduced to one.

The effect of reducing the amplitude of the evolution to $0.6A$ is shown in Fig. (4). Here we find the mean squared error of the adaptive filter predictor

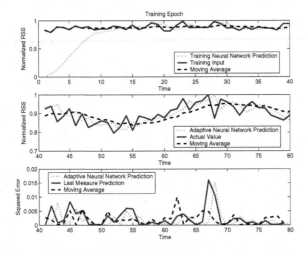

Fig. 4. Performance of the different prediction schemes with evolution period P and amplitude $0.6A$

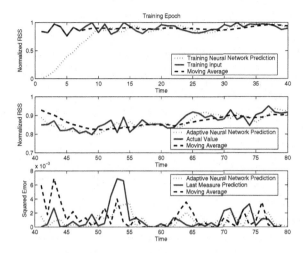

Fig. 5. Performance of the different prediction schemes with evolution period $2P$ and amplitude A

is 40% more efficient than that of the last-measure predictor, and 15% more efficient than the moving-average predictor. The adaptive predictor performs better in this case relative to its performance in the higher amplitude evolution.

To see the effect of increasing the time scale of the evolution, consider the results shown in Fig. (5) where a value of $2P$ is adopted for the period of the evolution. Here we find the mean squared error of the adaptive predictor is 25% more efficient than that of the last-measure predictor, and 50% more efficient than the moving-average predictor.

In order to probe the direct effect the signal-to-noise of the RSS has on the algorithms we repeat the last simulation with an order of magnitude increase in the RSS signal-to-noise (average of 100 simulated RSS values used). The results of this are shown in Fig. (6). Here we find the mean squared error of the adaptive predictor is only 10% efficient than that of the last-measure predictor, but 400% more efficient than the moving-average predictor. Clearly in this case the moving-average predictor is a relatively poor predictor in high signal-to-noise. The quality of the measurements is such that any evolution is better tracked by predictors weighted with the most recent measurement.

Finally, to see the impact of a bad learning rate consider the results of Fig. (7), where the learning rate is set at 0.2. Here the adaptive filter produces poor performance - not even matching the last-measure predictor estimate. This underlies the importance of testing and tuning the filter to the appropriate evolving environment.

2.3   An Entire QoS Map

Thus far we have considered the prediction algorithms as applied to a particular zone only. In practice of course we must apply the prediction algorithm to every zone of the QoS Map. Assuming the evolution of the QoS Map is the same in every zone, the same parameters of the filter (learning rate, order) can be used. Fig. (8) displays the performance over a QoS Map with 100 different zones. Here the period is set to 2P and the vertical axis shows the QoS Map Improvement factor, which is simply the ratio of the mean square error of the last-measure predictor to the mean square error of the adaptive filter predictor. For Fig. (8) the value of the QoS Map Improvement factor averaged over all the QoS Map area is 1.3.

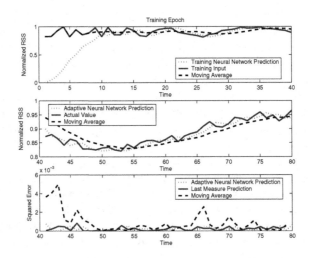

Fig. 6. Performance of the different prediction schemes with evolution period $2P$ and amplitude A, but with increased signal-to-noise

This result should be compared with that shown in Fig. (9), where the mean RSS value in each zone is set to a random value in the range -75 dBm to -50 dBm. Also, the period is varied randomly between P and $2P$ in each zone. Here, the QoS Map Improvement factor averaged over all the QoS Map area is 1.1, a significant reduction relative to the case of global evolution and constant mean RSS. Additional simulations show that the overall performance gain for the QoS Map can be increased back up to higher levels if the adaptive filter is localized. By this we mean the filter settings are different in different zones. Put another way, we find that use of a global adaptive filter in the more realistic scenario of varying mean RSS values and local evolution, will be at the cost of significant performance losses. Thus, although inconvenient, QoS map construction should be based on use of adaptive filters tuned to specific zones. Part of our ongoing work is to address the issue of how to dynamically optimize the local filters, by detecting the presence of temporal trends in the incoming QoS metrics reported for each zone.

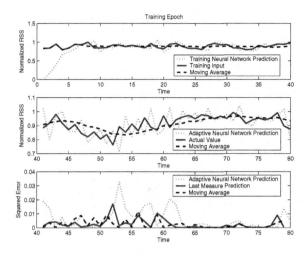

Fig. 7. Performance of the different prediction schemes with evolution period P and amplitude A, but with an increased learning rate of 0.2

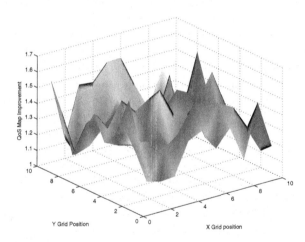

Fig. 8. The overall improvement in the predicted QoS for an adaptive filter relative to a last-measure predictor. A global evolution of period $2P$ and amplitude A is assumed to be present over the entire map area, and the mean RSS in every zone is -60 dBm

2.4 Prediction Algorithms and Experimental Data

It is perhaps worthwhile to conclude by looking at some actual experimental data collected during some of the experiments reported in [1]. The largest amount of data reported on in [1] was based on measurements of the link speed (data throughput) reported by the WLAN cards mounted on our GPS equipped laptops. Access points can degrade the design link speed as the quality of the RSS degrades - and report the new design link speed to the receiver. Normally the card can report data throughput as 11mbs, 5.5mbs, 2mbs, and 1mbs. Just under 2 hours of data have been obtained on a QoS Map zone whose reported link speed was observed to be highly variable. The results of applying the prediction

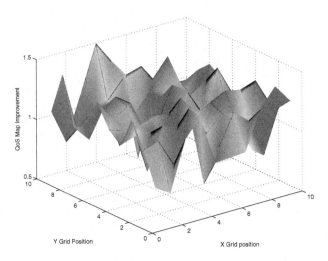

Fig. 9. The overall improvement in the predicted QoS for an adaptive filter relative to a last-measure predictor. Here local evolution is assumed to be present over the entire map area, and the mean RSS in every zone is randomly set in range -75 dBm to -50 dBm

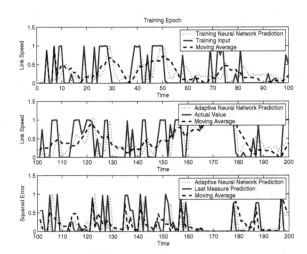

Fig. 10. Prediction tests on actual QoS Seeker link speed data

algorithms are shown in Fig. (10), where the time units corresponds to a 30 second interval. Here we find the mean squared error of the adaptive predictor to be 17% more efficient than that of the simple last-measure predictor, and 14% more efficient than the moving-average predictor. So that we see that even in real data with no coherent evolution of the radio environment present, small performance gains can be achieved.

3   Conclusions

QoS Maps, constructed from dynamic measurements collected from the roaming users of wireless networks, can be used to enhance the QoS performance of wireless applications. Empowering users of portable devices with the information on where in their vicinity the best communication connections are, has clear benefits for them and the efficiency of the network as a whole.

In this report we have specifically investigated the use of adaptive linear filters as a means to predict future QoS Maps from historical QoS Maps. By using the RSS as the QoS metric, we show that local adaptive filters can deliver significant performance gains relative to last-measure and moving-average predictors. Even under the mild evolution conditions we have investigated here, we have found up to 70% performance gains. Such gains have important impact on the functionality of QoS Maps.

We have also investigated the trade off between using global adaptive filters over all the QoS Map space, relative to local filters individually optimized for their specific zone. We find that a relative gain of order 20% is achieved by using local filters.

Many more simulations, covering wider evolution conditions and different filter settings, have been carried out. Although not reported on, these simulations show similar trends to those shown here. All our results show that when evolution of a radio environment is expected, adaptive prediction techniques have a significant role to play in the QoS Map construction.

Acknowledgment

This work has been supported by funding from National ICT Australia.

References

1. R. Malaney, E. Exposito, X. Wei, and D. Nghia, "Seeking VoIP QoS in Physical Space," NICTA-RAM-01-2004, 2004.
2. A Neskovic, N Neskovic, and D Paunovic, "Macrocell Electric Field Strength Prediction Model Based Upon Artificial Neural Networks," IEEE JOURNAL ON SELECTED AREAS IN COMMUNICATIONS, VOL. 20, NO. 6, AUGUST 2002, pp 1170 - 1177.

3. K. E. Stocker, B. E. Gschwendtner, and F. M. Landstorfer, "Neural network approach to prediction of terrestrial wave propagation for mobile radio," Proc. Inst. Elect. Eng. -H, vol. 140, pp. 315 - 320, Aug. 1993.
4. Qualcom-SnapTrack, http://www.snaptrack.com
5. G. Wolfle and F. M. Landstorfer, "Adaptive propagation modeling based on neural network techniques," in Proc. IEEE 46th Vehicular Technology Conf. (VTC), Atlanta, GA, 1996, pp. 623 - 626.

Location-Based Services for Scientists in NRENs

Stefan Winter[1] and Thomas Engel[2]

[1] Fondation RESTENA, 6, rue Coudenhove-Kalergi, L-1359 Luxembourg,
Luxembourg,
stefan.winter@restena.lu,
WWW home page: http://www.restena.lu.
[2] University of Luxembourg, 6, rue Coudenhove-Kalergi, L-1359 Luxembourg,
Luxembourg,
thomas.engel@uni.lu,
WWW home page: http://www.ist.lu./users/thomas.engel/

Abstract. This document describes the current status and future visions of roaming and locating users in science networks. At first it is described how roaming on network level can be accomplished by establishing a RADIUS-based roaming hierarchy, with a special focus on possible transport mechanisms for a user's location information. Afterwards, possibilities to build location-aware services for scientists on top of this existing roaming infrastructure are outlined.

1 Introduction

Scientific networks in Europe (often called "NRENs", National Research and Education Networks) are heterogent in respect to their physical layout (some offering wireless LAN, others providing "docking points" to their wired backbone), network admission techniques (VPN-based network admission control, web-based logins or IEEE 802.1X authentication) and their local usage policies. In order to provide a roaming service for scientific users that lets them travel throughout Europe and still log in with their usual credentials, a solution that integrates all these differences into a broader solution has to be used.

The need for such a solution was seen by the European NRENs several years ago. So, in parallel to the development of the well-known European Research Network GÉANT[1] a roaming service named "The European RADIUS hierarchy"[2] (often also called "EduRoam") was developed that aims to integrate the various different national solutions. This service was set up by the TERENA Task-Force Mobility[3] and is based on the authentication and authorisation protocol "RADIUS" (for Remote Authentication Dial-In User Service)[4]. The use of this flexible and standards-compliant authentication backend made it possible to integrate the different national authentication mechanisms and enabled users to travel freely between Points of Presence (PoPs) of participating NRENs. One aspect of roaming which is not yet available in EduRoam is the ability to provide location-dependent services at each of the PoPs.

T. Strang and C. Linnhoff-Popien (Eds.): LoCA 2005, LNCS 3479, pp. 341–349, 2005.
© Springer-Verlag Berlin Heidelberg 2005

In order to provide such services the location of the user should be known to both ends of the endwork, i.e. to the network where the user currently is (which is trivial) and to the network he is originating from. This, however, requires that the current location information is transported to the home location, where it can be evaluated and the location-based services can be generated. When transporting this information, special care should be taken that no unnecessary data is transferred in order to protect the privacy of the roaming user.

2   Related Work

The development of the European roaming service is conducted in the GÉANT2 research project, specifically in the Joint Research Area 5 (JRA5)[5]. The current research focus lies in the evolution of EduRoam, with the aim of replacing the current static RADIUS hierarchy by a more dynamic solution. A second aspect of JRA5 is the development of application-level roaming to make the access to a given resource independet of the location. The work in JRA5 is closely related to the activities of TERENA's Task-Force Mobility[3] (network-level roaming) and AACE Task-Force[6] (application-level roaming). The development of the protocols involved (RADIUS, Diameter) is taking place in the "Authentication, Authorization and Accounting (AAA) working group"[7] of the IETF. Another IETF working group that does closely related work to the topic of this paper is the "Geographic Location/Privacy (geopriv)" working group[8].

3   Roaming Technologies

A universal single-sign-on with roaming across Europe as described above can be split in two parts[1]. For one, setting up connectivity for registered users – and only for these – is an action that has to be done on a very low level on the network stack. Current national concepts regulate network admission either on ISO/OSI layer 2 (IEEE 802.1X as an example) or on layer 3 (Virtual Private Networks or web-based authentication).

The other part, location-independent access to services, must take place after network-level roaming is successfully completed. Therefore, location information that has implicitly become available through the roaming process can be used on application level to present personalized, location-based services and infor- mation. EduRoam is a network admission control infrastructure that operates both on ISO/OSI layer 2 (using the IEEE 802.1X protocol) and layer 3 (using either web-based redirects or VPN connections). It utilises a hierarchy of RA-

---

[1] It should be noted that roaming in the EduRoam sense is not a seamless roaming without loss of connectivity as in, for example, GSM networks. It is rather a means of travelling between NREN Points of Presence and having connectivity at these PoPs.

DIUS authentication servers to decide if a given user is allowed to enter the local network. It can integrate VPN- or web-based solutions because these other solutions can usually be configured to authenticate against a RADIUS backend. Users are identified via usernames that include their home domain so that their origin can easily be determined. Such usernames are similar to e-mail addresses since they use the @ character to seperate username and home domain, e.g. username@domain.tld. The NRENs that participate in EduRoam form a federation, which means that they trust each other's admission decision. That way, a user's admission request can simply be forwarded to his home domain which then authenticates the user with his home credentials and advises the foreign network to either allow access to the network or not, depending on the outcome of the authentication process.

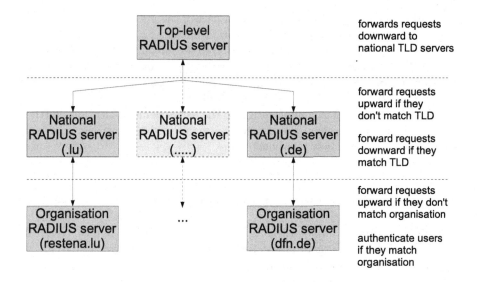

Fig. 1. The European RADIUS hierarchy (from [9])

Figure 1 shows a simplified structure of the EduRoam RADIUS hierarchy. A typical way that an authentication request from user stefan@restena.lu takes through this hierarchy when he is at a location within the DFN network in Germany could be:

1. The user enters his credentials (which is his username plus a piece of information that authenticates him, for example a password or a certificate) into an Authenticator at the DFN location.
2. These credentials are sent encrypted to the RADIUS server for dfn.de, which discovers that he is not responsible for users in restena.lu.
3. The dfn.de RADIUS server forwards the encrypted credentials to the RADIUS server responsible for the .de domain, which in turn discovers that he

is not responsible and forwards them to the European top-level RADIUS server.

4. The top-level server knows about all national servers and decides to forward the credentials to the .lu national RADIUS server.
5. Finally, the .lu server (knowing all organisational servers in .lu) forwards the request to the restena.lu RADIUS server which decrypts and evaluates the credentials and produces an answer, either an Access-Accept or an Access-Reject message.
6. Since each intermediate server keeps track of the pending request, the answer traverses the hierarchy in reverse order of the original request until it reaches the dfn.de domain.
7. The dfn.de RADIUS server instructs the Authenticator to which the user is connected to either permit or deny Layer-2 traffic.

Having the user's origin domain as integral part of the authentication procedure is a unique situation with regards to location-based services: usually, location-based services only take the current location of a user into account. But within the EduRoam environment it is also possible to provide services that use the user's origin, or even a combination of current location and origin. The next sections deal with the technical challenges of determining the current location of a user and transporting this information to the user's home institution. Furthermore, some possible location-based services are outlined that use the knowledge of both user locations, home and current.

4   Obtaining the Current Location

Obtaining information about the origin location of a user is almost trivial for both ends of the network infrastructure – his home institution is part of the user name. The other part, knowing about the user's current location is more difficult. Usually, only the network which the user is logged into knows the exact whereabouts of the network equipment. But it is not advisable to leave the administrative burden of providing location-dependent services solely to the visited location because it would lead to duplicate work in every such location (think of a "Welcome" page localized into several languages depending on the origin of the user). Furthermore, there are services that only the home location of the user can provide because it is based on information that is only locally available. On the other hand, having only the home location administer the services is not a viable solution as well since of course there is location-dependent information that is only relevant at and available from the visited network (think of the location of network services like printers or the current local traffic situation).

So, in order to provide a decent service both the visited network and the home network of the user should know about both locations and both should be able to deliver personalised content.

4.1   Transport via the NAS-Identifier Attribute

One way of transporting the location information from the visited network to the user's origin network is by using RADIUS attributes. Such attributes are tagged to a RADIUS message. They are defined in [4]. One special attribute, "NAS-Identifier", can be used to transport the required information. NAS is short for "Network Access Server" and is the device the user is directly connected to, e.g. a switch or a wireless access point. Current implementations of RADIUS-capable NAS often simply send the fully qualified domain name (in short: FQDN) of the NAS during RADIUS authentication requests. From a certain point of view, this can be seen as location information, but it may be very cumbersome to retrieve actual geographical information from the FQDN. Especially when the user belongs to an institution that resides under a gTLD, such as .org or .net, his domain name can not at all be used to derive his current location. However, the definiton of this attribute need not at all be the FQDN, as stated on page 52 of [4]:

> The String field is one or more octets, and should be unique to the NAS within the scope of the RADIUS server. [...] The actual format of the information is site or application specific, and a robust implementation SHOULD support the field as undistinguished octets.

So the attribute can be used to transport fine-grained information in a textual form as long as the identifier is unique to the device. One possible format could be the exact street address followed by the fully qualified domain name of the device (to ensure the uniqueness of the attribute in the case of two co-located devices) , as in the following example:

Example 1.

NAS-Identifier = "LU;Luxembourg;rue Coudenhove-Kalergi;6;

Fondation RESTENA; < FQDN > "

Setting the attribute to such a value would be a one-time administrative process at the NAS device. The content of this attribute will then be transported via the RADIUS infrastructure to the home location of the user. Then the appropriate organisational RADIUS server can evaluate the content of the attribute and send the current location of the user to the application-level, where the actual location-dependent service can be set up for the user (for example a web-browser can be instructed to prepare a web page with location-specific content that is to be displayed to the user). Many of the currently available RADIUS servers allow such a pass-over procedure to the application-level[2].

---

[2] For example, the popular server "Radiator" can be configured to execute an arbitrary program after a successful authentication (the so-called "Post-Auth-Hook" configuration directive).

4.2   Transport via VSAs

Using the NAS-Identifier provides the necessary possibilities to transport the desired information. Anyhow, a more flexible and dedicated solution will need to be developed to maintain scalability in the future. One approach is to use so-called *vendor-specific attributes* (VSA).

VSAs provide a way of defining new RADIUS attributes that need not be standardised by the IETF. They are placed in one specific IETF-standardised RADIUS attribute, and this container attribute carries all vendor-specific attributes and associated values. So, if any location attributes are to be invented, it is only necessary to treat EduRoam as a vendor and have the NAS fill these VSAs with the appropriate content[3]. That way, there is no need to use a specially formatted string in one attribute. Instead of this, a set of attributes can be defined and used to carry pieces of information. The information from example 1 could then be carried in the attributes shown in example 2, where all attributes preceded with "Location-" are VSAs.

Example 2.
NAS-Identifier = <FQDN>
Location-Country = LU
Location-City = Luxembourg
Location-Street = rue Richard Coudenhove-Kalergi
Location-Street-Number = 6
Location-Description = Fondation RESTENA

4.3   Granularity

The location cannot be determined any better than on a per-device basis when RADIUS attributes are used. This leads to a maximum granularity of several hundred meters because the NAS is either a wireless access point, which has a usual coverage of 100+ meters, or a network switch or hub, which may have a cabling length of 100+ meters as well (depending on the type of layer-2 equipment used).

4.4   Context Retrieval

Context retrieval in the EduRoam environment can be achieved by gathering statistical data about the use of the service. Every participating site has the ability to track how many users are currently using the service and how much time these users spend in that service. This data can be collected by using RADIUS to send accounting messages from the NAS to the RADIUS server. The server can then generate accumulated statistics about the service usage.

This information can, for instance, be used to detect if conferences or big meetings are currently taking place at a given location. An unusually high

---

[3] This means that a vendor ID (enterprise number) has to be registered with IANA because VSAs should begin with a unique vendor identifier.

amount of guest users that are logged in for a long time at the same NAS or adjecent NASes is a good indicator for meetings or conferences.

Currently, within the GÉANT development staff investigations are underway that evaluate the use of other VSA attributes to provide even more information about the users that are logging into the service. This information can include the *role* of a person within a network. Among other things, this can be used to seperate classes of service usage, e.g. a class "student", "professor" or "staff". Given that class information, the context retrieval could be improved, for example to seperate student conventions from academic conferences.

5   Privacy Considerations

The information about the user that is transmitted through the hierarchy is visible to every participating RADIUS server. Parts of the RADIUS server-to-server connections are encrypted via a shared secret, but this applies only to user credentials, not attributes like NAS-Identifier. So, since a RADIUS hierarchy like EduRoam involves several intermediate hops the user's information is exposed to these intermediate servers unnecessarily. This can be addressed in two ways.

The first option is to filter the delivered information so that only the required minimum is transferred over the entire hierarchy. For example, the first RADIUS server might get all the detailed information about the user but could decide (according to an agreed policy) to only forward country and city to the next hops because the distant servers probably don't need as fine-grained location information as the first server may need.

The second possibility is to establish a means of direct peer-to-peer connection between the two ends of the hierarchy, so that there are no intermediate servers that have to pass any sensitive information. This is, however, not provided in the RADIUS protocol, so protocol modifications or switching to a more advanced protocol like Diameter[10] would be necessary. Investigations in that direction are currently underway in the GÉANT2 project.

6   Location Services

Having the combination of current and home location at hand offers a new variety of services.

The home organisation can serve any information about the current location that it has gathered in the past. Such information may include

- Hotel discounts that the home organisation has negotiated at the visited location (of which the visited organisation may not be aware of)
- Links to some important officials or representatives who speak the visitor's native language
- Experiences from earlier visitors (possibly in the form of a blog or wiki)
- A list of local scientific contacts that are in touch with the home institution

All of this information can be presented in the user's home language, because the information originates from the home organisation.

Another location-based service that uses the user's home information is the personalised use of IP phones. Given the technical availability of IP phones that support the RADIUS protocol, the telephone number at the home institution of a user could be transferred to the phone in a RADIUS attribute and the user can dial with his home number from the visited location (if telephony regulations allow this in the visited country)[4].

The visited location can also make good use of the home location information. It could present the user with a start page that describes what kind of local services are available to him (which might depend on the user's origin if there are different service levels) and that makes him acquainted with the differences in network usage policies between the home location and the current location. Furthermore, it could provide an "Issue" page to the user with caveats that only apply to guest users from a specific institution, e.g that the user's home SMTP server will not be reachable through the visited network because of restrictive firewall settings. In that case, the visited institution could provide instant help by pointing the user to an own SMTP server.

7   Conclusion

Establishing a roaming service that spans entire Europe is a challenging task. The current state-of-the-art with a RADIUS hierarchy is already a viable and flexible approach that supports a wide variety of underlying authentication mechanisms. Users can authenticate over wired as well as wireless LANs and use a broad range of login credentials (X.509 certificates, passwords, Smart Cards etc.). It is possible to add more value to the existing authentication process by adding location information and classifying users into groups. This additional information can be evaluated and used to offer highly personalised new services to the users of this authentication system.

8   Future Work

As the evolution of the roaming service within GÉANT2 continues, new concepts for the roaming architecture are evaluated within JRA5. One of these concepts is the use of DNSSec to establish the connection between visited location and home location. This would obsolete the strict hierarchy of RADIUS servers and would also eliminate a "single point of failure" device, namely the RADIUS root server. By using mutiple DNS servers that are working as peers a higher resiliency

---

[4] As of December 2004, at least one hardware platform is available (Texas Instruments), though no end-user products have been built upon this platform yet.

compared to the single RADIUS root server can be achieved without any impact on the use of RADIUS attributes and the transported location information.

Another concept to enhance the roaming service is the use of the new Diameter protocol[10]. This protocol addresses several shortcomings of RADIUS. Since the attributes of Diameter are different to those in RADIUS, this would imply that changes in the way location information is transmitted have to be made.

Acknowledgements

This work is supported by the European Commission in the project GÉANT2, Fondation RESTENA Luxembourg and the SECAN-Lab at the University of Luxembourg.

References

1. The GÉANT homepage http://www.geant.net.
2. The EduRoam homepage http://www.eduroam.org.
3. The TERENA Task-Force Mobility homepage
   http://www.terena.nl./tech/task-forces/tf-mobility/
4. C. Rigney et. al., Request for Comments 2865: Remote Authentication Dial In User Service (RADIUS)
   ftp://ftp.rfc-editor.org./in-notes/rfc2865.txt
5. GÉANT2 – Joint Research Activity 5
   http://www.geant2.net./server/show/nav.00d00a005
6. The TERENA Task-Force Authentication, Authorisation Coordination for Europe
   http://www.terena.nl./tech/task-forces/tf-aace/
7. Internet Engineering Task Force. Authentication, Authorization and Accounting (aaa) working group homepage
   http://www.ietf.org./html.charters/aaa-charter.html
8. Internet Engineering Task Force. Geographic Location/Privacy (geopriv) working group homepage
   http://www.ietf.org/html.charters/geopriv-charter.html
9. Erik Dobbelsteijn et. al.:
   Mobility Task Force Deliverable D – Inventory of 802.1X-based solutions for inter-NRENs roaming (Version 1.2)
   http://www.terena.nl./tech/task-forces/tf-mobility/firsttofdels/delanddoc.html
10. P. Calhoun et. al., Request for Comments 3588: Diameter Base Protocol
    ftp://ftp.rfc-editor.org./in-notes/rfc3588.txt

Towards Smart Surroundings: Enabling Techniques and Technologies for Localization

Kavitha Muthukrishnan, Maria Lijding, and Paul Havinga

Fac. of Computer Science, University of Twente,
P.O.Box 217, 7500AE Enschede, The Netherlands
{k.muthukrishnan, m.e.m.lijding, p.j.m.havinga}@ewi.utwente.nl

Abstract. In this paper we identify the common techniques and technologies that are enabling location identification in a ubiquitous computing environment. We also address the important parameters for evaluating such systems. Through this survey, we explore the current trends in
commercial products and research in the area of localization. Although
localization is an old concept, further research is needed to make it really
usable for ubiquitous computing. Therefore, we indicate future research
directions and address localization in the framework of our Smart Surroundings project.

1 Introduction

Recent advances in wireless communication devices, sensors, hardware (MEMS)
technology make it possible to envision large scale dense ad-hoc networks acting
as high resolution eyes and ears of the surrounding physical space, making the
vision of Mark Weiser [36] into a reality. This calm technology promises various
applications and spans a wide range of fields ranging from medical and fitness,
security and safety, work, learning and leisure. Ad-hoc networks play a vital role
in modeling these future pervasive networks.

As we approach the level of ubiquitous network connectivity and pervasive
mobile devices, the enticing new category of *context-aware applications* has been
proposed. By definition [5] 'context' is any information that can be used to characterize the situation of an entity. An entity is a person, place or object that
is considered relevant to the interaction between a user and an application, including the user and applications themselves. A system is 'context-aware' if it
uses contexts to provide relevant information and services to the user, where rele-
vancy depends on the user's tasks. One of the important dimensions of context is
location. The proliferation of wireless technology, mobile computing devices and
Internet has fostered a growing interest in location-aware systems and services. It
is useful for accomplishing emergency services, E911, follow me services,finding
the nearest resources such as printer, habitat monitoring, patient tracking, asset
monitoring, buddy finder or a product finder, etc. *Localization* acts as a bridge
between the virtual and physical world [8].

T. Strang and C. Linnhoff-Popien (Eds.): LoCA 2005, LNCS 3479, pp. 350–362, 2005.

A mobile or a static node could answer the question 'Where am I?' in several ways. It might be relative to a map or relative to another node, or global coordinate system. This is referred to as *localization* or *location identification*.

Networked applications are often implemented in a layered architecture or protocol stack, and several layers of the protocol stack can benefit from localization [2]. Knowing the objects location not only promote context awareness at the application level, but also bumps up low level functionalities such as routing, service discovery and resource management.

Localization is by nature an interdisciplinary problem involving research in several areas of computer science and many kind of engineering. Consequently, research has proceeded in both the systems and the algorithmic fronts. In this paper we discuss the basic principles of localization, and the developments and advances made in the field. Finally, we show why localization is a corner stone in the future of pervasive computing.

1.1   Performance Parametric Measures of Localization

The key metric for evaluating a localization technique is the *accuracy* defined as, how much is the estimated position deviated from the true position. The accuracy is denoted by an accuracy value and *precision* value (e.g. 15 cm accuracy over 95% of the time). The precision indicates how often we expect to get at least the given accuracy. The accuracy of a location sensing system is often used to determine whether the chosen system is applicable for certain applications.

Calibration also plays a very important role. The uncalibrated ranging readings are always greater than the true distance and are highly erroneous due to transmit and receive delays [37]. Device calibration is the process of forcing a device to conform to a given input/output mapping.

Responsiveness is defined as how quickly the location system outputs the location information. It is an important parameter, especially when dealing with mobility. However, this parameter is mostly ignored in the description of the existing systems.

Scalability is also significant, as the proposed design should be scalable for large networks. Also of great importance is *self-organization* as it is infeasible to manually configure the location determination processes for a large number of mobile devices in random configurations with random environmental characteristics.

Cost, which includes the cost of installation, deployment, infrastructure and maintenance, is also a crucial parameter. An important cost factor when running the system in a real environment is *power consumption*. When scaling to thousands or millions of autonomous small devices it is clearly not feasible to change or recharge batteries very often, thus *energy efficiency* should be a goal of any localization mechanism meant for a large scale system.

In our view *privacy* is an important parameter of localization systems and it should form part of the architecture since its conception. Using localization it is very easy to create a Big Brother infrastructure that track users movements and allow to deduce patterns of behavior. However, this issue is being generally

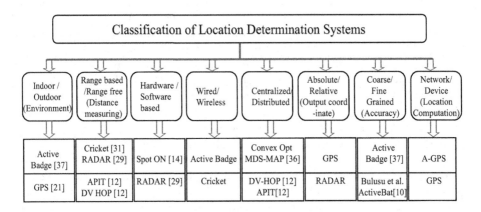

Fig. 1. Taxonomy of the existing location systems

overlooked in the design of systems and considered as an after thought only. Centralized systems are particularly weak with regard to privacy.

2   Taxonomy of Existing Location Systems

Though localization is not a new field of research,it has gained additional credit due to the advent of ubiquitous computing research and it is still an evolving research area. Figure 1 shows a taxonomy of localization technology.

3   Enabling Techniques and Technologies for Localization in Wireless Ad-Hoc Networks

Localization is defined as a mechanism to find the spatial relationship between objects [2]. An assumption in most of the localization system is the availability of *anchor nodes* or *landmarks* [29], while some other uses beacon nodes. Langendoen et al [19] differentiates anchor nodes and *beacon nodes*. They define anchor nodes as nodes having a priori knowledge of their own position with respect to some global coordinate system, while beacons (access points) are nodes based on external infrastructure (e.g. GPS-less, Cricket). Beacons have the same capabilities (processing, communication, energy consumption) as other nodes in the network. Fundamentally speaking, location systems needs some kind of input, for example it can be sensor reading originating from a sensor, or information from an access point as signal strength, for getting a symbolic representation. This information is then combined using a given technique to derive the location, either absolute or relative, of an object or set of objects. An absolute location system may use a shared reference grid for all located objects. For example, all GPS receivers use latitude, longitude and altitude for reporting the location, where as a relative location system has its own frame of reference. The following subsections give an overview about the existing technologies and techniques that enable localization.

3.1   Signal Technologies

The different types of signal technologies are tabulated in Table 1. Depending on the required range, propagation speed, cost, precision, bandwidth, etc. one can choose the required technology for a specific application. As you can see from Table 1, there are infrared based, ultrasonic based, electromagnetic based, inertial based, optical based, and radio frequency based systems. Depending on the type of frequency range used, Radio Frequency can be categorized into RFID (Radio Frequency IDentification), WLAN (IEEE 802.11b), Bluetooth (IEEE 802.15), wide area cellular and UWB. UWB is based on sending ultra short pulses (typically $< 1ns$). For location identification, UWB uses Time of Arrival measurement. The very short pulses lead to high accuracy and low power consumption.

3.2   Distance Measuring Techniques

The *ranging technology* forms the heart of any range based localization system. There are several range-based techniques such as Time-of-arrival (TOA), Time difference of Arrival (TDOA), Angle-of-arrival (AOA), and Received Signal Strength Indication (RSSI). The TOA and TDOA make use of signal propagation time for finding the range of distance. To augment and complement TOA and TDOA, AOA was proposed. AOA allows nodes to estimate and map relative angles between the neighbors. However, this approach requires costly antenna arrays on each node [30]. RSSI makes either theoretical or empirical calcula-

Table 1. Enabling signal technologies

Technology	Merits	Remarks
Infrared	Inexpensive (due to ubiquitous deployment) Compact Low power	Typical range is upto 5m Restriction to line of sight conditions Unusable in direct sunlight
Ultrasound	Relatively slow propagation (speed of sound) Allows for precise measurement at low clock rates, making the system simpler and inexpensive	Typical range is 3- 10m have been reported Environmental factors have substantial effects
Radio Frequency	Better than IR in terms of bandwidth, cost and speed	No proper propagation model exists Affected by multipath Typical range of bluetooth is 10-15m Typical range of WLAN is 50-100m Typical range of RFID is 1-10m Typical range of cellular systems is 100 -150m
DC Electromagnetic	High precision High signal propagation speed	Typical range is 1-3m Signals are sensitive to environments Precision calibration required,hence expensive Difficult to install
Optical	High precision Compact Low power	Typical range is upto 5m Restriction to line of sight conditions Unusable in direct sunlight
Inertial		Errors accumulate over time Calibration
UWB radio	High precision and accuracy Less affected by multipath than the traditional RF systems	Expensive Higher receiver density than the conventional RF systems nevertheless its easier to install

tions to convert the signal strength measurements to distance estimates.There are also other ranging techniques like E-OTD(Enhanced Observed Time Difference) and A-GPS(Assisted GPS)which are typically used in cellular location determination.

On the other hand range free schemes make use of some algorithm that calculates the distance in terms of hop count to anchor nodes [12]. Some algorithms that use range free schemes are Centroid algorithm, DV-Hop, Amorphous, Point-In-Test (PIT) and Approximate Point-In-Test (APIT) [12]. In the DV-hop method, the anchor nodes are placed at known position and they transmit broadcast messages that are flooded throughout the network containing the location of the anchor node, and the distance between the anchors is obtained by using hop count as a metric. Later, any of the below discussed position estimation method is used to estimate the node location. The other range free schemes are explained in Section 4.2.

3.3   Location Estimation Techniques

The next step after the determination of the distance is the *location estimation*. There are different methods for estimating the location. *Triangulation* uses the geometric properties of triangles to compute object locations. It is a positioning procedure that relies on angle measurements with respect to the known landmarks. Triangulation is subcategorized into *lateration*—using distance measurements— and *angulation*—using angle or bearing measurements [13]. *Trilateration* uses ranges to at least three known node position to find the coordinates of unknown nodes. The trilateration [24] procedure starts with an a priori estimated position that is later corrected towards the true position. There are many other types of lateration such as Atomic multilateration and Collaborative multilateration, which are addressed in detail in [31].

Proximity measures the nearness to a known set of points. The objects presence is sensed using the physical phenomenon with limited range. Scene analysis examines a view from a particular vantage point to draw conclusions about the observer's location [13][17]. The scene itself can contain visual images, such as frames captured by a wearable camera or any other measurable physical phenomena, such as electromagnetic characteristics that occur when an object is at a particular position and orientation.

When the localization problem is addressed in wireless sensor networks, there is yet another type of location estimation method called *min-max* algorithm [19]. The main idea is to construct a bounding box for each anchor using its position and distance estimate, and then to determine the intersection of these boxes. The intersection of the bounding boxes is computed by taking the maximum of all coordinate minimums and the minimum of all maximums. Hence it is named as min-max. The estimated position by min-max is found to match closely with the true position computed through lateration.

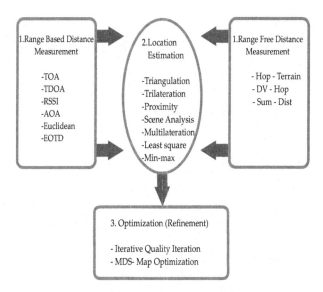

Fig. 2. Steps in localization

The least square algorithm [30] [9]is also used to derive position estimation from collection of reference points and their associated ranges.

Figure 2 shows that there are 3 steps involved in any localization system namely, distance measurement to anchors or beacons, position estimation and position refinement, which is an optional step [19]. The choice of the techniques and technology affects the granularity and accuracy of the location information.

4   State of the Art

4.1   Implemented Systems

Loran was the first navigation system, launched before the World War II to employ time difference of arrival of radio signals, developed by MIT radiation lab [26]. It was also the first true all weather position finding system, but it is only 2 dimensional. Transit was the first operational satellite based navigational system launched in the year 1959. Transit users determine their position on earth by measuring the Doppler shift of signals transmitted by the satellites. Global Positioning Systems (GPS) [26] is one of the oldest location technologies that provided the location of the users in 3D. GPS works well in urban outdoor environments, but it is quirky, unreliable and its accuracy degrades when the device is indoors and has limited line-of-sights to the satellites. It is not ubiquitously available and hence not suitable for under water and cluttered urban environments [2]. Additionally, it cannot be used for the future calm technology, as it is not practically implementable on low power devices and also size and the cost

Table 2. A comparison of several radio navigation methods

System	Method	Coverage	Dimensions	Accuracy
Loran	Hyperbolic	Cont	2D	250m
Transit	Doppler shift	Global	2D	25m
Omega	Hyperbolic	Global/cont	2D	2-4km
GPS	Spherical	GLobal/Cont	3D	5-10m

pose harsh limitations. Table 2 shows the comparison of various outdoor geo-location systems.

Bulusu has designed a *GPS-less system* for outdoor locations suited for very small, low cost devices [4]. The system uses a RF-based signal technology and proximity based position estimation and got satisfying results. In principle the system can be used indoors, but in this case the accuracy is bad.

The *active badge* [35] is the one of the early centralized indoor personal location system making use of IR technology. Each person in the office wears a badge, which emits a unique IR signal that is then gathered by the network of sensors and collected by a master server. The information is then relayed, to the visual display location manager. This was mainly used as an aid for the telephone receptionists to direct the phone calls to appropriate persons during working time. However its range was limited and obstacles such as walls and windows made it difficult for the IR signal to propagate. Nevertheless it works well for the intended application of routing telephone calls.

However, the Active Badge is not able to provide fine-grained 3D location information, which is needed by many applications. So the *Active Bat* [10] system was developed, with a primary focus on low power, low cost and accuracy. The bat is attached to the objects or persons whose location has to be determined. These bat transmitters emit ultrasound pulse, which are received by the receiver that is mounted on the ceiling. A central controller coordinates the transmitters and receivers. To locate a particular bat, the controller sends the unique ID over the radio channel. When a bat detects its ID, it sends an ultrasound pulse, which is picked by receivers in the ceiling. From the time-of-flight measurements, the system can calculate the 3D position of a bat to an accuracy of 3cm. Figure 3 shows the Active Bat System.[1]

Cricket location [28] support systems make use of proximity based lateration techniques for providing location information. Many beacons installed at known locations advertise the identity of that space with the use of some character string and every device in the network has a listener attached to it. The Listeners use some inference algorithm to determine the space in which they are currently located by listening to the beacon announcements. Each beacon sends two signals, an RF signal carrying the location data and an ultrasound carrying a narrow pulse. Based on the difference of arrival times, the device finds the absolute distance between the beacon and the listener.

---

[1] Graphics copied from [10].

Fig. 3. Active Bat System,developed by AT&T and Cambridge University

RADAR [1] makes use of RF signal for finding the user location indoors. A centralized system gathers signal strength information from multiple receivers and performs triangulation to compute the location of the user.

Pinpoint 3D-iD [27] is similar to RADAR, but expensive. Proprietary base stations and tag hardware are used to measure radio time of flight. It uses an installed array of antennas at known positions to perform multilateration. Pinpoint's accuracy is roughly 1 to 3 meters.

In the *SpotOn system* [14], special tags use radio signal attenuation to estimate the distance between tags. The aim in SpotOn is to localize a wireless device relative to one another, rather than to fixed base stations, allowing for ad-hoc localization.

HP Labs *SmartLOCUS* [16] uses synchronized RF and ultrasound differential time-of-flight measurements to determine the inter-nodal range between any two nodes. It yields an accuracy of 2-15 cm.

Georgia Institute of Technology's *Smart floor* [25] identifies people based on their footsteps. The goal of the system is to create an accurate system for recognizing a user's identity. However, this technologie's negative side is the huge installation cost and infrastructure cost.

UBISENSE [34] uses UWB technology to locate people and objects to an accuracy of 15cm. Sensors are mounted in the area to be monitored. Ubitags attached to objects or carried by people are then automatically tracked to provide accuracy and reliability.

There are also systems that make use of *Radio Frequency IDentification (RFID)* technology, for locating objects inside buildings.RFID is a radical means of enhancing the supply chain, especially for Returnable Transport Items [20]. RFID includes RFID readers, RFID tags and communication between them.One example is the *LANDMARC* systems [22].

Computer vision has also been used in localization. Microsoft's *Easy Living* [17] uses Digiclops real-time 3D cameras to provide stereovision-positioning capability in a home environment. However the dependence on infrastructural processing power, can limit the scalability of such systems for many applications.

Ekahau Positioning system is a software based positioning solution that can continuously pinpoint and track the location of mobile computing devices with an accuracy of 1-2 meters in indoor and campus environments. Unlike the competing solutions, Ekahau technology does not require any additional wireless infrastructure on top of the standard Wi-Fi network [7].

All the systems described above are already implemented. However, there are many other theoretical methods available to solve the location identification problem. We describe these methods in the next section.

4.2   Theoretical Methods

An often ignored issue in ongoing research is the impact of beacon density and the placement of the beacons. Self-configuring localization systems consider beacon density as an important parameter in characterizing the localization quality. Two algorithms were developed HEAP and STROBE [2] depending on the density of beacons in the network.

In the *convex optimization* [6] approach, the positional information is inferred from connectivity imposed proximity constraints. Few nodes have known locations, called the anchor nodes, and the remaining nodes infer their position from the knowledge about communication links.

MDS-MAP [32] is a method that makes use of connectivity information to provide locations in a network with or without beacons (known co-ordinates). The advantage of MDS-MAP is that it has a wide range of applicability, having the ability to work with both simple connectivity and range measurements to provide both absolute and relative positioning [32] [23]. Both convex optimization and MDS-MAP require centralized computation.

Whilst lot of research was initially based on range-based schemes, many developments in range free schemes also came in. Range free schemes make no assumption about validity of distance or angle information like the range based schemes. Some examples to quote are *Centroid algorithm, APIT, amorphous localization* and *DV-Hop algorithm*. In the centroid method [4], each node estimates its location by calculating the center of the locations of all seeds (or anchors) it hears. If seeds are well positioned, the location error can be reduced [3], but this is not possible in ad-hoc deployments. The APIT method [12] isolates the environment into triangular regions between beaconing nodes, and uses a grid algorithm to calculate the maximum area in which a node will likely reside. DV-based positioning algorithms are localized, distributed, hop by hop positioning algorithms [23] [24]. They work as an extension of both distance vector routing and GPS positioning in order to provide approximate positions for all nodes in a network where only a limited fraction of nodes have self positioning capabilities. They use the same principle of GPS, with the difference that the landmarks are contacted by hop-by-hop fashion rather than a direct connection and similar to distance vector each node at any time can communicate only with its neighbors. The amorphous method [21] is similar to DV-hop as the coordinates of the seeds are flooded throughout the network so each node can maintain a hop count to that seed. Nodes calculate their position based on the received seed locations and corresponding hop count. From an extensive study on range free schemes and on benchmarking various range free schemes, Tian He et al. [12] concluded that range free schemes offers cost effective solutions.

Table 3. Comparison of range free schemes

	Centroid	DV Hop	Amorphous	APIT
Accuracy	Fair	Good	Good	Good
Node Density	>10	>8	>8	>6
Anchors Heard	>10	>8	>8	>10
ANR	>0	>0	>0	>3
DOI	Good	Good	Fair	Good
GPSError	Good	Good	Fair	Good
Overhead	Smallest	Largest	Large	Small

Table 4. Summary of existing localization systems. Accuracy as reported in [11]

Technology	Accuracy	Location Estimation	Example
Infrared based	5-10m	Proximity	Active Badge
Ultrasound based	1-10cm	TOF-lateration	Active Bat
Vision based	1cm-1m	Scene Analysis	Easy Living
RF-UWB based	6-10cm	TOF-triangulation	Ubisense
RF-Bluetooth based	2-10m	Proximity, Triangulation	
RF-WLAN based	2-100m	Triangu.,Proximity,Scene An.	Radar
Satellite based	5-10m	Triangulation	GPS
RF cellular based	50-10m	Triangulation,Proximity	GSM localization
RFID based	5cm-5m	Proximity	Landmarc

Table 3 lists the comparison between various range free schemes. Tian He et al. experiment with several parameters such as node density (ND), anchors heard (AH), anchor to node range ratio (ANR), anchor percentage (AP), degree of irregularity (DOI), GPS error and placement of node and anchors.

Recently research on localization is focused on incorporating the mobility model. Although mobility makes the analysis more difficult, more accuracy is obtained. In [15], Lingxuan Hu etal. use a *sequential Monte Carlo Localization* method and argues that they exploited mobility to improve accuracy and precision of localization. Probabilistic techniques, such as *Markov modeling, Kalman filtering* and *Bayesian analysis* can also be used to determine the absolute location of a mobile node [18].

Table 4 gives a global view of localization techniques classified by achievable accuracy and the type of location estimation used for various technologies.

5   Conclusions and Future Work

Designing a location system for a particular environment presents difficulties when the system is applied to other environments. Despite the plethora of established location technology, there is no single location technology that may be relied upon in all environments to provide accurate location information. Clearly

"No one size fits all", there may not be a single best technology. However each of the techniques has its own pros and cons.

Ubiquitous computing is the *wave of the future*. Recent advancement in the various related technologies are paving way for the design and implementation of the future ubiquitous computing. *Location identification* is an important research area that had gained additional credit since the epoch of pervasive computing. This paper provided an overview of the existing location systems/algorithms and also highlighted the limitations of the existing technologies. In order to improve the existing techniques in future, we would focus on developing new distributed localization algorithms for resource poor ambient systems. They need to address specific issues like ease of deployment, scalability, and automatic configuration. Additionally, they need to provide easy adaptability to different types of environment, self-calibration, responsiveness, accuracy, be tolerant to node failures and range errors. Also of great importance is self-organization as it is infeasible to manually configure the location determination processes for a large number of mobile devices in random configurations with random environmental characteristics. Last but not least, we need to keep an eye on the cost factor—computational power, resources needed and money.

While there is plenty of research going on in developing new systems or algorithms, yet another available solution on hand is to make use of the existing systems/algorithms and choose the best in each case and fuse the location information reported by several technologies to get more meaningful results. It is not just sufficient to have technology development, also of foremost importance is to use the existing systems and have a means to have them integrated so that seamless transition between the available systems is achieved. Ideally the localization system should provide a framework to integrate location reading from all these sensor types into one seamless environment. The future research in the project Smart Surroundings [33] addresses all these issues and will provide an open platform for supporting new architectures and frameworks for the future ambient systems.

Acknowledgements

We thank Stefan Dulman,Faculty of Computer Science ,University of Twente for his comments and suggestions. The work presented in this paper is funded by the Ministry of Economic Affairs of the Netherlands through the project *Smart Surroundings* under the contract no. 03060

References

1. P. Bahl and V. Padmanabhan. RADAR: An inbuilding RF based user location and tracking system. In *Proceedings of IEEE Infocom*, volume 2, pages 775–784, Mar. 2000.
2. N. Bulusu. *Self-Configuring Location Systems*. PhD thesis, University of California, Los Angeles, 2002.

3. N. Bulusu, D. Estrin, L. Girod, and J. Heidemann. Scalable coordination for wireless sensor networks: Self-configuring localization systems. In *Proceedings of the Sixth International Symposium on Communication Theory and Applications (ISCTA '01)*, July 2001.

4. N. Bulusu, J. Heidenmann, and D. Estrin. GPS-less low cost outdoor localization for very small devices. *IEEE Personal Communications Magazine*, 7(5):28–34, Oct. 2000.

5. A. K. Dey, D. Salber, and G. D. Abowd. A conceptual framework and a toolkit for supporting the rapid prototyping of context-aware applications. *Human-Computer Interaction Journal*, 16(2–4):97–166, 2001.

6. L. Doherty, K. Pister, and L. Ghaoui. Convex position estimation in wireless sensor networks. In *IEEE INFOCOM*, Apr. 2001.

7. Ekahau positioning system. http://www.ekahau.com.

8. D. Estrin, D. Culler, K. Pister, and G. Sukhatme. Connecting the physical world with pervasive networks. *Pervasive Computing*, 1(1):59–69, 2002.

9. L. Evers, S. Dulman, and P. Havinga. A distributed precision based localization algorithm for ad-hoc networks. In *Proceedings of the Second International Conference on Pervasive Computing (PERVASIVE 2004)*, volume 3001 of *Lecture Notes in Computer Science*, pages 269–286. Springer Verlag, Berlin, Apr. 2004.

10. A. Harter and A. Hopper. A distributed location system for the active office. *IEEE Network*, 8(1):62–70, Feb. 1994.

11. M. Hazas, J. Scott, and J. Krumm. Location-aware computing comes of age. *IEEE Computer Magazine*, 37(2):95–97, Feb. 2004.

12. T. He, C. Huang, B. M. Blum, J. A. Stankovic, and T. Abdelzaher. Range-free localization schemes for large scale sensor networks. In *Proceedings of the 9th annual international conference on Mobile computing and networking (MobiCom'03)*, pages 81–95. ACM Press, 2003.

13. J. Hightower and G. Borriello. Location systems for ubiquitous computing. *Computer*, 34(8):57–66, 2001.

14. J. Hightower, R. Want, and G.Borriello. SpotON: An indoor 3D location sensing technology based on RF signal strength. Technical Report UW-CSE 00-02-02, University of Washington, Seattle, Feb. 2000.

15. L. Hu and D. Evans. Localization for mobile sensor networks. In *Proceedings of the Tenth Annual International Conference on Mobile Computing and Networking (MobiCom 2004)*, Sept. 2004.

16. A. Jiminez, F. Seco, R. Cere, and L. Calderon. Absolute locatilization using active beacons :a survey and IAI-CSIC contribution.

17. J. Krumm, S. Harris, B. Meyers, B. Brumitt, M. Hale, and S. Shafer. Multi-camera multi-person tracking for easyliving. In *Proceedings of the Third IEEE International Workshop on Visual Surveillance (VS'2000)*, pages 3–10. IEEE Computer Society, 2000.

18. A. M. Ladd, K. E. Bekris, A. Rudys, L. E. Kavraki, D. S. Wallach, and G. Marceau. Robotics-based location sensing using wireless ethernet. In *MobiCom '02: Proceedings of the 8th annual international conference on Mobile computing and networking*, pages 227–238. ACM Press, 2002.

19. K. Langendoen and N. Reijers. Distributed localization in wireless sensor networks: a quantitative comparison. *Comput. Networks*, 43(4):499–518, 2003.

20. LogicaCMG. Making waves: RFID adoption in returnable packaging, 2004.

21. R. Nagpal, H. Shrobe, and J. Bachrach. Organizing a global coordinate system from local information on an ad hoc sensor network. In *Proceedings of the 2nd International Workshop on Information Processing in Sensor Networks (IPSN '03)*, volume 2634 of *Lecture Notes in Computer Science*. Springer Verlag, Berlin, Apr. 2003.

22. L. M. Ni, Y. Liu, Y. C. Lau, and A. P. Patil. LANDMARC: Indoor location sensing using active RFID. *Wireless Networks. Special Issue on Pervasive Computing and Communications*, 10(6):701–710, 2004.

23. D. Niculescu and B. Nath. DV based positioning in ad hoc networks. *Telecommunication Systems*, 22(1-4):267–280, 2003.

24. D. Niculescu and B. Nath. Position and orientation in ad hoc networks. *Elsevier journal of Ad Hoc Networks*, 2(2):133–151, Apr. 2004.

25. R. J. Orr and G. D. Abowd. The smart floor: a mechanism for natural user identification and tracking. In *CHI '00 extended abstracts on Human factors in computing systems*, pages 275–276. ACM Press, 2000.

26. S. Pace, G. Frost, I. Lachow, D. Frelinger, D. Fossum, D. K. Wassem, and M. Pinto. *The Global Positioning System*, chapter GPS history, chronology and budgets, pages 237–270. RAND Coorporation, 1995.

27. Pinpoint 3D positioning system. http://www.pinpointco.com.

28. N. B. Priyantha, A. Chakraborty, and H. Balakrishnan. The cricket location-support system. In *Proceedings of the 6th annual international conference on Mobile computing and networking (MobiCom'00)*, pages 32–43. ACM Press, 2000.

29. C. Savarese, J. Rabaey, and J. Beutel. Locationing in distributed ad-hoc wireless sensor networks. In *ICASSP 2001*, 2001.

30. C. Savarese, J. M. Rabaey, and K. Langendoen. Robust positioning algorithms for distributed ad-hoc wireless sensor networks. In *Proceedings of the General Track: 2002 USENIX Annual Technical Conference*, pages 317–327. USENIX Association, 2002.

31. A. Savvides, C.-C. Han, and M. B. Strivastava. Dynamic fine-grained localization in ad-hoc networks of sensors. In *Proceedings of the 7th annual international conference on Mobile computing and networking (MobiCom'01)*, pages 166–179. ACM Press, 2001.

32. Y. Shang and W. Ruml. Improved mds-based localization. In *Proceedings of the 23rd Conference of the IEEE Communicatons Society (Infocom 2004)*, 2004.

33. Smart surroundings project. http://www.smart-surroundings.org.

34. P. Steggles and J. Cadman. A comparison of RF tag location products for real world applications. Technical report, Ubisense limited, Mar. 2004.

35. R. Want, A. Hopper, V. Falcao, and J. Gibbons. The active badge location system. *ACM Trans. Inf. Syst.*, 10(1):91–102, 1992.

36. M. Weiser. The computer for the twenty-first century. *Scientific American*, pages 94–104, Sept. 1991.

37. K. Whitehouse. The Design of Calamari: an Ad-hoc Localization System for Sensor Networks. Master's thesis, University of California at Berkeley, 2002.

Proximation: Location-Awareness Though Sensed Proximity and GSM Estimation

Aaron Quigley[1] and David West[2]

[1] Department of Computer Science,
University College Dublin,
Belfield, Dublin 4, Ireland
aquigley@ucd.ie
[2] Smart Internet Technology Research Group,
School of Information Technologies,
Sydney University, Australia
dwest@it.usyd.edu.au

Abstract. The realisation of ubiquitous in- and out-door location awareness needs the exploration of scaleable hybrid solutions that can utilize existing infrastructures in novel and complimentary ways. Our hybrid solution (BlueStar) incorporates mobile terminals (GSM smart phones) with a two-phase approach to location awareness, using existing infrastructure. The first phase relies on a network based signal measurement (timing advance) and cell id. In the second phase the mobile terminal "sniffs" for the identification of local wireless devices, which act as "beacons", in the environment. The mobile terminal does not connect to the beacons; it simple detects their presence. The aim is to offer a privacy enhanced yet flexible indoor/outdoor location management scheme, which allows for only the end-user to be aware of their fine-grained location data. A working example of our BlueStar system is presented along with a preliminary user study of "InfoHoard" a BlueStar game, in an indoor testing environment.

1 Introduction

Intelligent environment research is attempting to take the current *context* of the human activity into account when interacting with the individual or group [13,23,24]. Context includes information from the sensed environment (environmental state) and computational environment (computational state) that can be provided to alter an applications behavior, or is an application state, which is of interest to the user. One aspect of context aware computing is *location-awareness* for both indoor and outdoor applications [3,6,9,16]. Location-aware applications include finding services such as printing or telephones that are close by, tracking individuals, goods and resources, locating friends and colleagues, and localised information and guides [2,5,8,11,17].

This paper reports on the system implementation, user study and evaluation of a user-centric, location-awareness method based on "location sniffing" from a fixed infrastructure that preserves privacy. Our patented BlueStar *location sniffing* method is based on a combination of GSM timing-advance knowledge and simply being able to

T. Strang and C. Linnhoff-Popien (Eds.): LoCA 2005, LNCS 3479, pp. 363–376, 2005.

find a piece of local wireless infrastructure for which the location is known [1]. This notion of proximation does not support all classes of location-aware applications but rather applications that do not involve tracking or locating other people [16,19,25].

The rest of this paper is organized as follows. Section 2 provides a brief overview of both previous approaches to this problem and related work. Section 3 describes our system and the development of a test-bed infrastructure for the experiments. Section 4 describes our user study. Section 5 presents our user study results and evaluation. Finally, Section 6 concludes this paper.

2 Background

The increasing trend in mobile hand-held personal computers and mobile phones (mobile terminals) has seen the development of many different location-aware systems and technologies. Traditionally, location-awareness is synonymous with "tracking" and its associated social, control and big-brother implications. Tracking systems typically rely on a fixed infrastructure to determine the device's location [15,18,20]. However, it is now generally considered that location-aware applications can be based on location information, which is calculated either by the mobile terminal, the infrastructure, or a combination or hybrid approach [7, 14, 18,19]. Where the device has information, which allows it to locate itself, the applications, which reside on it, are self location-aware i.e. they do not require the infrastructure to provide the location calculation [19]. We now briefly review the key indoor location systems.

One of the earliest location-aware systems was PARCTAB from Xerox Palo Alto Research Centre (PARC) [21]. PARTAB was created to gather location data from special ID badges worn by each person. The system, once deployed, faced two major challenges: what are useful applications (follow-me phone, employee locator, and ad-hoc meeting planner) and some employees who participated in the study objected to having their every move tracked and recorded (the big-brother effect).

Research during the late 90's at AT&T Research lab in Cambridge, UK saw the development of a local -aware system called the Active Bat [12]. This system consists of a controller, a fixed node receiver infrastructure and a number of active bat tags. The system operates using a combination of RF and ultrasound, and calculates time-of-flight to estimate each tags location.

Predating the bat system was the Active Badge system based of the use on a diffuse infrared technology. The system operates by flooding the area with infrared light, similar to a light bulb. The infrared signals from the badge are bounced off the walls from the floor or ceiling for the receiving units to collect. The IR signal consists of an ID that is emitted on intervals or on demand [17].

The Cricket system from MIT uses a combination of RF and ultrasound technologies allowing a small "listener" device, which can be carried or attached to equipment, to estimate its distance to the closet beacon [19]. The infrastructure is based on a number of non-networked ceiling-mounted beacons fixed throughout a building.

The RADAR system from Microsoft Research is an RF in-building user location and tracking system using 802.11b wireless LAN technology [10]. This approach is analogous to numerours other research efforts which include Pelican [24] and Aura[7]. These approaches use the signal strength to measure the distance between Access Point (AP) and the mobile terminal. These distances, in conjunction with an estimated signal propagation model or one obtained from a site survey is used to compute the 2D position by triangulation.

The Global WiFi Positioning System from the PlaceLab project uses client received 802.11 access point MAC identifiers and signal strengths to privately compute a position [22]. This approach achieves scale and ubiquity through a community-oriented approach to beacon mapping and signal strength measurement.

3 BlueStar

BlueStar aims to allow only the end user to be aware of their location while indoors, akin to the Cricket system, rather than a system such as the Active Bat that tracks the user. The BlueStar sniffing method is a two-phase approach using a large-scale cellular network in conjunction with low cost local networks. Much of the existing research in this area has focused on improving the accuracy of either the network positioning or indoor location tracking, rather than addressing the need for a scalable user-centric complete solution. Further, many indoor tracking systems rely on special purpose receivers (badges) and transmitters in conjunction with a costly site "radio survey". The accuracy of the location computation is then typically a function of the resolution of the radio survey.

3.1 Deducing Location

In BlueStar the indoor location information (proximation) is deduced by handset/PDA-resident (mobile terminal) applications that have two sources of information:

- Evidence from passive sniffing of existing wireless infrastructure (Bluetooth or 802.11b) or low-cost beacons.
- Details about the local wireless infrastructure that are provided based on system knowledge of the user's approximate location from the GSM network positioning system.

The advantages of our approach include:

- 10m accuracy (using class 2 Bluetooth beacons) sufficient for a range of location aware applications such as mapping applications.
- No special purpose/expensive infrastructure.
- No special purpose tags or locator devices.
- No need to authenticate or formally introduce transmitter and receiver.
- More privacy protection as the system can only track at the GSM resolution.

In BlueStar the more privacy someone wants, the more abstract or redundant data they must be willing to accept to ensure a system cannot deduce where they are. Further, by including a network-centric approach to the delivery of the mapping, local information and local wireless network information, the proposed system can be deployed and tested on as large a scale as current global GSM network coverage.

The primary research component is to develop a scalable delivery and data encoding mechanism for both the location mapping and location-tied data. The primary development component of this project is to integrate the approximate network location technologies currently available with low-cost localised location-awareness. Typically, the handset-resident application will query for its approximate location using GSM mechanisms, which will then trigger the delivery of high level mapping, local information and existing local wireless network information, which we refer to as MIW (Mapping, Information, Wireless) data.

This refinement approach ensures the user's privacy and that the GSM network isn't being tasked with keeping track of individuals on a micro level.

3.2 System Architecture

Our architecture consists of a number of server-side components with which the BlueStar server communicates, as shown in Figure 1. The handset-resident application transmits to only the BlueStar server using HTTP over a GPRS connection. All replies from the server (including maps) are cached on the local device for a specified period, hence minimizing GPRS bandwidth usage and cost to the user.

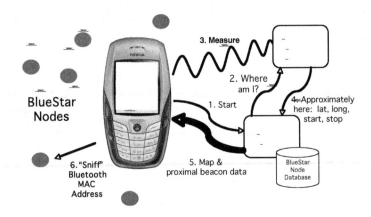

Fig. 1. BlueStar client-server-beacon application architecture

When the BlueStar application is started the first step is to determine the approximate location of the device. GSM location can be determined by querying the network for the approximate location in the form of a series of line segments (latitudes/longitudes) that define an enclosed region in space (using Cell ID plus TA). Once the handset-resident location module makes its initial inquiry, the BlueStar server contacts the location gateway, typically a GSM mobile positioning center (MPC), to determine the handset's approximately location. This approximate location element is passed to the world model gateway which converts it into a specific

BlueStar location ID (or IDs) used to efficiently query the information gateway, the wireless infrastructure gateway, and for a map-based application, the mapping gateway. A BlueStar location ID maps to a discrete area, such as a shopping centre for example. A GSM determined approximate location may map to more than one BlueStar location ID in a case where the GSM location ambiguously crosses two (or more) discreet locations.

Alternatively, approximate GSM location can be determined by querying the phone operating system directly, to obtain Cell ID, and perhaps Timing Advance information if available. The local device may then query the network to convert the Cell ID and TA information into a BlueStar location ID (or IDs). In addition the BlueStar server will return additional Cell IDs (and relevant TA information) corresponding to this location ID. This method of determining approximate location has two distinct advantages over the previous method.

- When entering an area covered by a discrete BlueStar location ID, location information need only be queried from the server once. This information is then cached on the local device and the network need not be contacted again, saving GPRS bandwidth, and cost for the user.
- Reduced queries for a user's approximate location actually increase privacy. In normal GSM networks the exact location of a phone within a location area is unknown unless the network pages the device (e.g. to locate it for an incoming phone call). By the phone caching location information, and determining its location directly, the network is not required to page the device.

Rather than caching all mapping, information and wireless details for the world, the approximate location allows BlueStar to extract a smaller portion of the world model as needed. This portion of the world model is delivered to the handset-resident application and is still large enough to ensure the system cannot micro-locate the end user.

The world model information consists of enough data for the handset-resident application to locate itself within the indoor setting, using wireless beacons. A location described by a discreet location ID may consist of more than one maps in the case of a map based application. For example, a shopping centre may consist of different maps for different floors. In addition, the information associated with a location (such as special offers for the shopping centre scenario) may be subdivided into smaller areas. The local device must query the BlueStar server for these maps and information. Here there is a tradeoff between the quantity of data that must be downloaded and cost to the user, versus the privacy of the user. To maintain complete privacy the device should query and cache the mapping and information data for the entire area at the granularity of the given GSM location. In this case redundant information may be downloaded however, as a user may never visit a particular location. To increase efficiency by reducing bandwidth consumption, and to lower cost, a user may specify that information and maps are only downloaded as required. So for example, when the user walks to a shopping centre floor for which they have no mapping data, and a known wireless device from that floor is 'sniffed', the map will be downloaded on-demand using HTTP over GPRS. This has the disadvantage of allowing the system to trace the location of the user to the granularity of a particular map. This could typically allow the system to tell that the user is in a particular building, on a particular floor, for example. Alternatively the user could specify an intermediate granularity where a number of maps within a certain distance are requested at once.

3.3 Implementation

In our current deployment we use existing Bluetooth infrastructure along with low-cost wireless Bluetooth beacons that we have developed. These beacons provide no data access functionality but instead "act" like wireless access points. These simplified devices require no network connectivity and have been built into the form factor of a power plug.

Our BlueStar server consists of an Apache web server running PHP, supported by a backend MYSQL database storing the MIW data. BlueStar clients make requests to the BlueStar server using a defined protocol over HTTP over GPRS.

We have developed two versions of our BlueStar client software for smart phones. The first consists of an implementation for Symbian series 60 devices, supporting phones such as the Nokia 6600. It is possible to obtain the Cell ID directly from the phone operating system. There is currently no published API for obtaining Timing Advance information. This implementation queries the BlueStar server for location IDs corresponding to the current Cell ID. This location information is cached on the phone. The user has the option of setting the application to download all mapping and information data on entering a new GSM location, or downloading the mapping data on demand. The application continually searches for discoverable Bluetooth devices to determine the local location. The length of time between discovery cycles is configurable to trade off phone battery consumption versus speed of detecting new in-range beacons.

In addition, we have developed a J2ME version, supporting MIDP 2.0 phones with JSR 82 (the Java Bluetooth specification). J2ME MIDP 2.0 does not currently support a way to obtain Cell ID or other GSM location information, so this implementation queries for the location ID directly from the network using a GPRS connection. This has the disadvantages outlined in the previous section.

4   User Study

We conducted an initial user study based on an "InfoHoard" game concept, to determine user acceptance of the system; to evaluate its ease of use; and to illicit privacy concerns of users of potential and existing navigation guidance systems. Twelve users were asked to engage in a simple treasure hunt scenario, using the BlueStar system running on a Nokia 6600. Six of the users were from an IT background the remaining six users were from a non-IT background. Four of the participants were female; eight of the participants were male. All the participants owned mobile phones.

To gauge the familiarity of our participants with data services on their mobile phones, they were asked to indicate the frequency with which they sent Short Messaging Service (SMS) messages, the frequency with which they used mobile Internet

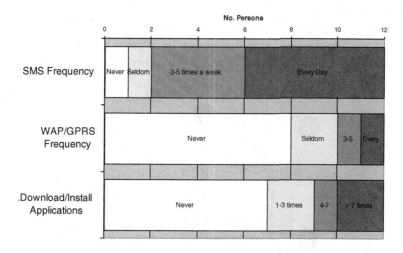

Fig. 2. Participant familiarity with data services on mobile phones

access (WAP/GPRS services) on their phone, and the number of times (if any) they had downloaded and installed application on their phones. The results are presented in Figure 2. All but two users sent SMS messages at least 3-5 times a week. One user sent SMSs less frequently than this, and one user never sent SMSs. Eight users had never used their phone to access the Internet, and seven users had never downloaded an application to install on their phone.

4.1 Treasure Hunt Scenario

To evaluate the system, participants were asked to engage in a simple treasure hunt task within an office building. They were given a Nokia 6600 running the BlueStar software. They were given a brief introduction to the system, consisting of an explanation of the symbols (blue stars representing nearby beacons, blue Xs representing target destinations, and blue halos representing off-screen target destinations [26]), as shown in Figure 3. They were shown how to use the options menu to search for the next location. Participants were given the first location to find. On reaching a target location, a message appeared on the phone asking a simple question, as shown in Figure 3 (1), (e.g. how many buttons are on the coffee machine?). The answer to the question could be found in the environment. Additionally, the message contained the next location for the user to search for where they would uncover the next message. Messages were only revealed in order at the various locations, so all users followed the same path. Users visited four distinct locations, answering seven questions (i.e. they visited some locations twice). The blue arc (halo) in Figure 3 (3) represents an off-screen location to the user. The target location is at the centre of the circle formed by the given arc.

The system logged the times for each user to arrive at each location and receive their next question. When the users completed the treasure hunt, they were asked to complete a questionnaire. The questionnaire consisted of 22 multiple choice questions, and six open-ended questions. The questionnaire was designed to illicit the users' pattern of use of mobile phones (as already discussed), their impressions of the ease of use and intuitiveness of the various aspects of the system, their likelihood of using such a guidance system, and questions relating to privacy of such systems.

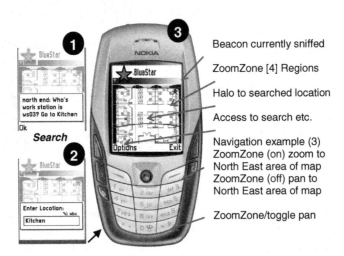

Fig. 3. "InfoHoard". (1) Question for current location & next location. (2) User search for next location. (3) UI after search

5   Results and Evaluation

5.1   Time to Task Completion

Figure 4 shows the duration between the user receiving successive treasure hunt questions 2 to 7. Part of task 1 involved explaining the operation of the messages to the participants, so the duration from the first question to the second question has been discounted.

The locations that participants would have to visit to answer questions were randomised such that the path participants took to answer all the questions involved them doubling back on themselves several times. In this way the time to completion of tasks would not be affected by participants anticipating the next location in advance. It can be observed from Figure 4 that participants completed the later stage tasks more rapidly than the earlier tasks. This would seem to indicate that as participants became more familiar with the system, their time to complete the individual tasks improved.

It can be observed that all participants completed the last three tasks in similar times: the shortest time being participant 2 completing the tasks in 3 minutes and 19 seconds, and the longest being participant 10 completing the task in 5 minutes and 40 seconds. This is true for participants that took significantly longer to complete the first tasks, such as participants 3, 6 and 10. This would indicate that even those

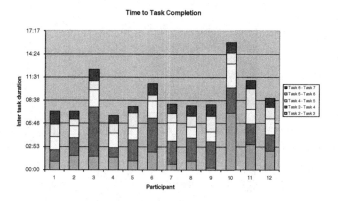

Fig. 4. Inter-task completion times (task 2-7)

participants who had some trouble with the system initially, converged significantly towards the group average times towards the end of the treasure hunt.

5.2 User Acceptance

Participants were asked to evaluate the ease of use of the following features: scrolling of the map, zooming of the map, searching for new locations, identifying current location using the map, and the general ease of use of the system. The questions were presented on a semantic differential scale of 1-6, 1 meaning the feature was easy to use or understand, 6 meaning the feature was difficult to use or understand results are presented in Figure 5.

Eight of the twelve users rated the ease of use of the system as 1 or 2, indicating most users found the system overall quite easy to use, their difficulty following the map notwithstanding. Three users found the system somewhat difficult to use, rating it 4 or 5. Most users found scrolling of the map, searching for new locations, and identifying the approximate distance and direction to off-screen target locations as relatively easy, with 8 or more users in each case having a difficulty level of 1 or 2.

One user noticed that the screen scrolled automatically as he encountered new beacons, and so they did not use or rate the scrolling feature. Two users indicated that they did not notice the arc mechanism for determining the location of off-screen locations. The majority of users (9) did not use the zoom feature [4], which was of little use for the Treasure Hunt task given.

Users found identifying their location on the map to be a difficult task generally, with all but three users indicating a difficulty level of 4 or higher. Much of this difficulty can probably be accounted for by the poor quality of the map we had available to us. The level of detail was perhaps too high for the task, and the names of locations were not marked clearly on the map.

The likelihood that participants would use a guidance system such as BlueStar to navigate in areas they knew well; in areas they were somewhat familiar with, and areas they were very familiar with, is shown in Figure 6. Not surprisingly, all but one

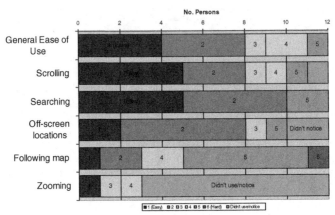

Fig. 5. Perceived ease-of-use of the interface elements

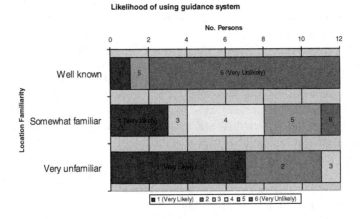

Fig. 6. User likelihood of using a guidance system in different types of location

user indicated they would be very unlikely to use the system in an area that was well known to them. All users indicated a likelihood level of 3 or above that they would use such a system in an area that was very unfamiliar to them. 7 users indicated they would be very likely, with likelihood score of 1, to use such a system. These figures indicate a high acceptance rate of the system by our study participants.

5.3 Privacy

We made the following claim in the questionnaire:

"There is a type of positioning used in this game. Only the smart phone you were using during the game knew your 'actual' position (it wasn't disclosed to anyone else). In addition, our game system running on a computer in Camperdown only knew you were in an approximately 500 metre square area within Sydney".

Participants were then asked whether they were likely to believe this claim coming from (a) someone they knew (e.g. the researchers running the study), or (b) someone they didn't know (e.g. a third party company). Figure 7 indicates the score indicated by our participants on a 6 point semantic differential scale.

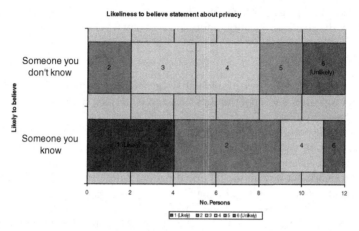

Fig. 7. User likelihood of belief in claims of resolution and self-positioning

As can be seen from Figure 7, user responses were fairly evenly distributed over the scale for someone they didn't know. No participant indicated they would completely trust a third party source (by giving a score of 1) however. Users were more inclined to trust someone they knew, with 9 users giving a score of 1 or 2, however a significant number of users still would not trust a source they knew to make such a claim about their privacy. This indicates a level of difficulty for the global deployment of a BlueStar system in assuring users that their location is not being tracked by the system.

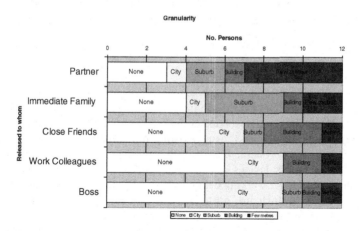

Fig. 8. Granularity at which users would be willing to expose their location to others

We also asked users if they would be willing to release their location information to others (their partner, immediate family, close friends, work colleagues, and their boss), and to what granularity if so (within a few metres, the name of a building, suburb, city, and none). As can be seen from Figure 8, a great range of answers was given by participants for this question. Some participants indicated they would be willing to divulge their location to within a few metres to all the individuals given. Others would not divulge their location information to any level of granularity.

5.4 Qualitative Questions

Our user study contained a number of qualitative questions that provide some insight in to the problems, issues and opportunities faced in deploying a passive rather than active location based system.

In terms of what aspects our test participants liked the most, users commented that "the circular navigation system was very good – gives a good indication of where to go given my current location", "The pop up questions when I came close", "It was generally easy to work out how to use", "Zoom feature where I am and where I need to be", and "Target location indicated on map".

In terms of perceived weakness in the interface and the application itself, users commented that the map was too detailed for the task at hand, and that the map should auto-orient itself and that the search function was not automated for the next location. Users commented that, "No auto rotate on the map, given low accuracy it's quite easy to walk a fair distance in the wrong direction first before noticing".

In broad terms many users commented on the strengths and weakness of the privacy that our system offers along with their general concerns for system monitoring. Users who had low privacy concerns commented that "As long as I can turn it off whenever I want, I'm fine with it", "City/Suburb information is quite innocuous and I imagine not a privacy concern to most people", "Very good for honest truthful people – but not for 'hidden' people" and "good for boss and assistant during office hours only". Users who were more concerned with their privacy commented that, "People don't want anyone to know where they are – Big Brother", "I enjoy my anonymity and dislike anything reminiscent of 1984" and "People are concerned with the 'wrong' people knowing where they are".

6   Conclusions

We have presented our scaleable hybrid solution for location awareness that uses existing wireless infrastructures in a novel and complimentary manner. By taking a two-phase approach to location awareness, using existing infrastructure, we have realised a large scale location aware system. Based on the mobile terminal sniffing for the identification of local wireless devices, which act as "beacons", in the environment.

We have further presented a working example of our BlueStar system along with a preliminary user study of "InfoHoard" a BlueStar game, in an indoor environment.

Location-aware services are slated to be the "killer application" for the next generation of mobile phones that incorporate large displays, more memory and substantially more processing power and battery. Along with the natural evolution of mobile phones there is now a confluence of PDA and mobile functionality into more power-ful and flexible computing devices with "always on" capabilities.

By focusing on a complete and privacy-centric approach to location-awareness the techniques and patented methods presented and evaluated here can be readily de-ployed by a telecommunications provider.

Acknowledgement

The authors would like to acknowledge the financial support of the Smart Internet Technology CRC Australia for this research project.

References

1. Quigley A., Ward B., Cutting D., Ottrey C. and Kummerfeld R, "BlueStar, a privacy cen-tric location aware system", IEEE Position, Location and Navigation Symposium 2004, Monterey, USA April 26-29 2004
2. Salber, D., Dey, A. K., Orr, R. J., and Abowd, G. D. "The Context Toolkit: aiding the de-velopment of context-enabled applications". In Proceedings of the 1999 Conference on Human Factors in Computing Systems, pages 434-441, Pittsburgh, PA.
3. Andreas Butz, Jorg Baus, Antonio Kruger, "Different Views on Location-Awareness", Technical Report University of Saarbrucken.
4. Daniel Robbins, Edward Cutrell, Raman Sarin, Eric Horvitz, ZoneZoom: Map Navigation for Smartphones with Recursive View Segmentation. In Proceedings of AVI 2004, Gal-lipoli, Italy, May 2004, pp. 233-240.
5. George W. Fitzmaurice. "Situated information spaces and spatially aware palmtop com-puters". CACM, 36(7):38–49, July 1993.
6. M. van Steen, F. J. Hauck, G. Ballintijn, and A. S. Tanenbaum. "Algorithmic Design of the Globe Wide-Area Location Service". The Computer Journal, 41(5):297–310, 1998.
7. Ousa S. and Garlan D., "Aura: An architectural framework for user mobility in ubiquitous computing environments". In Proceedings of the International Conference on Architecture of Computing Systems (ARCS'02) (Karlsruhe, Germany, April 2002), pp. 7–20.
8. Joseph F. McCarthy, Eric S. Meidel, "ACTIVEMAP: A Visualization Tool for Location-Awareness to Support Informal Interactions", Proceedings of the International Symposium on Handheld and Ubiquitous Computing (HUC '99)
9. Mike Spreitzer and Marvin Theimer. 1993. "Providing Location Information in a Ubiqui-tous Computing Environment". In Proceedings of the 14th ACM Symposium on Operating Systems Principles (SIGOPS '93). 270-283.
10. P. Bahl, N. Padmanabhan: "RADAR: An In-Building RF-based User Location and Track-ing System", in Proceedings of IEEE INFOCOM 2000, Vol. 2, Tel-Aviv, Israel (March 2000).
11. Pountain, Dick, "Track People with Active Badges", BYTE, Dec. 1993, pp. 57,58,62,64.
12. R. Want, A. Hopper, V. Falcao and J. Gibbons, "The Active Badge Location System", ACM Transactions on Information Systems, pp. 91-102, Jan. 1992.
13. Schmidt et al.; "Advanced Interaction in Context"; Lecture Notes in Science 1707, Sep. 1999. pp. 89-101.

14. Spreitzer, M. and M. Theimer, "Providing Location Information in a Ubiquitous Computing Environment", Proc. 14th Symposium on Operating System Principles, ACM Press, December 1993, pages 270-283.
15. "Ultrasonic Location System", The Oliveti & Oracle Research Laboratory, Oct. 16, 1998.
16. Ulf Leonhardt, "Supporting Location-Awareness in Open Distributed Systems", PhD thesis, Department of Computing, Imperial College of Science, Technology and Medicine, University of London, May 1998.
17. Want, Roy, et al., "Active Badges And Personal Interactive Computing Objects", IEEE Transactions, Feb. 1992, pp. 10-20.
18. Ward, A. Jones and A. Hopper, "A New Location Technique for the Active Office", IEEE Personal Communications, vol. 4, No. 5, pp. 42-47, Oct. 1997.
19. N. Priyantha, A. Chakraborthy and H. Balakrishnan, "The Cricket Location-Support System", Proceedings of International Conference on Mobile Computing and Networking, pp. 32-43, August 6-11, 2000, Boston, MA
20. N. Bulusu, J. Heidemann and D. Estrin, "GPS-less Low Cost Outdoor Localization For Very Small Devices", IEEE Personal Communications Magazine, Special Issue on Networking the Physical World, August 2000.
21. Want R., Schilit B. N., Adams N. I., Gold R., Petersen K., Goldberg D., Ellis J. R. and Weiser M.. "The PARCTAB Ubiquitous Computing Experiment". In Mobile Computing, Korth H. F. and Imielinski T., eds., Kluwer Academic Press, 1996.
22. Schilit, W.N., LaMarca, A., Borriello, G., Griswold, W., McDonald, D., Lazowska, E., Balachandran, A., Hong, J., and Iverson, V. Ubiquitous Location-Aware Computing and the "Place Lab" Initiative. The First ACM International Workshop on Wireless Mobile Applications and Services on WLAN, September 19, 2003, San Diego, CA, USA.
23. Andy Harter, Andy Hopper, Pete Steggles, Andy Ward, and PaulWebster. "The anatomy of a context-aware application". In MOBICOM 1999, pages 59–68, August 1999.
24. Bob Kummerfeld, Aaron Quigley, Chris Johnson, Rene Hexel, "Merino:Towards an intelligent environment architecture for multi-granularity context description", Workshop on User Modeling for Ubiquitous Computing, UM June 2003 Pittsburgh, USA.
25. Schmidt, A. and Beigl, M. "There is more to context than location" Environment sensing technologies for adaptive mobile user interfaces, 1998.
26. Baudisch, P., Halo: supporting spatial cognition on small screens., In UIST'03 posters and demos booklet (demo paper), Vancouver Canada, November 2003.

Author Index

Lecture Notes in Computer Science

For information about Vols. 1–3391

please contact your bookseller or Springer